D1192538

Learning to
Live with
STATISTICS

40722715

QA
276
.A82
2008

Learning to
Live with
STATISTICS

From Concept to Practice

David Asquith

LYNNE
RIENNER
PUBLISHERS

BOULDER
LONDON

Published in the United States of America in 2008 by
Lynne Rienner Publishers, Inc.
1800 30th Street, Boulder, Colorado 80301
www.rienner.com

and in the United Kingdom by
Lynne Rienner Publishers, Inc.
3 Henrietta Street, Covent Garden, London WC2E 8LU

© 2008 by Lynne Rienner Publishers, Inc. All rights reserved

Library of Congress Cataloging-in-Publication Data
Asquith, David, 1942–
    Learning to live with statistics: from concept to practice / David Asquith.
    Includes index.
    ISBN 978-1-58826-524-1 (hardcover : alk. paper)
    ISBN 978-1-58826-549-4 (pbk. : alk. paper)
    1. Mathematical statistics. I. Title.
QA276.A82 2007
519.5—dc22   2007031084

British Cataloguing in Publication Data
A Cataloguing in Publication record for this book
is available from the British Library.

Printed and bound in the United States of America

       The paper used in this publication meets the requirements
 ∞   of the American National Standard for Permanence of
       Paper for Printed Library Materials Z39.48-1992.

5 4 3 2 1

To Debbie, Heather, and Adam

*Who gave me time and
who knew just when to nag*

With love

Contents

A Note to the Beginning Statistics Student   xi

1 Beginning Concepts   1

A Preview: Text Overview 2
The Level of Measurement: Using the Right Tools 5
  Creating Categories 7 ▪ *Comparing Ranks 9* ▪
  When the Numbers Count 14
To What End: Description or Inference? 18
Exercises 19

2 Getting Started: Descriptive Statistics   21

Central Tendency: Typical Events 21
  The Most Common Case: The Mode 22 ▪ *Finding the Middle Rank:*
  The Median 23 ▪ *Counting the Numbers: The Average 25*
Diversity and Variation 29
  Highs and Lows: The Range 30 ▪ *Away from the Average:*
  Measuring Spread 31 ▪ *Using Differences from the Average 31* ▪
  Using the Original X Scores 36
Exercises 42

3 Probability: A Foundation for Statistical Decisions   45

Sorting Out Probabilities: Some Practical Distinctions 46
  Two Outcomes or More: Binomial vs. Random Outcomes 46 ▪
  Another Consideration: Discrete vs. Continuous Outcomes 47

Continuous Random Outcomes and the Normal Curve 48
*The Normal Curve and Probability 48 ▪ Keys to the Curve:
z Scores 50 ▪ Using z Scores and the Normal Curve Table 52*
Continuous Binomial Outcomes and the Normal Curve 61
Probabilities for Discrete Binomial Outcomes 71
*The Probability of Exactly X Successes 71 ▪ Combinations:
Different Ways to Get X Successes 73 ▪ Solving for Complements
78 ▪ A Handy Table of Binomial Probabilities 79*
Summary 82
Exercises 82
Continuous Random Probabilities 82 ▪ Binomial Probabilities 86

4 Describing a Population: Estimation with a Single Sample   **91**

The Theory We Need 92
*Curves of Many Sample Means 92 ▪ Curves of Many Sample
Proportions 95*
Working with Small Samples: t and Degrees of Freedom 97
When and Why Do We Use t*? 97 ▪ The* t *Table and Degrees of
Freedom 99 ▪ Only* z*: The Unusual Case of Proportions 101*
A Correction for Small Populations 103
Estimating Population Averages 105
Estimating Population Proportions 112
Sample Sizes: How Many Cases Do We Need? 118
*Estimating a Population Average 118 ▪ Estimating a Population
Proportion 123*
Summary 126
Exercises 126

5 Testing a Hypothesis: Is Your Sample a Rare Case?   **133**

The Theory Behind the Test 133
*Many Samples and the Rare Case 134 ▪ Null and Alternative
Hypotheses 138 ▪ Alternative Hypotheses and One- or Two-Tailed
Tests 139 ▪ Always Test the Null 140 ▪ Wrong Decisions? 142*
Nuts and Bolts: Conducting the Test 148
Options: z, t, *or a Critical Value 148 ▪ One-Tailed Tests:
Directional Differences? 152 ▪ Two-Tailed Tests: Any Differences?
155 ▪ Coping with Type II Beta Errors: More Power 159*
Summary 168
Exercises 168

6 Estimates and Tests with Two Samples: Identifying Differences **177**

Independent Samples 178
Many Samples and Usual Differences 178 ▪ *Estimating Differences Between Two Populations 181* ▪ *Testing Differences Between Two Samples 190*
Related or Paired Samples 200
Many Sample Pairs and Usual Differences 200 ▪ *Estimating Differences Between Related Populations 203* ▪ *Testing Differences Between Related Samples 212*
Summary 219
Exercises 220

7 Exploring Ranks and Categories **227**

Tests for Ranks 228
Large Samples: Wilcoxon's Rank-Sum Test 228 ▪ *Smaller Samples: The Mann-Whitney U Test 231*
Frequencies, Random Chance, and Chi Square 234
Chi Square with One Variable 235
A Goodness-of-Fit Test: Are the Data Normal? 240
Chi's Test of Independence: Are Two Variables Related? 243
Nuances: Chi's Variations 247
A Correction for Small Cases 247 ▪ *Measures of Association: Shades of Gray 253*
Summary 255
Exercises 255

8 Analysis of Variance: Do Multiple Samples Differ? **265**

The Logic of ANOVA 265
Do Samples Differ? The F Ratio 267
Which Averages Differ? Tukey's Test 269
Exercises 276

9 X and Y Together: Correlation and Prediction **281**

Numbers: Pearson's Correlation (r_{XY}) 283
Is r_{XY} Significant? 287
Using X to Predict Y ($\hat{Y}$) 292
A Confidence Interval for $\hat{Y}$ 295

Does *X* Explain *Y*? A Coefficient of Determination 297
Ranks: Spearman's Correlation (r_S) 304
Is r_S Significant? 304
Summary 309
Exercises 309

Appendix 1: Glossary   **315**

Appendix 2: Reference Tables   **333**

A. Areas of the Normal Distribution 334
B. Selected Binomial Probabilities up to $n = 15$ 338
C. Critical Values for the *t* Distributions 343
D. Critical Values for the Mann-Whitney U Test 344
E. Critical Values for the Chi Square Tests 346
F. ANOVA: Critical Values for the *F* Ratio 348
G. Tukey's HSD: Critical Values for the Studentized Range (Q)
   Statistic 351
H. Critical Values for Pearson's r_{XY} 353
I. Critical Values for Spearman's r_S 354

Appendix 3: Answers and Hints for Selected Exercises   **355**

Index   371
About the Book   379

A Note to the Beginning Statistics Student

IF YOU ARE READING THIS, YOU are most likely taking an introductory statistics course for which this book is required. The book will have been your instructor's choice, but I hope it is a choice you will eventually appreciate.

The text aims to present statistical concepts and procedures in as simple and straightforward a manner as possible. Nevertheless, it does cover a topic that students often fear. Even though I have not found any magic words, phrases, or reassurances that will make all apprehensions disappear, I would like to offer two basic thoughts.

First, my overriding impression is that when students do have trouble with statistics despite good efforts, in almost all cases it is because they are telling themselves (or have convinced themselves) that the material is beyond them. The excuses and explanations are legion: "I just can't understand numbers." "My whole family is poor at math." "My seventh-grade teacher said I was terrible at math, and he was right." "I liked it as a kid, but now it's too difficult." "I can't even balance my checkbook." "I want to work with people, not numbers. I don't see why I need this stuff." And on and on it goes. Sad to say, it sometimes becomes a self-fulfilling prophecy: If you are convinced you will do poorly, chances are you will do poorly. However, as I think many former students would confirm, the material is probably much less difficult than you have led yourself to believe. If you do not give up, if you tell yourself you can do it and have confidence in yourself, and if you are willing to work, then you *can* do it. I cannot put it any more directly than this: statistics are not impossibly mysterious or difficult.

Second, and even if you never become especially fond of statistics, a knowledge of basic statistical methods can be valuable. Statistical tools allow us to summarize and better understand what might otherwise be masses of unappreciated data. They help us to plot trends, estimate probabilities for certain outcomes (election victories, for example), evaluate products, test theories, perhaps uncover the unexpected, and occasionally challenge obvious or conventional wisdom. Even if the career for which you are aiming will not require you to produce your own statistical analyses, you will certainly be expected to understand and use numbers to some degree. Statistics are part of virtually all occupations, professions, and even world events today. When was the last time you read a daily newspaper or caught a radio or TV news broadcast completely devoid of numbers or statistics? Feeling comfortable with statistics and using them, or at least not recoiling from them, will almost certainly prove beneficial, if not essential, in your future.

Lesson number one, then, is: Hang in there—and possibly even enjoy it.

*   *   *

Despite my name being on the cover, this text reflects the input of various people. First and foremost, I am indebted to former students and especially to those who used previous versions of the book as works-in-progress and offered suggestions. Their insights and critiques forced me to clarify my thoughts and intent and made the writing easier. I have learned a great deal from their contributions and remain very grateful. Colleagues also have been generous with their time. One used earlier versions of the manuscript in his classes, others offered reviews and the benefit of their experiences and expertise. From a wider circle, I gained from conversations about statistics and about teaching them to new and sometimes wary students. For these insights, I would like to thank Yoko Baba, José Bautista, Miriam Donoho, Gary Eggert, Bob Gliner, Chris Hebert, Lawrence Lapin, James Lee, Ray Lou, Kim Ménard, Paul Munro, Max Nelson-Kilger, Wendy Ng, Mick Putney, Carol Ray, Rameshwar Singh, David Smith, Bob Thamm, and Geoff Tootell. Finally, I am indebted to Jean Shiota of SJSU's Center for Faculty Development for her helpfulness and skill with Adobe Illustrator and graphics; to the Sociology Department's Joan Block, who gave her typically invaluable administrative assistance; to Lynne Rienner and her staff for their support, encouragement, and patience; and to colleagues at San José State University and the administrators who granted me a sabbatical leave for the text's completion.

I would be most grateful for responses from those reading or using this text. I may be reached at dgasquith@aol.com. Thank you, and good reading.

—David Asquith

CHAPTER 1

Beginning Concepts

In this chapter, you will learn how to:

- Recognize nominal, ordinal, and interval-ratio levels of measurement
- Explain why the difference between levels of measurement is important in statistics
- Read frequency distributions
- Distinguish between descriptive and inferential statistical analyses

BEFORE GETTING INTO ANY DETAILS, STATISTICS, or formulas, it is worthwhile to consider both the scope of the text and some introductory concepts. This general overview introduces common questions and procedures in statistics and presents the sequence of topics in the text.

Statistics have options. First, there are different kinds of data or information with which we work, and they call for different statistical procedures. **Data*** consist of the measurements and numbers we summarize and analyze, but how do we decide which statistical methods are best? Which statistical procedures are appropriate and which are not? An important consideration in answering these questions is the **level or scale of measurement.** This refers to whether we have actual numbers or merely categories as data. Numbers naturally refer to actual quantities as data: the number of miles you drive per year, your age, how much money you earned last year, your exam scores, and so on. If you work for an immigration lawyer who wishes to know the typical length of time clients have spent in the United States before applying for

---

*Terms presented in bold face in the text appear in the glossary, which begins on page 315.

citizenship, you would have measurements or data points that consist of the numbers of months clients had been in the United States before submitting citizenship applications. Based on those numbers, you could calculate a typical or **average** time to applying for citizenship, and that would not be a complicated procedure. Sometimes, however, your data or measurements will not consist of numbers.

At a different level or scale of measurement, your information may consist simply of designations or categorizations people have made. Examples include checked boxes to indicate sex or gender, academic major, religious preference, and so on. These measurements are clearly not numerical. They are merely descriptive categories: female or male? Major in social science, chemistry, business, Spanish, etc.? Catholic, Buddhist, Protestant, Muslim, Jewish, and so on? These sorts of measurements or data would require different statistical treatments than would numerical information. Therefore, whether we have numerical or categorical data influences any decision as to what statistical procedures are acceptable. Simply put, the typical statistical procedures we use with a set of ages, for instance, do not work if we are asked to analyze data involving sex or religion, and vice versa. A further consideration is whether our purpose is descriptive or inferential statistical analysis.

Briefly, the distinction between **descriptive** and **inferential** statistical analyses refers to how broadly we wish to generalize our statistical results and the subsequent conclusions. If 200 people leaving the voting booth tell us their selections, is this a valid indicator of how the overall election may go? Well, maybe and maybe not. We may do statistical analyses on *any* set of data and simply *describe* that sample of cases. Summarizing how our 200 people voted would be easy enough. However, may we legitimately *infer* something about a whole population of voters from this 200? Are they representative of all voters in that precinct? If we wish to make *inferences* about larger populations, we must be especially careful to analyze truly representative and randomly selected samples from those populations. This is crucial; therefore, probability statements always accompany our inferences. What is the probability our inferences are correct? Incorrect? That makes inferential analyses different from a mere description. We will consider such issues in more depth later. For now, we turn to a text overview.

A Preview: Text Overview

As the text progresses, we will build upon more elementary concepts. We start with material that is no doubt familiar. Chapter 2 starts with statistics that tell us the **central tendency** in a set of data. These statistics give us a sense of the typical cases or central themes in sets of numbers. **Measures of variation** follow; they tell us how much numbers in a set tend to vary from each other. Are they spread over a wide range or, conversely, do they tend to cluster near the average? As used in Chapter 2, measures of central tendency and of variation are descriptive statistics. They simply summarize sets of available data.

The next topic, covered in Chapter 3, is **probability.** Probability forms a bridge between descriptive statistics and inferential statistics. All statistical inferences include information as to the probability they are correct. Probability, however, is a varied topic in itself. Chapter 3 provides an overview of the field and looks at two types of measurement commonly used in statistics, continuous and discrete binomial variables. **Continuous** variables may be measured in fractions (i.e., in less than just whole units). For example, time or distance may be measured down to thousandths of a minute or a mile if we wish. In contrast, **discrete** measurements exist only in whole numbers. How many courses are you taking? How many TV sets in your home? These measurements, of necessity, are made in whole units or numbers. Moreover, as the prefix *bi* suggests, **binomial events** are those in which only two things or outcomes are possible. Whether a newborn is a girl or boy and whether a coin flip comes up heads or tails are examples of binomial situations, and we have unique procedures for determining such probabilities. Examples include the probability of a woman having two girls and then a boy or the probability of 7 heads in 10 flips of a coin. Whether our variables are continuous or discrete and binomial, however, we may take advantage of **probability distributions** to help us determine, for any data set, what numbers are the most or least likely to occur. One of these distributions is especially useful: the **normal distribution,** or bell-shaped curve. It is a prominent part of Chapter 3.

Following probability, Chapter 4 turns to inferential statistics. It introduces **sampling distributions,** which are special cases of normal distributions that form the theoretical foundation for making inferences. We use sampling distributions for **estimation,** that is, estimating unknown averages or **proportions** (percentages) for large populations based upon samples. For instance, how much does the average student spend on campus per week? What proportion, or percentage, of students favor an ethnic studies course being required for graduation?

Chapters 5 and 6 also use sampling distributions for **hypothesis testing.** Hypotheses are simply statements of what we expect to be true or what we expect our statistical results to show. Chapter 5 examines whether a single sample average or percentage differs from a corresponding population figure. We might test the hypothesis that the average time to graduation in a sample of college athletes does not differ significantly from the campus-wide or population average. Alternatively, we might test whether the percentage of people owning pets is significantly greater among people over age 65 than among adults in general. Chapter 6 looks at two-sample situations. Instead of comparing a sample to a population, we compare one sample to another and ask whether they differ significantly. At graduation, do the grade point averages in samples of transfer and nontransfer students differ? Among drivers under age 18, does a random sample of females receive significantly fewer tickets than a similar sample of males?

Next, while still dealing with hypothesis testing, Chapter 7 shifts gears somewhat. Chapter 7 differs from preceding topics in two respects. First, it compares sample data *not* to population data but rather to other criteria. How does our sample compare to what is expected by **random chance** or expected according to some other stated criterion? We now test hypotheses that our sample statistics do not differ significantly from random chance or from other presumed values. For example, if a campus bookstore sells sweatshirts in three school colors, is there a statistically significant preference for one color over the other two? Random chance says all three colors should be selected equally and that buyers pick their colors at random. But do actual sales vary significantly from what random chance suggests they should be?

Second, Chapter 7 also introduces hypothesis tests for **ranked** and for **categorical data.** Here, we use data that must first be ranked or are already broken into categories. We may compare ranked scores (e.g., from essay tests or opinion scales) in two samples. Categorized measurements may be related to gender (female/male), religious denomination (Catholic/Protestant/Jewish/Muslim/Buddhist), or type of vehicle driven (car/truck/van/motorcycle). In these situations, we look at data in categories and consider the number of cases expected to fall into each category by random chance versus the number of actual cases in that grouping. For vehicles driven, for instance, how many people would be expected to list "car" according to random chance versus how many people actually did list "car?" Are drivers in our sample more (or less) likely to use cars than random chance would suggest?

Chapter 7 also uses categorical data to establish correlations or associations between characteristics. Whereas a hypothesis test may tell us whether an association between two characteristics differs from random chance, other statistics allow us to calculate the approximate strength of that association. Various **measures of association** tell us how closely two characteristics are correlated.

Chapter 8 rounds out hypothesis testing by introducing situations involving three or more samples. For example, suppose we wish to compare the average weight losses for people on four different diet plans. Do the average losses differ significantly? Chapters 5, 6, and 7 deal with one- and two-sample cases. Somewhat different methods are needed when we have more than two samples, however, and these make up Chapter 8.

Finally, in Chapter 9, we return to correlations between sets of numbers. Do the numbers of hours studied per week correlate with grade point averages? Do more years of education translate into (correlate with) higher incomes? Besides such correlations, we will also consider how statistical predictions can be made. Suppose we knew that scores on a first statistics test correlated strongly with scores on the second test. We may then predict scores on that second test based upon students' scores on the first test and also get an idea of how accurate those predictions might be. This is known as **correlation and regression.**

Each chapter concludes with a set of practice exercises. Some are essay questions, but most call for you to diagnose the situations described, pick out the relevant bits of information, and solve the exercise by using that chapter's procedures. The exercises are based on real-world situations, and you are asked to translate those word problems into workable and complete solutions. The task is to identify the nature of a problem and to then use the correct statistical procedures in your analyses—without being told specifically which formulas to use. That is part of the learning process: diagnose a typical situation, consider what you are asked to do, and decide upon the appropriate statistical solution. Your instructor may assign selected questions as homework or as classroom exercises. You are strongly encouraged to try as many of these exercises as your time allows, even if they aren't assigned. There is no one single strategy more conducive to learning statistics than practice, practice, practice. The end-of-chapter problems are designed to cover all the procedures and possible alternatives introduced. If you can do the exercises, you have mastered the chapter.

With this general plan of the text in mind, the remainder of this chapter turns to two issues important in any statistical analysis. First, with which kind or level of data are we working? Second, is our analysis descriptive or inferential in nature?

The Level of Measurement: Using the Right Tools

In the research process, **measurement** is the first step, preceding any statistical analyses. Measurement is simply a matter of being able to reliably and validly assess and record the status or quantity of some characteristic. For students' academic levels, for instance, we simply record their statuses as freshmen, sophomores, juniors, seniors, or graduates. We often refer to the things we measure as **variables,** that is, characteristics we expect to vary from one person or **element** to the next. Conversely, if something is true of *every* person or element, it is a **constant.** For example, grade point averages (or GPAs) among history majors would certainly vary, but the designation "history major" would be a constant. Constants become parts of our definition for the **population** we are studying, such as all upper-division history majors or all commuting students, and we often do not measure them directly. We assume everyone measures the same on those characteristics. Our measurements of variables, however, require statistical analyses.

We use the upper-case letter X to denote any single score. For instance, if we are measuring TV viewing, and if respondent number 17 watches 25 hours of TV per week, then $X_{17} = 25$. We need this way of referring to individual measurements. We must have a way to write formulas and express how we are going to treat the individual scores or observations. If we wish to square each measurement, we write "X^2." If we wish to multiply each X score by its companion Y score, it is "XY," and so on. "X" will appear in many formulas throughout the text, and we will have numerous occasions to refer to the "X variable."

One other feature of variables should be briefly noted here. To statisticians, the variables they analyze are obviously measurable and recordable. They have their measurements or data with which to work. Sometimes, however, a researcher assumes the variables reflect more broad or abstract concepts and characteristics. Research often includes more general and theoretical factors that do not lend themselves easily to direct measurement, such as personality, socioeconomic status (someone's location in a prestige or lifestyle hierarchy), employee morale, marital happiness, and so on. In these cases, the actual measurable variables or data become **empirical indicators** for the abstract concepts. For instance, answers to items on personality inventories (e.g., "Would you rather go to a movie with friends or stay home and read a bestselling book?") serve as indicators for broader and more theoretical dimensions of personality makeup. People may be asked about their incomes, educational histories, or occupations rather than directly asking their socioeconomic class levels. Measuring morale or marital happiness may mean asking about recommending one's job to a friend or about whether one has thought about divorce or would remarry the same person. Variables, then, and the resulting measurements will sometimes be obvious and concrete (sex, age), whereas at other times they may be indicators or reflections of more abstract concepts (sociability, mental health). In either case, the statistical analyses measurements used on these variables are of different kinds or **levels.**

The level of measurement involves the quantitative precision of our variables. Some variables naturally lend themselves to precise numerical measurements (e.g., age and income), whereas others do not (e.g., gender, academic major, and political party preference). Still other variables fall between these two extremes. Often, variables have been somewhat subjectively or arbitrarily quantified (e.g., test scores or attitudinal/opinion scores). Generally speaking, the *higher* or more quantitatively precise the level of measurement, the more we can do with the data statistically. The most precise levels are the **interval** and the **ratio.** Interval or ratio level data consist of legitimate and precise numerical measurements. This allows us to choose from a very wide range of statistical tools or operations. **Nominal** or **categorical** data, at the other end of the continuum, are sometimes described as nonquantitative. We are simply labeling or categorizing things (e.g., Democrat versus Republican) and are not measuring quantities. We may still summarize our data or test for multivariate correlations and so on, but our statistical choices are more limited. The basic point is that, with the higher levels of measurement, the more statistical options or possibilities we have. As we proceed through the following chapters, one consideration will be the kinds of statistical treatments appropriate when we have certain levels of data. Or, to think of this the other way around: If we wish to use a certain statistical procedure, what sort of data must we have?

Creating Categories

The nonquantitative level of measurement is called the nominal level. This is the most simple and elementary level or scale of measurement. A nominal variable's categories or measurements are qualitatively different from each other. The categories—say, female and male—cannot be ranked or put in any natural order or sequence. Designating females as category 1 and males as category 2 would be no more legitimate or correct than making males 1 and females 2. Besides gender, other examples are racial or ethnic background, marital status, academic major, or religious denomination. Considering the latter, we might establish the following categories: Protestant, Catholic, Jewish, Buddhist, Muslim, Other. We would have six major categories, but that is all. We could not go an additional step and rank the categories from highest to lowest. That would be nonsensical; the categories of measurement have no natural or logical sequence to them. They make just a much sense in *any* order: Muslim, Jewish, Catholic, Buddhist, Protestant, Other. Or, alphabetically, we would have Buddhist, Catholic, Jewish, Muslim, Protestant, and finally Other. The same is true if we list major racial/ethnic categories. We could list them alphabetically as African American, Asian American, Caucasian, Latino, Other, but it would be just as valid to list them in a different sequence: Asian American, Latino, African American, Caucasian, Other. Our measurements or categories do not fall into any *one* order or sequence. The measurements differ *qualitatively* from each other, not *quantitatively*. We often refer to these measurements as qualitative or categorical variables. We are measuring differences of type, not differences of amount. For instance, a researcher might ask your racial or ethnic extraction; the researcher would not ask how *much* race or ethnicity you have. Race, ethnicity, and the other examples are simply nominal and not quantitative variables.

Statistical operations with nominal data use the category **counts** or tallies, also called category **frequencies.** For example, with a sample of 100 people ($n = 100$) that includes 53 females and 47 males, we use the 53 and the 47 in any statistical operations. Saying a variable is nonquantitative does not mean we cannot do *any* statistics with the data. It just means we may only use the frequencies, tallies, or counts when we do so. The categories of measurement themselves are nonnumerical, e.g., Latino or Asian American, or drivers of cars versus trucks. Even so, we may at least count *how many* people fall into the respective categories, and it is these latter numbers that we use in our statistical analyses. We will work with frequencies and nominal data when we look at cross-tabulations and measures of association.

Frequency distributions illustrate the different levels of measurement. **Frequency distributions** *show all the measurements recorded for a particular sample or population.* The figures typically include the actual numbers and percentages falling into each measurement category. Researchers typically look at such frequency distributions before doing anything else. The frequencies (or tallies or counts or percentages) represent the most elementary level of analysis and are purely descriptive.

We look first here at frequency distributions in the form of tables. Although not shown, pie charts, line graphs, or bar graphs may be used as well. Each chart or table shows the responses for a single variable, starting with nominal variables. Table 1.1 shows the distribution of marital statuses in a recent survey of students at a large state university. Participants were enrolled in randomly

Table 1.1 Marital Status Among a Sample of Students at a State University

Marital Status	Frequency	Percentage	Valid Percentage
Single, solo	336	45.3	46.2
Single, attached	280	37.7	38.5
Married	98	13.2	13.5
Separated	2	.3	.3
Divorced	10	1.3	1.4
Widowed	2	.3	.3
Total	728	98.1	100.0
Missing	14	1.9	
Total	742	100.0	

Notes: "Frequency" refers to the actual number of students giving each answer. Altogether 742 students participated in the survey. "Missing" tallies the unusable responses. Fourteen students either omitted the question or gave unreadable answers. "Percentage" simply expresses all category frequencies as proportions of the total sample, totaling to 100%. "Valid Percentage" discounts the missing cases and recalculates the category percentages based on just those who answered the question. That number, or *n,* comprises 728 actual responses.

selected classes.* Not surprisingly, most students were single, but a distinction is shown between singles who were unattached ("solo") and those who reported being in monogamous relationships ("attached"). The main point, however, is that Table 1.1 illustrates a frequency distribution for a nominal variable. Each category or status is qualitatively different from the others. We do not have more marriage or less marriage. We have a series of different relationship statuses. Moreover, although the categories "Single, Solo" through "Widowed" may appear to be in some sort of logical order, we could actually list them in any sequence we wished. Nominal categories have no inherent order or linear, unidimensional feature to them. They are not different degrees of one thing. They are different things or statuses. Similar attributes of nominal variables are also illustrated in Table 1.2.

Table 1.2 shows the distribution of religious denominations among a recent sample of college students. As in Table 1.1, this table shows the actual frequencies, percentages, and the adjusted valid percentages for each category or answer. As was the case with marital status, the categories here might have been listed in any order. Nominal categories have no inherent order or sequence. Finally, note also that Table 1.2 includes an "Other" category. It would be prohibitive to list all possible religions in our original question, so this option makes the choices exhaustive. Students may answer no matter what their religious denominations. In Table 1.1, however, the choices shown for marital status include all options, and no residual or "Other" category is necessary.

At the next level of measurement, the order or sequence of categories is important. In fact, this feature is reflected in the name, the **ordinal level** or **scale.**

Comparing Ranks

In contrast to nominal variables, ordinal measurements can be ranked. As the name suggests, there is a natural or logical *order* to them. Common examples of ordinal variables are social class or socioeconomic status (SES), religiosity (how religious one is), one's degree of ethnocentrism or prejudice, political liberalism or conservatism, and scores on essay exams. Not only may we categorize respondents as to their liberalism or conservatism, religiosity, or

---

*The tables in this chapter—and most in the text—derive from recent campus surveys conducted by the author and students in survey research classes. The tables were prepared using SPSS, a comprehensive statistical program used on virtually all college campuses. Originally developed to aid social science research, its full name was the Statistical Package for the Social Sciences. It proved extremely popular, however, and its use spread to many academic disciplines and to business and other venues. To reflect its broader applications but still retain the familiar acronym, the name was changed to Statistical Products and Service Solutions. To most people, however, it remains simply SPSS.

Table 1.2 Religious Denomination Among a Sample of Students at a State University

Religious Denomination	Frequency	Percentage	Valid Percentage
None/not applicable	204	27.5	29.8
Catholic	232	31.3	33.9
Other Christian	125	16.8	18.2
Buddhist	66	8.9	9.6
Hindu	13	1.8	1.9
Muslim (Islam)	10	1.3	1.5
Sikh	7	.9	1.0
Jewish	11	1.5	1.6
Other	17	2.3	2.5
Total	685	92.3	100.0
Missing	57	7.7	
Total	742	100.0	

test scores, we may also rank the resulting categories or measurements. There is a logical ranking to them, for example: Very Liberal, Liberal, Middle-of-the-Road or Centrist, Conservative, and Very Conservative. We could reverse the order, of course, and put Very Conservative first, but the overall set or list would not make sense in any other sequence. Whenever they are listed, the categories should proceed from one end of the continuum to the other. Similarly, when we display a set of religiosity categories or test scores, we would logically list them from most to least religious, highest to lowest, or vice versa. We do not have qualitative differences between the measurements but rather differences based upon more of something or less of something, that is, upon differences of amount rather than differences of kind or type.

The preceding examples also illustrate the two different types of ordinal measurements we might encounter. On the one hand, we may have a set of *rankable categories.* That is the case with the liberalism-conservatism variable above. Not only do we place people's responses in particular political categories, we may also legitimately *rank* the categories from the most liberal to the most conservative (or least liberal, if you like). We may cross-tabulate sets of rankable categorical measurements with other variables or use measures of association similar to those used with nominal measurements. Tables 1.3 and 1.4 illustrate frequency distributions for sets of ordinal categories.

Please notice two things about Tables 1.3 and 1.4. First, the measurement categories appear in logical sequences. Table 1.3 ranks answers along

Table 1.3 Responses of a Student Sample to the Question: "Have You Ever Told Minor 'White Lies' or Fibs?"

Told Minor Lies	Frequency	Percentage	Valid Percentage	Cumulative Percentage
Many times	172	23.2	24.3	24.3
Occasionally	361	48.7	51.0	75.3
Rarely	156	21.0	22.0	97.3
Never	19	2.6	2.7	100.0
Total	708	95.4	100.0	
Missing	34	4.6		
Total	742	100.0		

Note: Percentages may not total 100.0 due to rounding.

Table 1.4 Responses of a Student Sample to the Statement: "If Asked, It Would Be OK to Help Family Members or Friends with the Answers on Tests."

OK to Help Family or Friends on Tests	Frequency	Percentage	Valid Percentage	Cumulative Percentage
Strongly agree	17	2.3	2.3	2.3
Agree	166	22.4	22.4	24.7
DK/NS	151	20.4	20.4	45.1
Disagree	295	39.8	39.8	84.9
Strongly disagree	112	15.1	15.1	100.0
Total	741	99.9	100.0	
Missing	1	.1		
Total	742	100.0		

Notes: DK/NS stands for "Don't Know/Not Sure," i.e., generally ambivalent or undecided. Percentages may not total 100.0 due to rounding.

a frequency dimension, "Many Times" through "Never," and Table 1.4 does the same with agreement, "Strongly Agree" through "Strongly Disagree." Second, a new column on the right shows the **Cumulative Percentage.** It shows the previous column, the Valid Percentage, *cumulatively* as one proceeds down the sequence of categories. Percentage-wise, it presents a running total for all the cases as we proceed from the first category to the last. The cumulative percentage makes sense only when we have ordered or rankable measurements because we are counting down the categories and accumulating more and more cases as we proceed, *in sequence,* from one category or measurement or rank to the next. The cumulative percentage has no meaning if the order of our categories is arbitrary or discretionary, as it is with nominal measurements. It would change with every different order in which the categories could be listed, and so it would not tell us anything useful.

In contrast, ordinal data may consist of numerical scores or measurements, that is, not rankable categories, but actual numbers. These might result from, say, grading exams or administering approve/disapprove opinion scales. Each person now has his or her own individual score, and the scores may be ranked from highest to lowest. Such a distribution is illustrated in Table 1.5. In three separate questions, college students were asked to agree or disagree that TV covered local, national, and world events well. Each question was scored with the same 5-point, agree-disagree answer format as in Table 1.4. "Strongly Agree" responses were the most positive and scored as 1, whereas "Strongly Disagree" responses were negative and scored as 5. Adding the scores for the three items yielded cumulative values ranging from 3 to 15. We may put

Table 1.5 College Students' Scaled Scores Evaluating the Adequacy of TV News Coverage

		Frequency	Percentage	Valid Percentage	Cumulative Percentage
(+)	3	8	.6	.6	.6
	4	10	.8	.8	1.4
	5	27	2.0	2.0	3.4
	6	102	7.7	7.7	11.1
	7	102	7.7	7.7	18.9
	8	167	12.6	12.7	31.5
	9	119	8.9	9.0	40.5
	10	149	11.2	11.3	51.8
	11	146	11.0	11.1	62.9
	12	185	13.9	14.0	76.9
	13	114	8.6	8.6	85.5
	14	65	4.9	4.9	90.5
(−)	15	126	9.5	9.5	100.0
	Total	1320	99.2	100.0	
	Missing	10	.8		
	Total	1330	100.0		

these combined scores in order, or rank them, but we may not claim, for instance, that a scale score of 6 is exactly twice as positive about TV news coverage as a score of 12. Our measurements of the variables are not that precise. All we may say is that the lower the score, the more positive a student's view of TV news programming.

Another example of a numerical ordinal scale comes from a survey on lying. Regarding work assignments, students were asked whether they had ever called an employer and falsely claimed to be sick and, separately, whether they had ever lied to a professor about the reason for late or missing assignments. Each item was originally scored on a 1 to 4 ("Many Times" to "Never") scale. Adding each student's tallies on the two items yields a possible range of combined scores from 2 through 8 (Table 1.6). As in the previous table, we may not claim that a student scoring 3 is twice as likely to have lied as a student scoring 6, but we may at least rank the X values. Our survey, after all, asked students to generally estimate how often they had lied: Many Times, Occasionally, Rarely, or Never. We had not asked for actual and precise numbers of times. We may claim a set of ordered and rankable numerical scores, with the lower score meaning more lying about reasons for not meeting one's obligations, but we cannot claim to have measured those "lying histories" with true mathematical precision.

Ordinal data in numerical form are suitable for statistical procedures using ranks. Realizing the scores do not represent precise mathematical increments of the variable, we simply rank them from highest to lowest or vice versa, and

Table 1.6 College Students Scaled Responses About Lying to Employers and/or Professors

		Frequency	Percentage	Valid Percentage	Cumulative Percentage
Most	2	9	1.2	1.3	1.3
	3	21	2.8	2.9	4.2
	4	83	11.2	11.6	15.8
	5	112	15.1	15.6	31.4
	6	204	27.5	28.5	59.8
	7	170	22.9	23.7	83.5
Least	8	118	15.9	16.5	100.0
	Total	717	96.6	100.0	
	Missing	25	3.4		
	Total	742	100.0		

thereafter we ignore the original numbers and use the ranks in our statistical procedures. We will return to this concept in Chapters 7 and 9. For now, there are two remaining levels of measurement.

When the Numbers Count

The interval and ratio levels of measurement are legitimately accurate and precise measurements with no subjectivity or doubt. These are the kinds of measurements about which there is no ambiguity. An earlier example, age, is such a variable. Other examples would be the number of units you are taking this semester, how many units you have accumulated over your college career, the number of miles you traveled, or the number of blocks you typically walk to campus, how many people live in your household, how much money you earned last year, and so on.

Statistically, the difference between the interval and ratio levels of measurement does not matter. We may treat interval level measurements just as we would ratio level observations. However, there is an important difference between the two. Ratio scales have a true (or legitimate and meaningful) zero point, that is, a complete absence of whatever is being measured, and interval scales do not. Annual income, for example, would constitute a ratio scale. Someone could, at least theoretically, have absolutely no income or even a negative income, so therefore an $X = \$000$ measurement could be legitimate and valid. In contrast, the Fahrenheit temperature scale represents interval level measurements. A reading of zero degrees has no particular meaning because freezing occurs at 32° Fahrenheit. And yet both scales, income and Fahrenheit, are numerical, accurate, precise, and unambiguous. One has a meaningful zero point, however, and the other does not. This is the difference between interval and ratio scales, but we may treat interval and ratio data the same and ignore that difference. It does remain essential, however, to distinguish between nominal, ordinal, and interval-ratio measurements.

As examples, Tables 1.7 and 1.8 show frequency distributions for interval-ratio measurements. As for the previous tables, Table 1.7 shows data from a campus survey and clearly reflects a college population. Notice the comparatively large frequencies for people in their mid-twenties and just single-digit tallies at age 33 and above. This predominance of people under age 30 is also confirmed in the cumulative percentage column. We have accumulated or accounted for fully 89.1% of the sample when everyone up through age 30 is counted. This feature of cumulative percentages also has another name. We sometimes refer to it as the **percentile rank** of a number, defined as *the proportion or percentage of cases that fall at or below a certain point in a distribution*. With 89.1% of the cases falling at age 30 or below, someone exactly 30 would fall at about the 89th percentile rank in this distribution. We could

Table 1.7 Age at Last Birthday Among a Sample of College Students

Age	Frequency	Percentage	Valid Percentage	Cumulative Percentage
17	1	.1	.1	.1
18	30	4.0	4.2	4.4
19	49	6.6	6.9	11.3
20	62	8.4	8.8	20.1
21	96	12.9	13.6	33.7
22	115	15.5	16.3	49.9
23	104	14.0	14.7	64.6
24	48	6.5	6.8	71.4
25	47	6.3	6.6	78.1
26	24	3.2	3.4	81.5
27	16	2.2	2.3	83.7
28	20	2.7	2.8	86.6
29	8	1.1	1.1	87.7
30	10	1.3	1.4	89.1
31	10	1.3	1.4	90.5
32	13	1.8	1.8	92.4
33	5	.7	.7	93.1
34	6	.8	.8	93.9
35	2	.3	.3	94.2
36	2	.3	.3	94.5
37	1	.1	.1	94.6
38	1	.1	.1	94.8
39	5	.7	.7	95.5
40	3	.4	.4	95.9
42	3	.4	.4	96.3
43	3	.4	.4	96.7
44	2	.3	.3	97.0
45	3	.4	.4	97.5
46	3	.4	.4	97.9
47	3	.4	.4	98.3
48	1	.1	.1	98.4
50	2	.3	.3	98.7
51	2	.3	.3	99.0
54	1	.1	.1	99.2
55	1	.1	.1	99.3
57	1	.1	.1	99.4
59	1	.1	.1	99.6
67	1	.1	.1	99.7
77	1	.1	.1	99.9
82	1	.1	.1	100.0
Total	707	95.3	100.0	
Missing	35	4.7		
Total	742	100.0		

Note: Percentages may not total 100.0 due to rounding.

Table 1.8 If an Immigrant: Number of Years in US (College Student Sample)

Years in US	Frequency	Percentage	Valid Percentage	Cumulative Percentage
1	7	1.0	3.4	3.4
2	13	1.9	6.3	9.8
3	11	1.6	5.4	15.1
4	8	1.2	3.9	19.0
5	12	1.7	5.9	24.9
6	11	1.6	5.4	30.2
7	5	.7	2.4	32.7
8	7	1.0	3.4	36.1
9	9	1.3	4.4	40.5
10	19	2.8	9.3	49.8
11	12	1.7	5.9	55.6
12	14	2.0	6.8	62.4
13	5	.7	2.4	64.9
14	5	.7	2.4	67.3
15	7	1.0	3.4	70.7
16	8	1.2	3.9	74.6
17	6	.9	2.9	77.6
18	8	1.2	3.9	81.5
19	1	.1	.5	82.0
20	10	1.4	4.9	86.8
21	12	1.7	5.9	92.7
22	1	.1	.5	93.2
23	5	.7	2.4	95.6
24	2	.3	1.0	96.6
25	1	.1	.5	97.1
26	2	.3	1.0	98.0
30	1	.1	.5	98.5
31	1	.1	.5	99.0
32	1	.1	.5	99.5
45	1	.1	.5	100.0
Total	205	29.7	100.0	
Missing	485	70.3		
Total	690	100.0		

Note: Percentages may not total 100.0 due to rounding.

also say that a 40-year-old student would be at almost the 96th percentile rank. He or she would be as old or older than 96% of all students in the sample. We will use percentile ranks again in Chapter 3, when we discuss the bell-shaped curve.

Table 1.8 shows a similar distribution but from a different survey. It illustrates an interval-ratio variable and distribution from a survey on immigration.

In this case, a questionnaire asked immigrant students how many years they had been in the United States. The **range** of data is extensive, from a low of 1 year (rounded off) to a high of 45 years. According to the cumulative percentage column, about half the immigrant students (49.8%) had been in the United States 10 years or less. Moreover, notice the large number of *Missing* cases in Table 1.8. Most students ($n = 485$, or 70.3% of all respondents) were *not* immigrants and, of course, did not answer the question. For this question, they were coded as omits or "Missing." Still, fully 29.7% of students in the survey did answer as immigrants, no doubt reflecting recent and changing demographics among young adults in the United States.

Interval-ratio measurements allow us to have the utmost confidence in their mathematical accuracy and precision. Therefore—and this is a key point—any calculations may use the actual X scores or measurements. Unlike nominal or ordinal measurements, the data are not in categories nor must we convert the data values to ranks. To justify their use, however, the original scores must be absolutely reliable, unambiguous, and precise measurements of the variable in question.

When we do have such reliable data, we may use the actual X scores in our calculations. If we need to know the average or typical or usual response, we may add up all the X values and then divide by the total number of cases or values we have, or n. If we have a set of ages, the X variable being age, we may do this. If we have a set of essay exam scores, however, we have ordinal-level measurements only and should not calculate an average. We may not be sure how to *precisely* interpret each exam score, so we should not use them in any statistical procedure. Instead of an average, we have alternative statistics available, as discussed in Chapter 2.

Two final considerations are important. First, as noted earlier, when we do have interval-ratio data, we may use a broad range of statistical treatments. This text assumes we have such data for the most part. Therefore, we may look at a full range of introductory statistical tools. We will, however, also look at statistics specifically designed for ordinal and nominal data.

Second, sometimes we are working with more than one level of data at the same time. If we are correlating measurements on two variables, one nominal, say, and the other interval-ratio, what do we do then? A common rule is to use a statistic (or statistics, plural) appropriate for the lower level. If we have both nominal and interval-ratio data, for example, we use statistics suitable for the nominal level. The higher-level variable meets all the assumptions and criteria of the lower level of measurement, but the reverse is obviously not true. Interval-ratio measurements meet all the criteria of nominal measurements (we can distinguish between different categories or scores of the variable), but nominal data would certainly not meet the precise quantitative requirements of the interval-ratio level. Consequently, we should use statistics appropriate for the nominal level. That is, we choose a level of measurement we are sure is met by both our variables. As in most statistical situations, we would probably have *some* choice in exactly how to treat the data, but we must be prepared to sacrifice that higher level of measurement in one of our variables on occasion.

Beyond this, and no matter what level(s) of data we have, there is another matter to consider. What is the purpose of our analysis? Are we simply summarizing data, or do we wish to make inferences about a larger universe or population based upon seeing just a tiny fraction or sample of it? Do we wish to summarize what we have *or* do we want to make educated guesses beyond the data immediately available? These questions lead to the last part of this introduction: the difference between descriptive and inferential kinds of statistical analyses.

To What End: Description or Inference?

Statistical analysis has two very broad areas: the descriptive and the inferential. The former is the more basic and, as the name suggests, amounts to describing and summarizing data. Given a set of numbers, what is the average, do the numbers vary much or very little from that average, and what percentage of the cases fall below such-and-such a score? Examining the data at hand, descriptive statistics look at variables one at a time and simply summarize the data by calculating various statistical measures (e.g., averages, medians, standard deviations) and by showing frequency distributions. If we have 20 variables, we may easily summarize the scores or measurements for each one and provide a report. We are distilling a lot of raw or original data down into more comprehensible summary measures. Because this involves averages and the like, a good part of descriptive statistics and the material in the following chapter should be familiar to you (or will come back quickly). After that, we begin moving toward the second branch of statistics, inference.

Inferential statistics are a bit more involved and theoretical. The term *inferential* derives from the fact that we are making inferences about larger universes or populations based upon just sample data. And this, in turn, requires that we have **random samples.** Random samples (sometimes called probability samples) are those that give every member of the population a statistically equal chance to be selected. The procedures required for good random samples can be quite involved and are beyond the scope of this text. Nevertheless, we may justify making inferences about the population only if we have random and representative samples. Our discussions of inferential statistical procedures assume we are dealing with random samples.

This second branch of statistics also covers most of what we think of when we use the term "statistics": the normal or bell-shaped curve, procedures known as confidence interval estimates, hypothesis testing, correlation and regression, and so on. But, as noted before, when we look at inferential statistics, we must also consider probability. Statistical inference is based upon probability. Whenever we make that extrapolation or inferential leap from the sample to the population, there is always a chance we are wrong. What is the probability the population average is or is *not* what we have estimated? These probabilities must accompany any inferences we make.

Succeeding chapters look first at descriptive statistics, next at probability, and finally turn to inferential statistics. Before proceeding, however, we must be wary about assuming too much from this introduction: The level of measurement, on the one hand, and the distinction between descriptive and inferential statistics, on the other, are quite independent concepts. They do not necessarily correlate in any way. Descriptive analyses, being the more simple and elementary of the two, do not apply only to the nominal level of measurement. Inferential statistical methods, being more involved and complex, do not apply only to the higher levels of measurement. We may have descriptive statistical summaries involving *any* scale of measurement, nominal to interval-ratio. The same is true of inferential analyses, which may involve nominal, ordinal, or interval-ratio data. It is not a case of a certain level of measurement being appropriate for either descriptive *or* inferential methods. It is a matter of selecting a basic statistical treatment (descriptive, inferential, or both), depending on the study's purpose, and only thereafter tailoring the specific statistics used to the level(s) of measurement involved.

In Chapter 2, we turn to a few descriptive statistics, some of which will be familiar. We first consider averages, more formally known as measures of central tendency, and then look at a popular and valuable measure of variation.

Exercises

1. What is the level or scale of measurement, and why is it important in statistical analyses?

2. Describe the principal levels of measurement, including the characteristics of each and at least one example of each.

3. How do the levels of measurement differ regarding the statistical procedures possible with each?

4. What is the difference between numerical measurements at (1) the ordinal level and (2) the interval-ratio level?

5. How do descriptive and inferential statistical analyses differ?

6. What is the importance of random sampling to statistical analyses?

CHAPTER 2

Getting Started: Descriptive Statistics

In this chapter, you will learn how to:

- Determine the mode, median, and average
- Decide which measure(s) of central tendency work for your data
- Determine the range and calculate the standard deviation
- Interpret descriptive statistics

WE START WITH TWO SORTS OF DESCRIPTIVE analyses. First, there are measures of **central tendency,** or location. These measures summarize a set of data by showing the typical response or observation. You will almost certainly have done some of this before, such as when you calculated an average. Second, there are **measures of variation.** As the name suggests, these are indicators of how much difference exists in a set of data. Are the individual scores all pretty much alike or do they vary a lot? This latter material will probably be new to you if you have not studied statistics before. Here we will look at a very important statistic called the **standard deviation.**

Central Tendency: Typical Events

Measures of central tendency give us the average or typical observation. If we wish to portray the general nature of a set of numbers, to present a general idea of the main theme or tendency in the data, the place to start is with measures of central tendency. There are three measures of central tendency or location: the mode, the median, and the mean or average. Each reveals a different dimension of the overall data set.

The Most Common Case: The Mode

The **mode** is *the most frequently occurring score or observation.* What number or score (or category) occurs with the greatest frequency? What is the most common score or category for a variable or question of interest? As an example, a recent survey by the author asked how many miles (one-way, approximately) students commuted to campus. For $n = 20$ cases, randomly selected from a much larger probability sample, we have the following observations or X values expressed in miles:

Case No.: 1 2 3 4 5 6 7 8 9 10 11 12 13 14 15 16 17 18 19 20

Miles (X): 0 80 57 9 45 35 2 15 0 15 7 3 15 8 10 6 10 4 20 5

                         ↑   ↑   ↑

The distance of 15 miles occurs three times, which is more times than any other single value. Note that the mode is 15 miles, not 3. Three is the count, tally, or frequency (f) that identifies the mode. We would describe the modal distance as 15 miles.

Unlike the other two measures of central tendency (the median and the mean), it is possible to have more than one mode. Had another distance also occurred three times, we could have described this sample as **bimodal,** meaning *having two modes.* One mode would be 15 miles and the second mode would be whichever other distance also occurred three times. It is possible, of course, to have more than two modes. It depends on how our data are distributed, on what sorts of X values we get.

We sometimes talk of major and minor modes. The **major mode** is *the value or score with the highest frequency,* and the **minor mode** is *the value with the next highest frequency.* It may make sense to cite a major and minor mode, especially if these two values have frequencies obviously greater than those for remaining values or scores. In our data set, 15 miles is the major mode, but distances of 0 miles (live on or next door to campus) and 10 miles each occur twice. We might therefore describe 0 miles and 10 miles as minor modes. Then again, we have to use good judgment in citing minor modes. In appearing twice, $X = 0$ and $X = 10$ miles occur only once more than all the other scores. Are they really prominent modes or "spikes" in the data? We would probably be better off here to simply cite the mode as 15 miles and ignore the 0- and 10-mile distances. Nevertheless, when appropriate, it is permissible to record more than one mode.

We may also cite modes for categorical data. In fact, the mode is the *only* measure of central tendency we may use for nominal data. In Chapter 1,

Tables 1.1 and 1.2 display frequency distributions for nominal variables. For a sample of college students, Table 1.1 depicts marital status and 1.2 shows religious denomination. Table 1.1 reveals the major modal status to be single and "solo," or unattached ($n = 336$), and the minor modal status is single and "attached" ($n = 280$). These two categories clearly dominated the sample of 740 students. Similarly, in Table 1.2, two categories stand out. The major modal denomination is Catholic ($n = 232$), and the minor modal preference is "no religion" or no particular denomination ($n = 204$). Still another example from the same campus survey is shown in Table 2.1, summarizing students' political party preferences. The Democratic party is the *major* modal choice ($n = 368$), and "Other/None" constitutes a *minor* mode ($n = 202$). Again, the mode is not the number 368, but the category "Democratic party."

The mode is the score or category with the highest count, but there may be data distributions for which we cite more than one mode. Other measures of central tendency go beyond simple counting, however.

Finding the Middle Rank: The Median

The **median** is the *middle ranking score or value.* As nearly as possible, 50% of all cases fall above the median, and 50% fall below. For example, for a data set with an odd number of values, such as 3, 5, 7, 20, 30, 100, and 110, the median is simply the middle number, 20, which has three values below it and an equal number of values above it.

Taking the earlier example of 20 students' commuting distances to campus, to find the median, we rearrange them in ascending order:

Miles (X):   0 0 2 3 4 5 6 7 8 9 10 10 15 15 15 20 35 45 57 80
                                            ↑

Table 2.1 Political Party Preferences among a Sample of College Students

Political Party Preference	Frequency	Percentage	Valid Percentage
Democratic	368	49.6	55.8
Republican	90	12.1	13.6
Other/none	202	27.2	30.6
Total	660	88.9	100.0
Missing	82	11.1	
Total	742	100.0	

Note that we could have arranged them in descending order as well. For an even number of data points, the median is the midpoint between the *two* middle-ranking scores. To get that midpoint, we add those two adjacent numbers and divide by two. Here, the exact midpoint or middle rank falls between X values of 9 miles and 10 miles, respectively, so the median distance is 9.5 miles. There are exactly 10 cases above that middle rank and 10 cases below it.

Accordingly, we may think of the median as the 50th **percentile rank.** Recall from Chapter 1 that the percentile rank is defined as the percentage of cases falling below a certain score. Since 50% fall below the median, by definition, the median is always the 50th percentile rank.

Another feature of the median is important. As illustrated with the commuting distances and unlike the mode, the median may be a value that does not actually occur in a data set. This is permissible for *derived* statistics. All statistics are derived from the raw data and, as such, often are values that do not, or even *could not,* occur in the original data. As an example, we often hear claims that the typical US family has 1.84 children. Obviously, actual, countable children exist in whole numbers, not fractions, so such a household is impossible. When we are talking about statistics summarizing the raw data, however, such figures are quite permissible and logical.

Moreover, unless we have an obvious middle point in the data, finding a median may require some interpolation in locating about where the 50th percentile rank would fall. Consider Tables 1.5 and 1.6, which measured, respectively, students' views of TV news coverage and their reported frequency of lying to employers or professors. In Table 1.5, for instance, what is the middle-ranking score on our scale of TV news opinions? For $X = 10$, the percentile rank is 51.8; meaning 51.8% of all cases fall at or below that score. That is too high a value for the median. For $X = 9$, the percentile rank is 40.5, too low for the median. The median therefore lies somewhere between scores of 9 and 10, and perhaps you have surmised that it would be closer to 10 than to 9. Cumulative percentages allow us to estimate where that point would be. **Cumulative percentages** *are the running totals of the proportion of percentage cases in ranked data as one proceeds from the lowest to the highest observations or vice versa.* The interval from $X = 9$ to $X = 10$ includes an increment of 11.3% of the distribution ($51.8 - 40.5 = 11.3$). From $X = 9$ to the median is a 9.5% increase ($50.0 - 40.5 = 9.5$). To get to the median, then, we must go 9.5/11.3, or about eight-tenths ($9.5/11.3 = .8$) of the way through the entire interval between 9 and 10. Our estimated median on the news rating scale is thus about 9.8 ($9 + .8 = 9.8$).

We may do the same with our scale of lying (Table 1.6). The table shows the cumulative percentage (percentile rank) for $X = 6$ to be 59.8, too high for the median. But for $X = 5$, the cumulative percentage is only 31.4, well below the median. The median score lies between the *X* values of 5 and 6, but where exactly? How far into the 5 → 6 interval must we go? The entire interval includes 28.4% of all scores (59.8 – 31.4 = 28.4). From *X* = 5 to the median, we must go 18.6% (50.0 – 31.4 = 18.6). To get to the exact 50th percentile rank, then, we must go about seven-tenths of the interval (18.6/28.4 = .7). Given our distribution, we estimate the median score to be 5.7 (5 + .7 = 5.7). Interpolation and estimation work for determining the medians in such cases, but they may not be necessary. In Table 1.7, for example, 49.9% of college students' ages fell at or below 22 years. That is very, very close to an exact 50%. Without interpolation, we could simply assert that the median age is 22 years.

Another point, as may be obvious from the discussion thus far, is that to determine a median, we must have at least rankable scores or observations. This means we must have at least ordinal-level or quantitative data. With rankable data (the ordinal and interval-ratio levels), we are able to derive a median. With nominal data, we cannot do so.

To illustrate this point, consider the following: What is the median marital status in Table 1.1, the median religious denomination in Table 1.2, or the median political preference in Table 2.1? We cannot put these nominal categories into any natural order or hierarchy. We could list them in the tables in any order we choose, but we cannot order or rank them the way we can a set of test scores. What would be the middle-ranking marital status, religion, or party? What categories rank at the respective 50th percentiles? The 50th percentile obviously depends upon the order or hierarchy in which we list the separate categories, and such a concept is nonsensical for these data.

Finally, unlike the mode, there is only one median. Only one spot or location or value falls at that middle rank. There is only *one* center, no alternative centers and no major ones versus minor ones. The median complements the mode and tells us something a little different, that is, not the most common data point but the most central. The third measure of central tendency, the average, is the most useful of all, however.

Counting the Numbers: The Average

The **average,** or **mean**—the *arithmetic balance point in a set of quantitative data*—is the most familiar (and statistically most useful) measure of central tendency. The magnitude or weight of the scores *above* the mean should be offset by (i.e., balanced out by or equaled by) the magnitude or weight of the scores *below* the mean. It is not the *number* of scores above and below the mean that are important. That consideration is important only for the median. For the mean, it is rather the values of the scores that count.

When we say the weight of the scores above the mean is balanced out by that below, we are really talking about the deviations from the mean. The total by which scores deviate or differ from the mean on the positive side (above) equals the total by which other scores deviate from the mean on the negative side (below).

Because we use it so much, we need a symbol for the mean. In fact, there are two symbols. They represent, respectively, the **sample mean** and the **population mean.** Recalling the discussion about inferential statistics, we deal with both samples and populations. The time will come when we will need to distinguish between **sample averages** and **population averages.** To make this easier, we use different symbols for the two. The sample mean is indicated by the symbol $\overline{X}$, read as "X bar." The Greek symbol μ, or "mu," stands for the population mean.

An example would be calculating an average for the previous data set of the number of miles a sample of students reportedly drives to campus.

Miles (X): 0 80 57 9 45 35 2 15 0 15 7 3 15 8 10 6 10 4 20 5

For the mean, we merely add up all the X values and divide the total by n, the number of observations in question. Expressing this operation for a **sample average,** we write:

(1)
$$\overline{X} = \frac{\Sigma X}{n}$$

The symbol Σ is the uppercase Greek letter sigma. It may be translated as "the summation of." You will encounter this symbol frequently in statistical formulas. It means to add up whatever follows. In this case, it means to add up all the X values. In our example, we get an answer of 17.30 miles* using equation 1:

$$X = \frac{\Sigma X}{n} = \frac{346}{20} = 17.30$$

The mean of 17.30 miles becomes the arithmetic balance point for this sample. Let's look at this mean as it relates to the individual data points. We write these 20 data points in ascending order, as we did for the median:

Miles (X): 0 0 2 3 4 5 6 7 8 9 10 10 15 15 15 20 35 45 57 80
                                                                     ↑

---

*It is customary to go to at least two decimal places in statistical work. A common practice is to do all the calculations to the third decimal place and then round the final answer down to two places. To make things a little easier, we will work to two decimal places.

Now calculate the deviations from the mean, as shown in Table 2.2. Note that the number of *X* values falling below the mean of 17.30 is not the same as the number above. Fifteen observations are smaller than the mean, whereas only five are larger. That does not matter as long as the weights or magnitudes of the deviations above and below the mean, respectively, are equal.

Calculating a sample average is therefore a straightforward process. We add up all the *X* values and divide the total by the number of cases *n* we have. It is the same when we calculate a population mean, but that formula includes population symbols:

$$(2) \qquad\qquad \mu = \frac{\Sigma X}{N}$$

We use *N* here rather than *n*. Capital *N* is used to denote the population size, whereas *n* indicates the number of observations in a sample.

Besides defining the mean and showing how it is calculated, *when* is it appropriate to use this measure? A mean should be calculated only for interval-ratio

Table 2.2

Miles (X)	Deviations from X, or $(X - \overline{X})$	
0	−17.3	
0	−17.3	
2	−15.3	
3	−14.3	
4	−13.3	
5	−12.3	
6	−11.3	
7	−10.3	
8	−9.3	
9	−8.3	
10	−7.3	
10	−7.3	
15	−2.3	
15	−2.3	
15	−2.3	Sum of the negative deviations = −150.50
20	2.7	
35	17.7	
45	27.7	
57	39.7	
80	62.7	Sum of the positive deviations = +150.50

data. It is not appropriate for nominal or ordinal data. Recall from Chapter 1 that only with interval-ratio data may our actual *X* values be considered precise and reflective of true values of the variable. This fact is important. In calculating a mean we use each of these actual numbers or *X* values. We add up all the *X* scores. To justify this, we have to be sure each of these *X* values is indeed a truly mathematical measurement, not an arbitrarily scored observation. While some ordinal data may be numerical or quantitative (see, for example, Tables 1.5 and 1.6), only interval-ratio measurements allow full confidence in their mathematical precision. For ordinal measurements, the median is adequate because it does not take each actual *X* value into account. It merely considers the observations' ranks, and the middle ranking *X* value (or hypothetical *X* value) is the median.

We may easily illustrate this difference between the median and the mean with data used earlier:

Miles (*X*): 0 0 2 3 4 5 6 7 8 9 10 10 15 15 15 20 35 45 57 80
                                                    ↑

As you may recall, we found the median commuting distance to be 9.5 miles, or simply the middle rank in the data. It fell at the midpoint between 9 and 10. The mean for the same data, calculated later, was 17.30 miles. Now, however, assume one *X* value is different. Assume the person commuting an already daunting 80 miles now commutes 100 miles instead. What happens to the median? Nothing. It is still 9.5 miles. The original ranks of the *X* values are still intact. The middle of the data still falls midway between 9 and 10 miles. If the 9 or the 10 had changed, of course, the median would have been affected because the middle rank would have shifted, but that is the *only* way to influence the median. Changing any other value does not alter this statistic because the median is not sensitive to *every* observation, only the middle one(s). Not so with the mean. The mean would go from 17.30 to 18.30 miles, reflecting the new and higher *X* value. Again, the mean *is* based on every actual value or number, not just the relative ranks. For this reason, we should calculate a mean only when we have the utmost confidence in each data point's validity and accuracy, and this requires interval-ratio level measurements. If we are sure of only ordinal or rankable data at best, only the mode and the median are appropriate.

The previous example brings up another point. Some data distributions are **skewed;** that is, they *include extreme scores at one end of the continuum or range.* Even if these are interval-ratio data, the median may be the preferred measure of central tendency. The average would be pulled in the direction of the extreme *X* values and thus may give a distorted view of the typical score. The median, however, remains the middle rank and is not affected by these unusual and outlying cases. A hypothetical example may illustrate the point.

Suppose a labor contract is up for renewal. The contract covers a variety of jobs, and its salary range is from about $55,000 per year to $125,000. The modal salary is $62,000 and the median a little higher, at $68,000. The average salary, on the other hand, is $77,000, inflated by a comparatively small number of workers earning more than $100,000 per year. During negotiations, the union emphasizes that the typical worker makes just $62,000, and half make $68,000 or less, conditions the new contract must address. Management counters that the average salary is already $77,000 annually, allowing little room for substantial wage increases. Who is right, and who is wrong? Well, both sides are right, of course, but which is the more representative or "true" measure of central tendency in this case? As a compromise and despite the dollar figures representing interval-ratio measurements, perhaps the median is more representative.

The relationship between the levels of measurement and the measures of central tendency is summarized in Table 2.3. An asterisk (*) indicates that a particular statistic is appropriate for a given level of measurement.

Each of the three measures of central tendency gives us a different but typical observation for various distributions. Care must be exercised when using these statistics; they are not universally applicable to all levels of data. As we move on to other statistics, however, it will become obvious that the mean is the most widely used measure of central location. This is apparent in the next section, as we turn to measures of variation.

Diversity and Variation

Variation involves how much the X values in a given sample or population differ from each other. Are they all fairly similar or do they differ widely? We will consider two measures of variation here. The first is the range, a very

Table 2.3 Appropriate Measures of Central Tendency for Given Levels of Measurement

	Measure of Central Tendency		
Level of Measurement	Mode	Median	Mean
Nominal	*		
Ordinal	*	*	
Interval-ratio	*	*	*

simple and no doubt familiar statistic. The second is the standard deviation, more complex but much more useful statistically.

Highs and Lows: The Range

The **range** is *the difference between the largest and smallest values in a quantitative data set.* Our familiar data set consists of 20 *X* values, the lowest value being 0 miles and the highest 80 miles. The range is 80 – 0, or 80 miles. We say the range or span covered by our sample is a distance or gap of 80 miles. Another example illustrates this concept.

A different survey on the author's campus asked students how many hours per week they were employed during the school year. A random sample of 10 cases was abstracted from a much larger sample. In ascending order, those *X* values were:

Hours per Week Employed (*X*): 0   0   0   0   4   16   20   27   36   40

There was a difference of 40 hours between the two most extreme work schedules, so the range of employment hours was 0 to 40, or 40 hours.

Some people calculate the range a bit differently. In some cases, the range is calculated as the difference between the highest and lowest scores *plus 1*. The rationale is that the full or true range includes not only the distance *between* the extreme scores but, technically, also includes those extreme scores as well. It is not just the distance from the lowest score to the highest, for instance; it should *include* the highest score as well. This means we go all the way to the true upper limit of the range, that is, all the way up through the highest number.

Calculating the range this way slightly changes the results for our examples. In the earlier case, with values of 0 to 80, we now have a full range of 80 – 0 + 1, or 81 instead of 80 miles. With the data for hours employed, our complete range becomes 40 – 0 + 1, or 41 instead of 40 years. Which way is better or preferred? Technically, this second way is more accurate and proper. However, adding the 1 is not going to make much difference, and there is something to be said for the ease of understanding and simple logic of the first method: high score minus low score. Either method for deriving the range seems acceptable today. Some recently published texts present one way, some the other. Whichever way you pick is less important than your realizing there are two ways, because, as you will see, such options are not uncommon in statistics.

The value of the range is its simplicity: it is easy to understand. The trouble is, it is not a very useful statistic. In statistical parlance, it is not a very *powerful* statistic; it does not use much of the information available. It gives us no indication whatever about how the data are distributed between the two extremes. It uses only the two most extreme X values, whereas there may be dozens, hundreds, or even thousands in a total sample. We do not know whether the X values tend to peak or cluster in the middle of the distribution, whether they are lumped together or skewed toward the upper or lower end of the range. Certainly, the distribution of employment hours presented here is skewed; fully 4 of the 10 observations fall at the very lowest extreme (0 hours). With the range, we simply have no idea what the very broad middle of a distribution looks like. Obviously, we need a statistic that uses more of the data and also conveys more information. This is the standard deviation.

Away from the Average: Measuring Spread

Unlike the range, the **standard deviation** *uses all the X values to measure the amount of spread in a distribution.* When measuring the variation or spread in a set of numbers, the obvious question becomes: Variation or spread from what? From the lowest score? From zero? In fact, as the arithmetic center or balance point for a distribution, we measure variation from the mean. We are asking: For all our X values, what is the typical (or standard) deviation from the average? Although it sounds awkward, in a way we are asking: What is the average deviation from the average? If we have a wide range and a lot of variation in our X values, then we will have a large standard deviation from the mean. In contrast, if our X values do not differ much and tend to cluster close to the average, our standard deviation from that average will be small.

One cautionary note: If we are measuring deviations from the average, we must have data for which it is appropriate to calculate an average in the first place. We must therefore be reasonably confident of having interval-ratio data. We should not calculate standard deviations for ordinal level data sets.

For the mechanics, two formulas are available to derive a standard deviation. It is easier to understand the concept and meaning of the standard deviation if we look first at the deviation formula. When doing the actual calculations, however, it is easier to use the raw score formula. The deviation formula is presented first so that you might get a better feel for this statistic, and then we will look at the quicker formula.

Using Differences from the Average

To consider an example, we start with the now-familiar data set of $n = 20$ commuting distances to campus. Step 1 is to calculate the mean ($\overline{X}$), which we earlier found to be 17.30 miles. Step 2 requires finding the deviations from the mean. We subtract the mean $\overline{X}$ from each X value, which we did in Table 2.2 when we looked at the average as the balance point between positive and negative deviations.

A first glance at Figure 2.4 might suggest adding up the second column, the $(X - \bar{X})$ values, to get a measure of variation around the mean. As we have seen in our discussion of Table 2.2, however, the sum of this column is always zero (allowing for rounding error). The positive deviations above the mean always total the same as the negative deviations below the mean and cancel each other out. As a result, the $(X - \bar{X})$ column by itself is useless as an index of variation around the mean.

While not providing a final answer, the deviation scores $(X - \bar{X})$ do make a good starting point in showing how the data vary from the mean. We must do something further to them, however, to come up with a meaningful index. There are two possibilities. First, we could take the absolute value of the deviation scores. *The* **absolute value** *of a number simply uses its numerical value as a positive quantity regardless of the number's original sign.* If we add up the numerical values of the $X - \bar{X}$ column without regard to the + or − signs in Table 2.2, we would get a total of 301.00. Dividing this number by n (20 in this case) yields a little-used statistic known as the **mean absolute deviation,** or **MAD.** Here, the MAD = 15.05 miles, which can be interpreted to mean that the typical observation differs from the mean by an absolute value of 15.05 miles.

So far, so good. Then why is the mean absolute deviation not sufficient? Why do we need to look for another measure at all? Remember something you learned in elementary school—if you do something to one number, you must do the same thing to each number in the set. Otherwise, the original ratio or relationship between the numbers is altered. For example, if you square one number, you must square the rest. That principle still holds, and it is the reason absolute values do not work well in statistics. With absolute values, we are changing or ignoring the signs of some values (negative numbers) but not others (positive numbers). Mathematically and statistically, negative and positive are important, and we cannot just wave away or ignore the differences. Statistically, then, because the MAD's absolute values are awkward and unusable, we avoid them.

Our second strategy for coping with the $(X - \bar{X})$ values works better. Instead of taking the absolute values, we *square* each $(X - \bar{X})$ quantity, as shown in column (3) of Table 2.4. This also gets rid of the minus signs. So for step 3, we square each $(X - \bar{X})$ quantity and add them. We now have a number that reflects variation from the mean, obviously will not equal zero, and is not based on absolute values. The greater the variation from the mean, the larger this total will be. For our data, $\Sigma(X - \bar{X})^2 = 8,472.20$.

There are two more steps to get a workable standard deviation, but an important aside before that has to do with the term we just calculated: $\Sigma(X - \bar{X})^2$.

Table 2.4

(1) Miles (X)	(2) Deviations from $\overline{X}$ ($X - \overline{X}$)	(3) Deviations from $\overline{X}$, Squared ($X - \overline{X})^2$
0	−17.3	299.29
0	−17.3	299.29
2	−15.3	234.09
3	−14.3	204.49
4	−13.3	176.89
5	−12.3	151.29
6	−11.3	127.69
7	−10.3	106.09
8	− 9.3	86.49
9	−8.3	68.89
10	−7.3	53.29
10	−7.3	53.29
15	−2.3	5.29
15	−2.3	5.29
15	−2.3	5.29
20	2.7	7.29
35	17.7	313.29
45	27.7	767.29
57	39.7	1576.09
80	62.7	3931.29
$\Sigma X = 346$	$\Sigma(X - \overline{X}) = 00.0$	$\Sigma(X - \overline{X})^2 = 8,472.20$

This is *the sum of the squared deviations from the mean.* That is a mouthful, so it is usually called the **sum of squares,** or **SS.** The important point, however, is that this sum of squares is a key concept; it crops up a lot in statistics. This term is a basic measure of **variation,** and you will see it again in Chapter 8. At this point, do not necessarily struggle to memorize its formula, although you should understand the sum of squares and be able to explain it in your own words. It involves variation, a key concept in statistics.

Continuing on, for step 4, since we want the average, typical, or *standard* deviation from the mean, we divide $\Sigma(X - \overline{X})^2$ by n. This step is similar to how we took an average before. Doing the math for our present example, we get

$$\frac{\Sigma(X - \bar{X})^2}{n} = \frac{8472.20}{20} = 423.61$$

At this point, we have a statistic called the **variance,** which is *the square of the standard deviation.* The variance will be very useful later on, but for now, we will discuss the standard deviation itself. So as a final step, step 5, we take the square root of the variance to get the standard deviation. Also, the symbol for a sample standard deviation is *s,* so we write:

(3)
$$s = \sqrt{\frac{\Sigma(X - \bar{X})^2}{n}}$$

Taking the square root makes sense because we had squared our deviations from the mean in step 3. Since we squared our original units of analysis, and we wish to express the standard deviation in the same units with which we started, the final step is logically to take the square root. For our example,

$$s = \sqrt{\frac{\Sigma(X - \bar{X})^2}{n}} = \sqrt{\frac{8472.20}{20}} = \sqrt{423.61} = 20.58 \text{ miles}$$

Note that this does not say anything about a positive or negative deviation. It merely says that, on the average, our observations deviate from the mean by 20.58 miles. Some deviate above the mean and some below. Overall, the typical or standard deviation is 20.58 miles, either positive or negative. As a tip, however, the standard deviation is always a positive number when calculated. If our calculations had resulted in –20.58, or any other negative value, we would have known immediately that a mistake had been made. After all, we have squared all of our deviation scores, so all our numbers *should* be positive. Besides, we are talking about a deviation from the mean per se, not a positive or negative deviation, and conceptually this deviation could never be a negative quantity.

Up to this point, you should have a better feel for this tool called the standard deviation. We have covered the basics, but before we get to the quick method of calculating *s,* there is a small adjustment we must make to equation 3, the formula for *s.*

Recall the discussion of inferential statistics. We use sample data to estimate things about populations. Most of the time, our sample standard deviation, *s,* is expected to serve as an estimator for the **population standard deviation,** or σ (lowercase Greek letter sigma). We rarely have an interest in just a sample itself, especially a random sample. Typically, the sample has value only insofar as it helps us to understand the population.

Further, intuitively, there is probably more variation in the whole population than we will find in any single sample. Populations are usually many times larger than the samples we use and tend to be slightly more varied. Using equation (3), we would get an accurate standard deviation for the sample, but we would probably *underestimate* the true population standard deviation. In order to make s an unbiased estimator of σ, we should amend equation (3) slightly. Instead of dividing by n, we divide by $n - 1$. This is called a **correction factor,** which is common in statistics. In this case, using $n - 1$ makes the denominator slightly smaller, which makes the overall result or answer slightly larger. We then have an unbiased estimator for σ. To rewrite our formula for the sample standard deviation, we get:

(4)
$$s = \sqrt{\frac{\Sigma(X - \bar{X})^2}{n - 1}}$$

To slightly revise and correct our example:

$$s = \sqrt{\frac{\Sigma(X - \bar{X})^2}{n - 1}} = \sqrt{\frac{8472.20}{19}} = \sqrt{445.91} = 21.12 \text{ miles}$$

This is a better, more accurate estimator of the population standard deviation than the original result without the correction factor.

Likewise, we may rewrite the formula for the sample variance, s^2, an unbiased estimator for the population variance, or σ^2:

$$s^2 = \frac{\Sigma(X - \bar{X})^2}{n - 1}$$

To complete the set of formulas, we must also consider the procedure for deriving the population standard deviation itself. It will sometimes occur, however rarely, that we do have data for a whole population. When this is the case, there is no need for any correction factor. We do not have to *estimate* σ; we may calculate the true figure directly. Using the appropriate symbols, the formula for the population standard deviation becomes:

$$\sigma^2 = \frac{\Sigma(X - \mu)^2}{N}$$

Similarly, the formula for the **population variance** is written as:

$$\sigma^2 = \frac{\Sigma(X - \mu)^2}{N}$$

To summarize our first example, we would say that the average commuting distance to campus is 17.30 miles. However, there is clearly a lot of variation around this figure. In fact, the typical student's commute tends to vary by as much as ±21.12 miles from this average. This may appear a bit confusing. If we go in a negative direction, that is, −21.12 miles or 21.12 miles below the average, we end up in negative numbers or minus miles. The average minus one standard deviation, 17.30 − 21.12, puts us at −3.82 miles, an obvious impossibility. That does sometimes happen, however, with variation and standard deviations. It is due to the fact that we have a skewed or lopsided sample here. The standard deviation is inflated by the very high scores (long commute distances) of a handful of respondents, one commuting as much as 80 miles each way. Naturally, in any report, we would have to alert the reader to the skew or imbalance in our data to explain this apparent anomaly.

Because the standard deviation is so important, a second example may be worthwhile—one without the skew evident in the miles commuted. A recent survey in Silicon Valley asked about acceptable uses of recycled water. Of interest were data regarding respondents' years of formal education. A smaller set of $n = 15$ was randomly abstracted from the original sample of 1000+ cases, and those data appear in the left-hand column of Table 2.5.

By following the steps outlined previously, we first get the average: $\Sigma X/n$ or 227/15 = 15.133 years of education in column (1). In column (2), we get our deviation scores by subtracting the average, 15.133, from each original X value. Recall that the sum of the deviation scores should equal zero, allowing for rounding. Finally, we get the squared deviations from the mean in column (3). We need that sum, the sum of squares, or SS, to determine the standard deviation, the typical deviation from the average.

$$s = \sqrt{\frac{\Sigma(X - \bar{X})^2}{n - 1}} = \sqrt{\frac{33.736}{14}} = \sqrt{2.410} = 1.552 \text{ years}$$

or ±1.552 years of education as the typical deviation from the average. Respondents in the recycled water survey, at least in the 15 abstracted cases, typically varied by approximately ±1.5 years from the average of slightly more than 15 years of education. We next see how these same results are easier to obtain with the **raw score** formula.

Using the Original X Scores

Given the preceding introduction to the concept of the standard deviation, we now consider a somewhat easier formula. It may not look it at first, but the

Table 2.5

(1) Years of Education (X)	(2) Deviations from $\overline{X}$ $(X - \overline{X})$	(3) Deviations from $\overline{X}$, Squared $(X - \overline{X})^2$
12	−3.133	9.816
13	−2.133	4.550
14	−1.133	1.284
14	−1.133	1.284
14	−1.133	1.284
15	−.133	.018
15	−.133	.018
15	−.133	.018
16	.867	.752
16	.867	.752
16	.867	.752
16	.867	.752
16	.867	.752
17	1.867	3.486
18	2.867	8.220
$\Sigma X = 227$	$\Sigma(X - \overline{X}) = 0.000$	$\Sigma(X - \overline{X})^2 = 33.736$

calculations are easier and involve fewer steps. Fewer steps or calculations mean less chance for error, and that is all to the good. Indeed, electronic calculators use this alternative procedure to derive standard deviations. It is called the **raw score** formula because we use the raw or original scores themselves. We work only with the actual X values and do not have to calculate any deviations from the mean. In fact, we do not have to determine the mean at all for this procedure. Equation 8 shows the raw score method for determining the sample standard deviation.

(8)
$$s = \sqrt{\frac{\Sigma X^2 - \dfrac{(\Sigma X)^2}{n}}{n - 1}}$$

The difference is in the numerator. We calculate the sum of the squared deviations from the mean (the sum of squares, or SS) differently. The denominator,

$n - 1$, remains the same. Comparing equations 4 and 8 for s, we note the following equality for the sum of squares:

$$\Sigma(X - \bar{X})^2 = \Sigma X^2 - \frac{(\Sigma X)^2}{n}$$

wish to calculate the sum of squares. We may illustrate the comparative ease of the raw score method, however, with the data sets used previously. First, for the sample of 20 commuting distances (compare Table 2.4), we now have one basic step, and then we plug numbers into equation 8. We have the original X values in the left-hand column. We square each of those values and then sum each column: ΣX and ΣX^2, as shown in Table 2.6.

Table 2.6

(1) Miles (X)	(2) (X^2)
0	0
0	0
2	4
3	9
4	16
5	25
6	36
7	49
8	64
9	81
10	100
10	100
15	225
15	225
15	225
20	400
35	1225
45	2025
57	3249
80	6400
$\Sigma X = 346$	$\Sigma X^2 = 14,458$

Next, inserting the column totals into the formula, we have:

$$s = \sqrt{\frac{\sum X^2 - \frac{(\sum X)^2}{n}}{n-1}} = \sqrt{\frac{14{,}458 - \frac{(346)^2}{20}}{19}} = \sqrt{\frac{14{,}458 - 5985.80}{19}} = \sqrt{\frac{8472.20}{19}}$$

$$= \sqrt{445.91} = 21.12 \text{ miles}$$

We see that $s = 21.12$ miles is the same result as before using equation 4, but this time we did not have to bother with deviation scores and negative numbers. Again, we may conclude that the typical commuting distance would deviate from the average by ±21.12 miles.

We may also review the sample of $n = 15$ cases of years of education in the sample of Silicon Valley residents. Our standard deviation using the original deviation formula (equation 4) was 1.552 years of education. We also get this with the raw score formula, equation 8.

Table 2.7

(1) Years of Education (X)	(2) (X^2)
12	144
13	169
14	196
14	196
14	196
15	225
15	225
15	225
16	256
16	256
16	256
16	256
16	256
17	284
18	324
$\sum X = 227$	$\sum X^2 = 3{,}469$

Plugging the column totals into equation 8, we get:

$$s = \sqrt{\frac{\Sigma X^2 - \frac{(\Sigma X)^2}{n}}{n-1}} \;=\; \sqrt{\frac{3469 - \frac{(227)^2}{15}}{14}} \;=\; \sqrt{\frac{3469 - 3435.267}{14}} \;=\; \sqrt{\frac{33.733}{14}}$$

$$= \sqrt{2.410} = 1.552 \text{ years}$$

Our answers are the same using either equation 4 or equation 8. Equation 8 contains fewer steps, and, again, that means less chance for error. However, it is certainly a matter of personal choice. If you are more comfortable using the deviation formula, by all means stick with that one. In either case, if our two methods of calculating the sum of squares are equivalent, we may rewrite the other formulas in terms of the raw score method. For the sample variance:

(9)
$$s^2 = \frac{\Sigma X^2 - \frac{(\Sigma X)^2}{n}}{n-1}$$

Similarly, the population standard deviation and variance, respectively, become:

(10)
$$\sigma = \sqrt{\frac{\Sigma X^2 - \frac{(\Sigma X)^2}{N}}{N}}$$

(11)
$$\sigma^2 = \frac{\Sigma X^2 - \frac{(\Sigma X)^2}{N}}{N}$$

Descriptive statistics include measures of central tendency and variation. On the one hand, we describe a set of data in terms of the average or typical observation. On the other hand, we may summarize and describe the amount of spread, difference, or variation in a sample or population. One point to remember is that we cannot apply these statistics indiscriminately. We must use good judgment and consider the level of measurement or whether we have sample or population data.

Before moving on to probability, two final topics deserve mention. One has to do with comparing standard deviations, and the other involves possible measures of variation for ordinal and nominal data. First, how do we compare standard deviations from different distributions? One measure, the **coefficient of variation (CV),** considers each standard deviation in relation to its mean. In population and then in sample terms, we have:

$$CV = 100 \left(\frac{\sigma}{\mu}\right) \text{ or } 100 \left(\frac{s}{\bar{X}}\right)$$

For our earlier examples of commuting distances and years of education, we had, respectively, averages of 17.30 miles and 15.13 years, and we got standard deviations of 21.12 miles and 1.55 years. To calculate the CVs:

$$CV_{MILES} = 100 \left(\frac{21.12}{17.30}\right) = 100(1.221) = 122.1$$

$$CV_{YEARS} = 100 \left(\frac{1.55}{15.13}\right) = 100(.102) = 10.2$$

Relative to their means, students' miles driven to campus show much more variation than do years of education among the urban public. The CV is a handy little statistic that allows us to compare variations in different units of analysis and distributions.

Second, although we encounter them very infrequently compared to the standard deviation and variance, there are measures of variation for ordinal and nominal measurements. With numerical ordinal measurements and the median, one sometimes sees reference to the **interquartile range,** which is simply *the difference between the 25th and 75th percentile ranks, that is, the range for the middle 50% of all cases.* It gives us an idea of how clustered or spread out ordinal measurements might be. One may also see this reported as the **semi-interquartile range,** or half the interquartile range. Alternatively, there exists the **index of dispersion** (otherwise known as the **index of qualitative variation**) for nominal and categorized ordinal data. Without going into detail, this statistic examines the actual number of paired differences in the data (comparing each respondent's answer to that of every other respondent) versus the total possible number of such differences, which is a function of sample size and the number of categories. How many varying responses are there compared to how many there could be? Although not covered here, references to the index of dispersion may be found in some statistics texts and online.

In the pages ahead, however, we will build upon two statistics in particular: the mean and the standard deviation. We turn now to probability, the foundation of inferential statistics, and particularly to the statistical model upon which many of these procedures are based. This is the normal or bell-shaped curve.

Exercises

For the data sets in exercises 1–10:

a. Find the modes, medians, means or averages, and standard deviations.

b. In brief paragraphs, interpret what your results mean for each case. (*Hint:* Actually interpret or explain your results. Do not merely repeat what the numbers were or what mathematical steps you went through to get them.)

1. Before designing an ad campaign aimed at college students, a chiropractor randomly sampled 10 college students and asked whether she could weigh their backpacks and book bags. Rounded to the nearest whole pound, her data are recorded below:

Weight in pounds (*X*):   14   20   13   8   16   15   17   13   9   12

2. As they were about to graduate, 10 randomly sampled college seniors were asked how many different jobs or positions they had held while in college. Their answers were as follows:

Jobs held (*X*):   3   5   4   0   10   1   4   8   2   0

3. College students often rent rooms near campus, but the rooms can be expensive. A random sample of newspaper ads revealed the following monthly rates for nearby rooms:

Monthly rate in US dollars (*X*):   485 535 495 550 510 625 575 500 525

4. A sample of high school teachers reported their respective years in the teaching profession:

Years teaching (*X*): 14   9   22   17   7   28   10   16   5   2   15

5. A children's baseball team played a 14-game season. The number of runs scored in each game, respectively, is shown below:

Runs scored (*X*):   6   2   0   17   4   5   3   8   1   13   4   7   2   4

6. A sample of US adults was asked how many days per week each read a daily newspaper (defined as reading not just the comics and at least one article besides sports).

Days read paper (*X*):   5   3   0   7   4   0   7   6   1   0   1   3   2   3   2

7. Doing a random survey in a college bookstore, a student found the following numbers of books required in selected courses:

Books required (X):   3   4   8   2   3   5   4   5   1   4   2   1   1   4

8. A sociologist asked a sample of US adults to estimate the percentage of their friends who belonged to ethnicities and races other than their own. He got the following answers:

Percentage of different race (X): 20 75 33 0 33 67 40 10 0 25 33 50

9. A sample of full-time college students estimated the number of hours per week they spent doing academic research online during the school year.

Hours per week online (X): 10 20 0 8 20 40 30 0 20 0 9 15 20

10. "More than 30 minutes, and it's free!" Dorm-bound student researchers sampled eight pizza places and, under similar conditions, recorded each one's delivery time.

Minutes until delivery (X): 18   25   23   17   29   36   25   42

11. How does the level or scale of measurement affect the kinds of statistical analyses possible with any given data set. (Be specific here, with examples.)

12. What is variation, dispersion, or deviation? Please explain this statistical concept in your own words.

13. What is the coefficient of determination? How does it allow us to compare different data sets?

14. If the mode for a particular data set was 12, the median 16, and the average 20, what could you conclude about the distribution?

15. Why would a claim that s = −1.33 be a mistake? (Explain in detail, not just superficially in one sentence.)

16. Even though s and σ are never negative, could μ or $\bar{X}$ ever be negative numbers? Why or why not?

17. In your own words, please define:

    a. Mode   c. Average   e. Range
    b. Median   d. Standard deviation   f. Sum of squares

18. Are students' grade point averages (GPAs) based upon original letter grades statistically justified? Why or why not?

CHAPTER 3

Probability: A Foundation for Statistical Decisions

In this chapter, you will learn how to:

- Recognize different probability situations
- Convert *X* values into z scores
- Use *z* scores to determine normal curve areas and probabilities
- Convert percentile ranks into *X* values
- Calculate discrete binomial probabilities

INFERENTIAL STATISTICS ALLOW US TO MAKE estimates and conclusions about populations based upon random samples by extrapolating our sample results and findings to the population. We must always state what level of confidence we have in our extrapolations, and we couch these statements in terms of probability. What is the probability we are correct? Conversely, what is the probability of error and that our estimates or conclusions are wrong? Therefore, our eventual estimates and conclusions are *always* accompanied by probability statements.

We rely upon statistical models or guidelines to make these probability statements. The most important and well-known of these is the normal or bell-shaped curve. We will use this model extensively in our discussion here. Normal distributions, or at least approximations of them, are found among many variables in the empirical (or real) world (e.g., height and weight, attitudinal or opinion scores, IQ scores, task completion times, waiting times, or how closely the *actual* sizes and weights of items fit manufacturers' specifications). The normal distribution, which describes what data patterns will emerge over large numbers of samples, is part of statistical theory. Accordingly, statisticians use the normal curve model to guide their analyses in numerous probability situations.

Sorting Out Probabilities: Some Practical Distinctions

Before taking a detailed look at the normal curve, a more general view of probability theory would be helpful. Besides probability being the basis of inferential statistical analyses, the study of uncertainty and the probability of given **outcomes** or **events** occurring constitutes a principal part of statistics. In the following discussion, we first consider probabilities for continuous random variables. We use the mean, standard deviation, measures called z scores, and areas or proportions of the normal curve to determine probabilities. This requires a normally distributed random variable; that is, the variable X may assume any one of a broad range of different values (it is a *random* variable) and the distribution of actual X scores is reasonably normal. The scores or cases in a normal distribution tend to peak around the average in the middle of the range and to progressively taper off toward the extremes of that range.

But what if we do not have a continuous and random variable? What do we do if the X variable is of the proportional or **binomial** kind—that is, *when each* X *score is either a 1 or a 0, Yes or No, Agree or Disagree, or a Success or Failure type of measurement?* For questions like these, we need other concepts and formulas. What if we wish to know the probability of events such as voters passing a bond issue? Running into snow while driving over the Rockies in December? Your flight being on time? Or a family having two girl babies in a row? To see how these questions fit into the overall probability picture, we may divide the field into segments by considering first how many outcomes, or results, are possible in a probability situation: only two or more than two? Second, are the possible outcomes discrete or continuous? That is, do they only exist in whole numbers, or may they occur in fractions?

Two Outcomes or More: Binomial vs. Random Outcomes

First, the **binomial probability experiment** (or situation) has *only two possible outcomes*. The simplest example is a coin flip. The outcome is heads or tails, one or the other. Or, we may ask a sample of voters whether they intend to vote for Candidate A. From the original choices of Yes, Not Sure, and No, the responses may be dichotomized into Yes and Other, which creates a binomial situation. Or, we may wonder about the probability of getting X number of successes out of so many attempt or trials. In fact, the two outcomes in any binomial situation or experiment are commonly referred to as **success** and **failure**. The decision as to which outcome constitutes a "success" and which a "failure" is sometimes arbitrary (e.g., heads or tails?), but questions or problems are typically formulated so that we are looking for the probability of a certain number of successes (X) out of a larger number of trials (n), given that each trial may result in either a success or a failure.

In contrast to binomials, **random variables** *may assume a variety of different values or outcomes, not just two.* Age and height and weight are examples here. Any survey would almost undoubtedly reveal a wide variety of adult ages, heights, and weights. With random variables, our X measurements may assume numerous different outcomes or values.

Another Consideration: Discrete vs. Continuous Outcomes

We also consider whether the X is discrete or continuous in nature. A **discrete variable** *may be measured only in whole numbers or units.* Examples are the number of television sets in your home or the number of children in your family. These variables and measurements are discrete. Each X outcome must be a whole number. You may have 0, 1, 2, 3, or however many TV sets in your home, but you can never have, say, .77 or 1.29 sets. You may report 0, 1, 2, 3, 4, or however many children, but cannot have 1.45 or 3.16 children. These are discrete variables—variables that may exist and be measured only in whole numbers of things or outcomes.

Binomial events or outcomes are always discrete, by definition. We always consider the probability of a certain number or proportion of successes. The outcomes for any trial or person are *always* discrete: success or failure, approve or disapprove, and so on. The outcomes do not exist in fractions of successes; they are only whole numbers. There are situations, however, for which we assume this is not entirely true, and we will use the continuous normal curve model to determine binomial probabilities.

Then there are variables that are legitimately and always continuous. **Continuous variables** *may exist and be measured in fractions or parts of wholes.* For instance, age, height, and weight, besides being random variables, are also continuous variables. Certainly, ages exist in fractions of years. One does not have to be exactly 20 or 21. One may be 20.52178 years old, or whatever. There are fine increments of the variable (and its measurement) existing between the whole numbers. There is a *continuous* scale or very small increments of the variable. The same is true of height and weight. Height does not exist only in whole inches nor weight in whole pounds. Each exists in a continuous scale of tiny increments, in millionths of inches or grams if we wish, limited only by our technological ability to measure them.

This gives us two factors to take into account when we approach any probability situation. First, we consider the number of different outcomes or X values possible: two or more than two? Do we have binomial outcomes or a random variable? Second, may the X variable be measured only in whole outcomes or, on the other hand, in tiny increments or fractions?

This chapter looks first at continuous random variables and then turns to probabilities involving continuous binomial situations. The chapter concludes with discrete binomial probabilities, probabilities for which there is no curve and where we must solve for the probabilities of each single (discrete) value of *X*.

Continuous Random Outcomes and the Normal Curve

The Normal Curve and Probability

The **normal distribution** (or normal curve) is a *theoretical distribution for continuous random measurements in which the observations tend to peak in the middle of the range and taper off toward both extremes* (see Figure 3.1). In terms of the vertical axis, the height of the curve represents the frequency or the **relative frequency** with which various *X* values occur. The higher the arc of the curve, the greater the frequency of scores at that point along the *X* axis. Under the arc, the **area** of the curve shows the relative frequency or proportion of cases falling into certain parts of the distribution. This area or relative frequency also translates into the probability of any *X* measurement falling into that part of the distribution. The horizontal or **X axis** shows the values for the *X* variable, and that variable will change, of course, from one problem and distribution to the next. The mean and standard deviations are marked off along the *X* axis. Finally, the normal distribution looks like a cross section of a bell, and hence its informal name: the bell-shaped curve.

As illustrated in Figure 3.1, the model itself has certain important characteristics:

- It is symmetrical around a central peak.
- Along the lower or horizontal axis (the *X* axis), the mean, median, and mode all fall at this central peak.
- A span of six standard deviations occupies almost the full range between the lower and upper tails of the curve: three standard deviations above the mean and three below the mean. The curve is *asymptotic,* meaning its tails never actually touch or terminate at the horizontal axis.
- The total area of the curve is said to equal 1.0000. Partial areas or segments of the curve may be found and expressed as proportions (e.g., .5000 would equal half the area of the curve).
- Areas of the curve are synonymous with probabilities. If part of the *X* axis falls under a curve area of .2166, for example, there is a 21.66% chance of those *X* values occurring. Similarly, for the part of the *X* axis (the *X* values) falling under the top 5% of the curve, there is a 5% probability of those *X* scores happening.
- Proceeding outward from the mean, certain set areas of the curve correspond to standard deviations along the horizontal axis. In round numbers:

Figure 3.1 The Normal Distribution

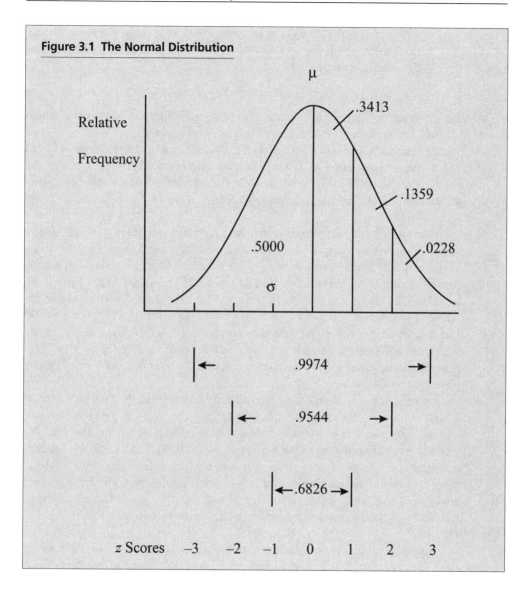

- 34% of the curve always lies between the mean and one standard deviation away along the *X* axis; 68% of the curve lies between ±1 standard deviation from μ;
- 48% of the curve always lies between the mean and two standard deviations away along the *X* axis; 95% of the curve lies between ±2 standard deviations from μ;

• 49% of the curve always lies between the mean and three standard deviations away along the *X* axis; 99% of the curve lies between ±3 standard deviations from μ.

Keys to the Curve: *z* Scores

We can imagine the normal curve as also being a distribution of *z* **scores,** which identify any spot on the horizontal axis (any *X* score) in terms of the *number of standard deviations away from the mean*. *z* scores always have a mean of 0.00 and a standard deviation of 1.00. If you are one standard deviation above the mean, your *z* score is 1.00. Being three standard deviations below the mean gives a *z* score of −3.00. Falling right on the average or mean, for instance, gives a *z* score of 0.00.

Our procedures will require translating *X* values into their corresponding *z* scores. Suppose we sampled full-time college students and found they spend an average of $40 around campus per week for meals and incidentals, with a standard deviation of $12. Then $\overline{X} = 40$, and $s = 12$. If you spend $52, your *z* score is 1.00 ($52 − 40 = +$12). You fall exactly 1.00 standard deviation *above* the mean. If you spend $28, your *z* score is −1.00 ($28 − 40 = −$12), exactly 1.00 standard deviation *below* the mean. In contrast, if *X* = $24, the *z* score equals −1.33, one and one-third standard deviations below the average ($24 − 40 = $16). If we know the mean and standard deviation of a distribution, any *X* value can be translated into a *z* score.

z scores have certain advantages over raw scores. First, they tell us where an *X* value stands in relation to the mean. Two features of a *z* score do this: the size and the sign. The size or magnitude of a *z* score tells us how many standard deviations away from the mean the *X* score lies (e.g., 1.00, 2.00, 1.28, .43 standard deviations). The sign (+ or −), then, tells us on which side of the mean *X* falls. *z* scores below the mean are always negative, and those falling above the mean are always positive, although we normally do not write the + signs. A *z* score thus tells us more than a raw score does by itself; it precisely locates *X* in relation to the mean.

Second, *z* scores are also **standardized scores,** which means *they make scores from different distributions comparable*. Percentages do this, too, but they do not convey as much information as do *z* scores. Assuming they represent interval-ratio measurements, consider an example with multiple choice test scores. Say you score 76 on an English exam and 28 points on a statistics test. Your *X* scores are 76 and 28, respectively. But how do your two scores compare? The *X* values do not tell us much by themselves because we do not know how many points each test was worth. If we express your scores as percentages, however, they become comparable. Assume your English exam netted you 76/100 points. You got 76%. Your statistics performance was 28/35 points, or 80%. Relatively speaking, you did slightly better on the statistics test. But how do you compare to others taking the exams? From just your percentages, we have no way of knowing.

Converting your scores to *z* scores allows them to be compared with each other and also with the average scores on each test. If the mean on the English exam is 70, with a standard deviation of 8 points, your 76 gives you a *z* score of .75, three-fourths of a standard deviation above the mean (+6 points above the mean divided by the standard deviation of 8 yields 6/8 = .75). Suppose the mean on the statistics test is 21, with a standard deviation of 5 points. Your *z* score in statistics is then 1.40. You are 1.40 standard deviations above the mean (+7 points above the mean divided by the standard deviation of 5 yields 7/5 = 1.40). We therefore see that your percentage scores alone do not tell nearly the whole story. They are very similar: 76% versus 80%. Your *z* scores, however, reveal that your *relative* performance is considerably better in statistics. You are higher above the average in statistics and have a higher percentile rank: more area of the curve falls below your statistics score than below your English score.

z scores therefore convey more information than raw scores themselves, and they also standardize scores from different distributions. We look next at how *z* scores are calculated and then at how they are used with the normal curve.

If we know a distribution's mean and the standard deviation, any *X* value may be translated into a *z* score. The formula for determining a *z* score is evident if one thinks about what it is. The *z* score expresses how many standard deviations above or below the mean a particular *X* value happens to lie. The first step in translating *X* into a *z* score is to see how many points that *X* value is from the mean: $X - \mu$, or $X - \overline{X}$ in sample terms. We then divide this difference by the standard deviation. This tells us how much the difference equals in standard deviations. Writing this as a formula and using population symbols, we have:

(13)
$$z = \frac{X - \mu}{\sigma}$$

or in sample terms:

(14)
$$z = \frac{X - \overline{X}}{s}$$

To reconsider previous examples, if the mean score on the English exam is 70, the standard deviation is 8, and *X* = 76, then:

$$z_{76} = \frac{76 - 70}{8} = \frac{6}{8} = .75$$

On the statistics exam: $\overline{X} = 21$, *s* = 5, and *X* = 28. Then:

$$z_{28} = \frac{28 - 21}{5} = \frac{7}{5} = 1.40$$

When X falls below the mean, the calculations will automatically result in a negative z score. Say your neighbor scored 19 on the statistics test ($X = 19$), two points below the mean:

$$z_{19} = \frac{19 - 21}{5} = \frac{-2}{5} = -.40$$

Using z Scores and the Normal Curve Table

How do z scores relate to the normal distribution? Basically, z scores are the key to using the normal curve. The only way we can make sense of, interpret, or use the normal curve is through z scores. The normal curve model comes with a table showing how much of the area falls above and below any given z score. We will use areas of the curve, recalling that areas of the curve correspond to probability. By looking up the area (or proportion) of the curve in the table, we will be able to tell the probability of getting a score above or below any given point on the X axis, that is, above or below any given X value.

We start with a series of X values that may be assumed to be reasonably normally distributed. We calculate a mean and a standard deviation and thus may convert any X score into a z score. By consulting the normal curve table, we finally determine the probability (area) of a score falling above or below this X value. We may also use the same general procedure in the other direction. Say we wish to know what X score falls at the 90th percentile rank. In using the table, we first look up the z score separating the lower 90% of the curve's area from the top 10%. This point is the 90th percentile rank. We then translate that z score into an X value or raw score.

Examples will help you to see how the normal curve table works. First, however, consider its general features. Table 3.1 excerpts part of Table A. Table A shows columns labeled A, B, and C. Column A lists z scores to two decimal places, every possible z score from 0.00 to 3.25 and a few beyond that. Column B shows the area of the curve lying between a z score (where that z score falls on the X axis) and the mean. That segment of the curve is given as a proportion of the curve's total area and is shown to four decimal places. Column C lists the proportion of the curve falling beyond or farther out from any z score, that is, the area that lies from z outward toward the tail of the curve. (Figure 3.2 illustrates these features of the curve and the table for a z score of 1.02.) In Table 3.1, notice that, for any given z score, columns B and C must always add up to .5000. Together, the area from z to the mean and the area in the opposite direction, from z toward the tail, must total exactly half the curve. It follows that, for the low z scores (those

Table 3.1 Excerpts from Table A: The Normal Distribution

z Score A	Area from z to the Mean B	Area from z Outward to the Tail C
.15	.0596	.4404
.16	.0636	.4364
.17	.0675	.4325
1.00	.3413	.1587
1.01	.3438	.1562
1.02	.3461	.1539
2.31	.4896	.0104
2.32	.4898	.0102
2.33	.4901	.0099

Figure 3.2 Tabled Areas of the Normal Curve for *z* = 1.02

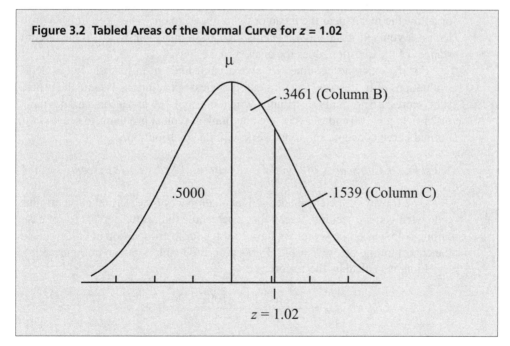

closer to 0.00), the areas in column C are larger than those in column B. These *z* scores are close to the mean, so there is much more of the curve going outward from *z* (to the tail) than between *z* and the mean. As the *z* scores get larger and move away from the mean, this ratio of columns B and C reverses itself. As we move away from the mean, the area in column B gets larger while that in column C gets smaller. Always, though, the two areas total .5000. These features of the curve are true of both the left (negative) and right (positive) halves.

About that left half of the curve—note that there are no negative z scores in the table. As we have seen, however, negative z scores are possible. They do not appear in the table because that would be redundant. We use z scores to read areas of the curve in columns B and C. The left half of the curve (negative z scores) is the mirror image of the right half (positive z scores), so the respective areas of the curve are identical for both a positive z score and its negative counterpart. We will now solve problems involving both positive and negative z scores so you can get a feel for this feature of the table. Indeed, a few examples may help you become familiar and comfortable with z scores, the table, areas of the curve, and how all these things apply to particular problems.

As a tip before starting, however, try to think of normal curve problems intuitively rather than just mechanically and in terms of formulas. Keep your wits about you as you read a problem. Visualize the curve, think of what information you are given, and think of where that X value or area of the curve falls—above or below the mean, near the mean or the tail, and so on. When you finish a problem, ask yourself if the answer seems logical. In other words, given the circumstances of the question, does the answer make sense?

For the example, assume we have a valid interval-ratio scale for students' evaluations of professors' teaching effectiveness. In addition, assume the professors' scores are normally distributed with a mean of 100 and a standard deviation of 25 (i.e., $\mu = 100$ and $\sigma = 25$). We may look at typical problems to see how the normal curve concepts and tools work with this distribution.

1. *Professor Ego scores 142 on the evaluations. What is Ego's percentile rank?*

First (always), sketch, shade, and label a model of the normal curve to illustrate what you are seeking. Recall that a percentile rank is the percentage of cases falling below a certain score. We have to solve for the proportion of cases (area of the curve) falling below $X = 142$. Ego's percentile rank is shown in Figure 3.3.

First, we calculate the z score:

$$z_{142} = \frac{X - \mu}{\sigma} = \frac{142 - 100}{25} = \frac{42}{25} = 1.68$$

From Table A, we see that the area from z to the mean (column B) = .4535. Our illustration may now be amended as shown in Figure 3.4: For the final answer, .4535 + .5000 (left half of curve) = .9535. Ego's position is therefore the 95.35th percentile rank.

Figure 3.3

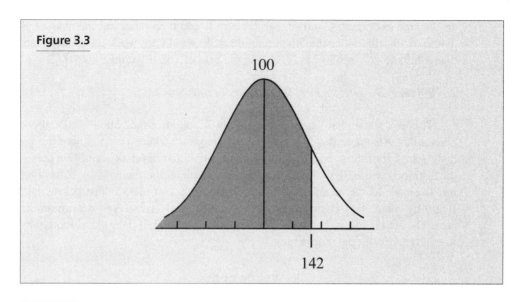

Figure 3.4

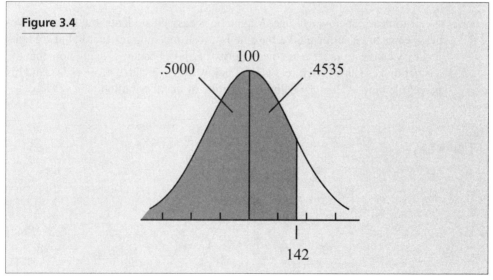

Not bad at all. If we can look past Ego's possible air of self-importance, her classes are no doubt worthwhile. Ego is rated in the top 5%. Note also that this answer seems correct and logical. Ego's score of 142 is well above the average of 100. She *should* have a high percentile rank. If we made the mistake of simply leaving the answer at .4535, about the 45th percentile rank and below the average, it would clearly be too low for her. Similarly, making the mistake of looking

up the area in column C rather than column B and then reporting the answer as
.0465, or slightly under the 5th percentile rank, would not work. Logic suggests
that would be way too low for Ego's high score of 142 if the mean is 100.

2. Professor Harmless scores 96. Where does he rank?

This is a somewhat vague question, but if you are asked about ranks, think
percentiles. Also note that $X = 96$ is below the mean. Make a mental note immediately that Harmless' z score should be negative and his rank should be below
the 50th percentile. Recall that the 50th percentile is the median; by definition,
it is the same as the mean in a normal distribution. Harmless is just below that.
Begin by calculating Harmless's z score, and then use Table A to determine the
area. The area below z (column C) = .4364 (see Figure 3.5). Harmless' score puts
him at the 43.64th percentile rank.

$$z_{96} = \frac{X - \mu}{\sigma} = \frac{96 - 100}{25} = \frac{-4}{25} = -.16$$

Harmless is below average. Maybe he is easy on students, but his effectiveness seems to be suffering. Perhaps he is a well-meaning instructor, but students
feel they can learn more from the majority of his colleagues. Again, note that the
answer makes sense. Harmless is a bit below average with a score of 96, and his
percentile rank is also just below the middle of the distribution.

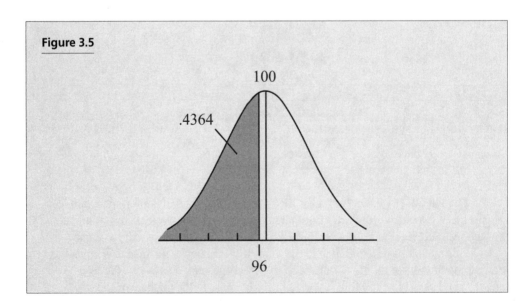

Figure 3.5

3. *In a particular department, Professor Wheeze scores the low of 80, while Professor Mumble rates the highest at 122. What proportion of the campus' professors score between these two extremes?*

Here we have two X scores: 80 and 122. One falls above the mean and one below, and the question calls for the proportion (area) between the two scores. We must get the z score for each X value, find the area between z and the mean for each (column B entries), and add up the two areas. This will give us the total area of the curve falling between the X values of 80 and 122 (see Figure 3.6). First, find the z scores for each:

$$z_{80} = \frac{X - \mu}{\sigma} = \frac{80 - 100}{25} = \frac{-20}{25} = -.80$$

$$z_{122} = \frac{122 - 100}{25} = \frac{22}{25} = .88$$

Then from Table A, find the corresponding areas. The area from the mean to z (column B) = .2881, and the area from the mean to z (column B) = .3106. The shaded area of the curve = .2881 + .3106 = .5987.

Therefore, 59.87% of the scores for all professors fall between these two extremes. Wheeze is poor, but Mumble, when you can tell what he is talking about, is somewhat above average. The department appears on par for this campus.

Figure 3.6

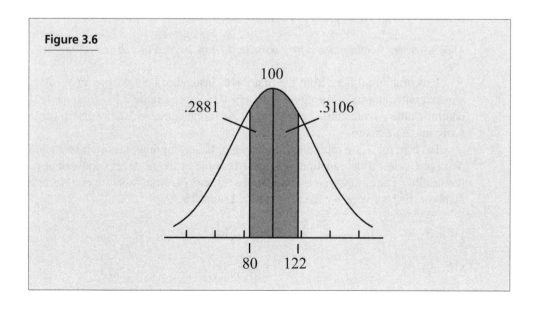

About 60% of all faculty fall between these two professors, close to the middle 60%, in fact. Not poor, but not great either.

This problem illustrates an important point. Although we may add or subtract areas of the curve, we can never add or subtract *z* scores. This is because what a *z* score is worth in terms of curve area depends upon where it is placed. One *z* unit (a *z* score of 1.00) near the mean corresponds to about 34% of the curve between it and the average in the center. The second *z* unit from the mean (between $z = 1.00$ and $z = 2.00$) is a much smaller area, about 13.5% of the curve. For the third *z* unit, that is, from $z = 2.00$ on out into the tail, the area is less than 2.5% of the curve (see Figure 3.1). These same percentages pertain to negative *z* values. As we go away from the central average, the areas of the curve taper off and get smaller as they progress over these *z* units. Hence, the relationship between *z* units along the horizontal axis and areas of the curve is not linear; it is not 1:1. In effect, adding or subtracting *z* scores would be like adding or subtracting the proverbial apples and oranges. We cannot do it. In the curve, *z* scores have meaning only insofar as they indicate areas of the curve. Since *z* units indicate different-sized areas, we cannot add or subtract them.

Areas of the curve are different, however. All areas of the curve are part of the same and unchanging 1.0000 unit area, the same 100%. We are dealing with similar units—proportions or percentages of the curve—no matter where those areas lie: near the tail, close to the mean, or wherever. We may add and subtract those areas of the curve as need be.

More example problems will help you to remember: areas of the curve may be added or subtracted; *z* scores may not. Back to the professors' ratings, with a mean of 100, and a standard deviation of 25.

4. *What score would place a professor in the top 10% of this distribution?*

This problem differs from the first three. Instead of being given an *X* value, we start with an area of the curve: "the top 10%" (see Figure 3.7). We must first determine the *z* score cutting off the top 10% of the curve and then convert this *z* score into an *X* score.

To solve for *X*, we must algebraically juggle our previous *z* formula to isolate *X* on one side of the equation. The new formula is the algebraic equivalent of the familiar *z* formulas given in equations 13 and 14. In population and sample symbols, respectively, we have equations 15 and 16:

(15) $$X = \mu + z\,(\sigma)$$

(16) $$X = \overline{X} + z\,(s)$$

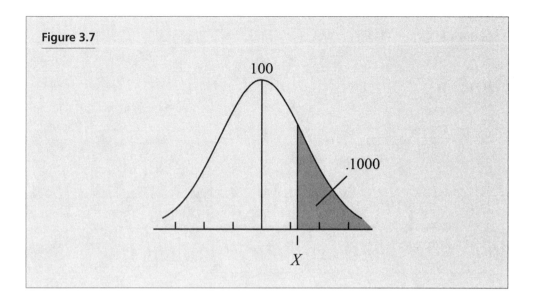

Figure 3.7

To solve for *X*, we first consult Table A to find the *z* score that cuts off the top 10%. Column C lists areas in the tail of the curve, so we need the *z* score corresponding to 10%, or .1000, in that column. We locate an entry of .1003 (close enough) in column C. Reading from column A, our *z* score at that point is 1.28. This tells is we must go 1.28 standard deviations above the mean to reach any normal distribution's top 10%. Then, using our given data,

$$X = \mu + z\,(\sigma) = 100 + (1.28)\,(25) = 100 + 32.00 = 132.00$$

Any professor scoring 132.00 or higher falls in the top 10%.

5. *What score places a professor into the most extreme 5% of all instructors?*

An interesting question: What is meant by "most extreme" 5%? It may be tempting to say it is the top 5%, but what about the lower tail? The lower tail is an extreme area also. Actually, we want the most extreme 5% *wherever* it falls in the curve. This means of the lowest 2.5% plus the uppermost 2.5%. We want the *X* values cutting off these two areas of the curve (see Figure 3.8).

We are given the areas of the curve with which to start. We must get the *z* scores first and then convert them into *X* values. Except for one being positive and the other negative, our two *z* values are the same. Each marks off 2.50% or .0250 in a tail of the curve.

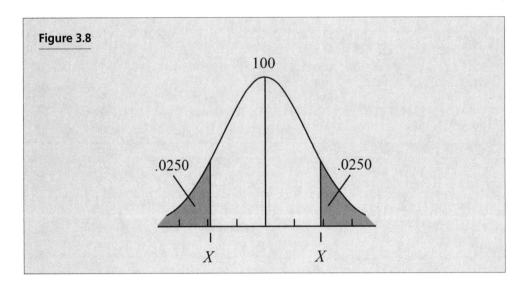

Figure 3.8

Column C of Table A shows an entry for .0250, corresponding to a z score (column A) of 1.96. The X values we seek lie 1.96 standard deviations from the mean. For the X value in the upper tail:

$$X = \mu + z\,(\sigma) = 100 + (1.96)\,(25) = 100 + 49.00 = 149.00$$

For the X value in the lower tail (remember, z is negative here):

$$X = \mu + z\,(\sigma) = 100 + (-1.96)\,(25) = 100 - 49.00 = 51.00$$

Any instructor scoring 149.00 or above or any scoring 51.00 or below falls into the most extreme 5%. Put another way, such professors fall *outside* the middle 95% of all professors. They are either terribly good or just plain terrible, one or the other.

We have now looked at two ways to use the normal curve model. First, given an X value, we may solve for the area of the curve falling above or below that point on the horizontal axis and interpret our answer as a probability figure. Second, given an area of the curve (or percentile rank) with which to start, we may solve for an X value. Summarized and diagrammed, the two procedures are shown in Table 3.2.

Notice that no matter which procedure we are using, the first step is always to get a z score—we must take what information is given and derive a z score. This fact illustrates the importance of z scores for the normal curve. We simply cannot use the curve without them.

Table 3.2

Given	Method	Answer
An X value	Convert the X value into a z score (equation 13 or 14)	Look up the area of the curve by using the z score and z table (Table A)
An area of the curve	Look up z score in the table	Translate the z score into an X value (equation 15 or 16)

Note that if we wanted the z score that cut off *exactly* the top 10% of the curve, there is no entry in Table A (column C) for *exactly* .1000. When we worked this problem, we claimed .1003 was close enough. That was, for our purposes. If we wish to be more exact, however, we must interpolate. For example, for z = 1.28, Table A (column C) shows the area to be .1003, and for z = 1.29, the area is .0985. The *exact z* score for .1000 lies somewhere between the z scores of 1.28 and 1.29. We could find that score through interpolation; your instructor will decide whether that is required for z. For this text, however, we will take the nearest z score, the Table A value, without interpolation. (There will, however, be two easy and obvious exceptions to this rule, but they will come later.)

So far, we have looked at the normal curve and shown how it may be used with continuous random variables, particularly as a probability distribution or model. We may use the same model to determine binomial probabilities, at least under certain conditions. These conditions exist when the binomial probability distribution appears continuous enough to approximate the normal curve.

Continuous Binomial Outcomes and the Normal Curve

Remember that binomials are yes-no, success-failure types of variables. To begin to use the normal curve with binomial outcomes, we must first consider a statistical contradiction. Binomials are inherently discrete, so on any single binomial trial or series of trials, the outcomes are always whole events (whole successes or failures) or *numbers* of events. But, as our sample size or number of trials increases, the **binomial distribution,** when graphed, looks less and less discrete and more and more continuous. It begins to look like the continuous normal curve. When this happens, we may use the normal distribution to estimate binomial probabilities.

To proceed, we need symbols to stand for the probability of "success" and of "failure." Lowercase p stands for the probability of success on any trial, and lowercase q stands for the probability of failure on any trial. Since success and failure, by definition, are the only two possible outcomes in a binomial situation, one or the other is *certain* to happen. Their respective probabilities must therefore total to 100%, 1.00, or certainty. Moreover, p and q remain constant over all binomial trials; they do not change. In other words, we say the binomial trials are **independent** of each other: *What happens on one trial* (i.e., what outcome occurs) *has no effect on what happens on any other trial.* The result of your first coin flip, for instance, has no effect on what you will get with your next flip.

To see the difference between discrete and continuous binomial distributions it helps to look at them in tabular or graph form. Figure 3.9 illustrates two binomial probability distributions. In each case, the horizontal axis lists the possible values X may assume, and the vertical axis represents the probability of X occurring. The heights of the respective bars or lines represent the probability of X occurring: the higher the bar or line, the greater the probability.

The two distributions clearly differ in shape and appearance. The first distribution (a), with $n = 5$ and $p = .5$, is symmetrical and peaks in the center, but it has a somewhat block-like and discrete appearance. We could not easily superimpose the smooth normal curve over graph (a). Each X value is represented by a quite discrete and distinct bar or column. In fact, it is somewhat misleading to depict the horizontal axis in (a) as an unbroken, continuous line. Technically, no X values can exist between the whole numbers, e.g., between 0 and 1 or between 1 and 2. In contrast, a smooth distribution or curve is more obvious in graph (b). The second distribution, with $n = 20$ and $p = .5$, appears smoother and more continuous. Even though each X value is still represented by a discrete bar or column, the overall appearance of the graph suggests smoothness and continuity *and takes on the appearance of the smooth normal curve.* When this happens, we may use the normal curve model to at least *estimate* binomial probabilities.

If a discrete binomial distribution appears smoother as n increases, the question becomes: At what value of n are the data less discrete and more continuous? Where is the magic threshold, as it were? Obviously, in Figure 3.9 when $p = .5$ and $q = .5$, it is somewhere between $n = 5$ and $n = 20$. But how large does our sample size or number of trials have to be before the distribution starts looking like Figure 3.9(b)? In other words, when may we start using the normal curve to estimate binomial probabilities? The **criteria for continuity** and use of the normal distributions require that both:

(17) $$np \geq 5 \text{ and } nq \geq 5$$

When asked to determine any binomial probability, it is best to go through a series of questions or steps. Think of them as steps in a diagnostic process; you are diagnosing the circumstances described in the question and, one by one, eliminating possible ways to set up or solve the question. The first step is to

Figure 3.9

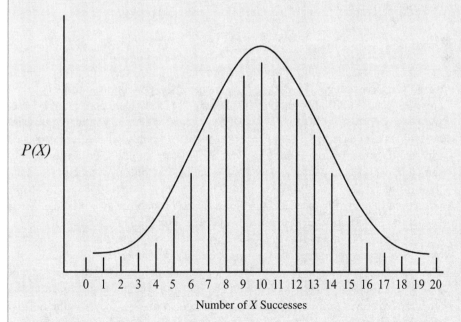

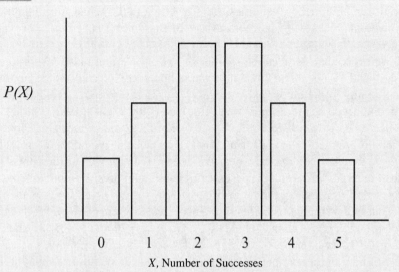

$P(X)$

X, Number of Successes

(a) Discrete distribution: $n = 5, p = .5, q = .5$

$P(X)$

Number of *X* Successes

(b) Continuous distribution: $n = 20, p = .5, q = .5$

determine whether the binomial distribution for the question approximates the normal curve. If *np* and *nq* are each at least 5, one may use the normal curve to estimate binomial probabilities. Before getting to the actual *z* formula used with binomials, however, one more aspect of the *np* and *nq* criteria for continuity needs to be addressed.

Notice that, besides the sample size (*n*), the criteria also take into account the *p* and *q* values. The function of the sample size is to contribute smoothness or continuity to the binomial distribution. As the sample size increases, the distribution becomes more smooth and continuous, or at least it takes on that appearance, as we saw in Figure 3.9(b). The role of the *p* and *q* values is to contribute a certain symmetry to the distribution. For instance, if *p* (and therefore *q*) is close to .5, the distribution peaks near the center. When *p* = .5 or thereabouts, about half the trials will result in successes and half in failures. *X* would assume a value close to the middle of the range of possible *X* values; no matter how many trials, about half of them should result in successes. In Figure 3.9(b), for instance, the distribution peaks at an *X* value of 10. Given *n* = 20 and *p* = .5, we should have about 10 successes out of 20 trials: half successes and half failures.

The average number of successes expected in a binomial situation is often referred to as the **expected value,** or *E*(*X*): *On the average, given p, how many successes should we expect over* n *trials?* The answer is *np*. For the mean of a binomial distribution:

(18) $$\mu = E(X) = np$$

To illustrate this point, if you flip a normal coin 10 times, how many heads would you expect to get? Five, right? Instantaneously, your brain said: Well, given only heads and tails, half should be heads, so that is 5. You are correct, of course, and you used equation 18 for *E*(*X*): 10(.5) = 5 heads, even if you weren't aware of it. In another case, if 15% of grocery carts have wobbly wheels, your next 20 trips to the supermarket should get you 20(.15), or an expected three wobbly carts. Or, for another example, say your usual bus or train is never early and arrives on schedule only 78% of the time. Over the next year and about 230 trips, you could expect it to be late 50.6 times. (The probability of its being late is 1 – .78, or .22. Therefore, you can expect 230(.22) = 50.6 late arrivals.)

Let's turn our attention back to Figure 3.9(b): Suppose that *p* = .8 instead of .5. We would then expect to get *np* = 20(.8) = 16 successes over 20 trials. The distribution peaks at *X* = 16; it no longer peaks exactly at the center, nor is it symmetrical. In fact, it looks like Figure 3.10(a), with a negative skew and an extended lower tail. Is it *symmetrical enough* so that we may use the normal curve? Well, in this case the answer is no (*np* = 16, but *nq* = 4, less than 5).

The general principle is that a large sample size will compensate for a lopsided or extreme *p* value (and *q* value). If we have a small sample for the normal curve

Figure 3.10 Binomial Probability Distributions

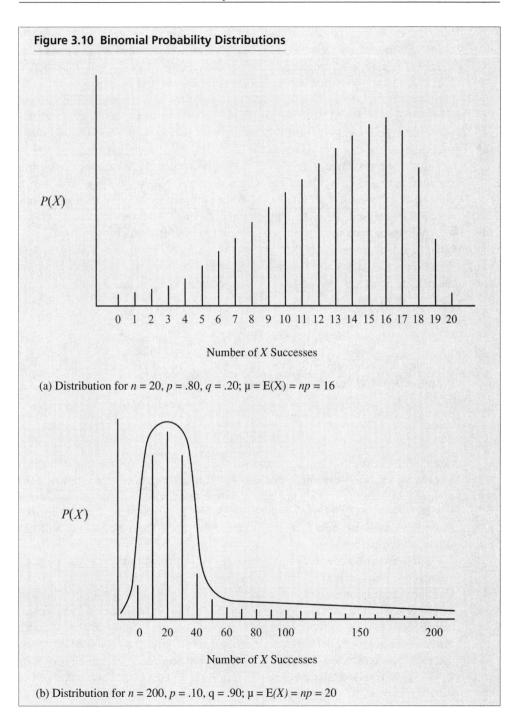

(a) Distribution for $n = 20, p = .80, q = .20; \mu = E(X) = np = 16$

(b) Distribution for $n = 200, p = .10, q = .90; \mu = E(X) = np = 20$

(i.e., if *n* is close to the borderline for using the *z* curve), we need *p* and *q* values close to .5 to produce the required normality in the binomial distribution. In contrast, if *n* is large, even extreme values of *p* and *q* (e.g., .1 or .9) will still produce binomial distributions that appear sufficiently smooth and symmetrical so as to permit the use of the *z* curve. This last fact is illustrated in Figure 3.10(b), where *n* = 200 and *p* = .10. Most of the distribution conforms to a generally normal arc, while small and negligible probabilities taper off into the upper tail. For all practical purposes, the distribution meets the requirement of normality. The general rule requires that the sample size (*n*) and the *p* and *q* values work together to ensure the continuity and the symmetry of continuous binomial distributions. One factor, either *n* or *p/q*, may compensate for the other to produce a normal-looking distribution. For this reason, the criteria for continuity include both factors, *n* and *p* or *q*.

With these criteria in mind, we now turn to the *z* formula for binomial distributions. It is fundamentally the same as the earlier *z* formulas (equations 13 and 14), except that terms and symbols appropriate for binomials are substituted for the original symbols. As we have seen in equation 18, the mean equals *np*. The standard deviation can also be rewritten for continuous binomial distributions:

(19) $$\sigma = \sqrt{npq}$$

Mimicking the original *z* formula (equation 13), we get *X* minus the mean in the numerator and the standard deviation in the denominator:

(20) $$z_x = \frac{X - np}{\sqrt{npq}}$$

Equation 20 works under many circumstances, but it often includes another term, a correction factor. Remember that we are using a continuous curve or model to approximate what is still a somewhat discrete distribution. We argue that, when *n* is less than 100, even though the binomial probability distribution looks like a continuous normal curve, it retains enough discrete features that a correction factor is necessary.

Furthermore, if we are using a continuous curve with discrete numbers (*X* values), we must consider where an *X* value truly begins or ends if it is part of a genuinely continuous distribution. This means using the **true limits** of *X* instead of just *X* itself. For instance, suppose we have *X* = 36 successes in a binomial distribution. In a smooth or continuous distribution, the whole number 36 really starts or begins at 35.5. That is, a value above 35.5 would be rounded up to the discrete, whole number 36. Thus, the **true lower limit** of 36 is actually 35.5, or *X* − .5. **True upper limits** are found similarly. If *X* must be a whole number but is rounded off, anything up to 36.49 becomes 36. The true upper limit of 36 is therefore 36.49, and presumably 36.50 is the lower limit of 37.

A question arises as to how finely detailed we need to be. How closely must we mark a number's true limits? To one decimal place? Two? For our work, we will simplify things by using $X \pm .5$ as the true limits of X. Our correction factor is $X - 5$ or $X + .5$. We use this correction with the normal curve only when n is less than 100, that is, when the binomial distribution approaches but does not quite meet complete continuity. Alternatively, when n is 100 or more, we assume the binomial probability distribution to be sufficiently continuous as to be indistinguishable from the true normal distribution. The correction factor is therefore no longer needed for $n \geq 100$.

For situations permitting use of the normal curve to estimate binomial probabilities, but when n is less than 100, we use equation 21 to calculate the z score and then use it to look up an area of the curve :

$$(21) \qquad z_X = \frac{(X \pm .5) - np}{\sqrt{npq}}$$

We use either $X - 5$ or $X + .5$, but not both at the same time. But when do we use the lower limit of X (or $X - .5$) versus its upper limit (or $X + .5$)? There is a guideline or rule to follow in deciding which one to use.

Say we wish to find the area (or probability) up to and including an X value. If $X = 36$, and we want the probability of X scores up to and including 36, or $P(X \leq 36)$, we consider where 36 really *ends* if it is a truly continuous number. It ends at its true *upper* limit, or 36.5, so we use the $X + .5$ version of the formula. From X, we are going to the left in the curve and including X values of 36 and below. We want the area of the curve extending from the lower tail all the way up to the true upper limit of 36.

Conversely, suppose we want $P(X \geq 36)$, which includes the area of the curve starting at $X = 36$, going to the right, and all the way into the upper tail. But in a continuous distribution, where does 36 really begin? Its true *lower* limit is 35.5. We solve for the area of the curve falling at $X = 35.5$ and above, and use the $X - .5$ version of equation 21.

The general rule for using the $X - .5$ or $X + .5$ correction is: When seeking an area above or to the right of X, use the true lower limit $(X - .5)$, and when seeking an area below or to the left of X, use the true upper limit $(X + .5)$. A student once put it a different way—a way that helped her remember it: "Do you mean we always slightly *expand* the area we're seeking?" The answer is, yes, we do, and this is a useful way to think of it. Whenever n is less than 100, the area we are seeking is always slightly enlarged by the correction factor. We are going all the way out (up or down) to the X value's true limit rather than just to that X score itself.

6. *In a recent college survey of 589 students, 200 answered Yes when asked if they routinely studied with peers from "other cultural or ethnic backgrounds." Based on these data, what is the probability that, of any 50 students, 20 or more regularly study with diverse peers?*

First, we know this is a binomial given the nature of the variable: one either studies with others of different cultural and ethnic roots or one does not. Pulling some initial numbers from the question: $n = 50$ students, $p = P(\text{Study with others of different backgrounds}) = 200/589 = .34$, and $q = 1 - p = 1.00 - .34 = .66$. To see whether a continuous or discrete binomial distribution applies, we do our continuity check: $np = 50(.34) = 17.00$, and $nq = 50(.66) = 33.00$. Each is greater than 5, so the criteria for a continuous binomial distribution are met.

Because n is less than 100, we must use the correction factor. The question asks about "20 or more" students, so we are going upwards or to the right in the curve from $X = 20$ (Figure 3.11). That means we start where 20 actually begins, that is, at its true lower limit. The z formula therefore includes $(X - .5)$ in the numerator:

$$z_X = \frac{(X-.5)-np}{\sqrt{npq}} = \frac{(20-.5)-50(.34)}{\sqrt{50(.34)(.66)}} = \frac{19.50-17.00}{\sqrt{11.22}} = \frac{2.50}{3.35} = .75$$

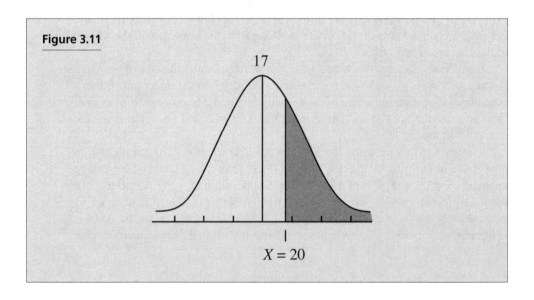

Figure 3.11

17

$X = 20$

Finally, for $z = .75$, column C in the normal curve table (Table A) reads .2266. Therefore, $P(X \geq 20)$ is estimated to be about 22.66%.

To show how the $X + .5$ correction works, we revise the problem slightly. Suppose the question asks about 20 *or fewer* students studying with a diversity of peers? All else remains the same, but now we seek the area to the left and toward the curve's lower tail (Figure 3.12). We want the probability of X equaling anything up to and including 20 students, and that means going to the true upper limit of 20. The z formula now becomes:

$$z_X = \frac{(X+.5)-np}{\sqrt{npq}} = \frac{(20+.5)-50(.34)}{\sqrt{50(.34)(.66)}} = \frac{20.50-17.00}{\sqrt{11.22}} = \frac{3.50}{3.35} = 1.05$$

When $z = 1.05$, column B of Table A reads .3531. Adding the lower half of the curve (.5000) gives the final answer of .8531. We estimate that, of any 50 students, there is an 85.31% chance that 20 or fewer study with others of diverse backgrounds.

Notice the tone of caution in the summaries of these two examples. The final percentages or probabilities are "estimates." This is because the binomial distribution *approximates* a normal, continuous distribution. They do not match perfectly, especially when n is less than 100 and we need a correction factor to justify using the curve at all. Binomial probabilities that we derive by using the normal curve, although generally acceptable, are not considered completely accurate. Acceptable, but not perfect. Note also that the two probabilities, those of "20 or more" and "20 or fewer" students, will not add up to 100% of the students because the case of exactly 20 students is included twice, once in each category.

Figure 3.12

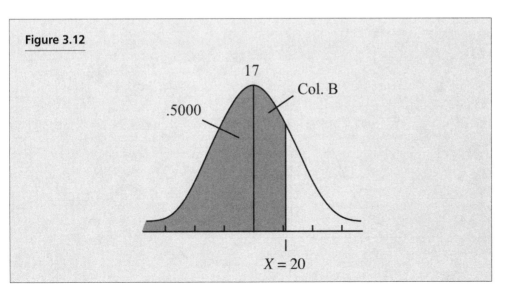

7. *In a survey of college students, 536 out of 738 sampled agreed with the statement: "It is sometimes OK to lie to avoid hurting someone's feelings." Of any 100 students, what is the probability at least three-fourths agree with this proposition?*

To set this problem up, we realize first that it is a binomial. One either agrees with the statement or one does not. Then actual numbers narrow the problem down further: $n = 100$ students, $p = P(\text{Agree}) = 536/738 = .73$, and $q = 1 - p = 1.00 - .73 = .27$. We do our continuity check to see whether our binomial distribution is continuous or discrete: $np = 100(.73) = 73.00$, and $nq = 100(.27) = 27.00$. Each is at least 5, so the criteria for a continuous binomial distribution are met, and the graph can be sketched as in Figure 3.13.

We note two additional bits of information. Realizing that $n = 100$, we do not need the $X \pm .5$ continuity correction. Also, we are asked for the probability of "at least three-fourths" agreeing. Three-fourths of 100 gives us $X = 75$. We therefore want $P(X \geq 75 \mid n = 100$ and $p = .73)$, read as "The probability that X is equal to or more than .75, given that $n = 100$ and $p = .73$." Note that the vertical line or "$\mid$" is read as "given."

For z, we get:

$$z_X = \frac{X - np}{\sqrt{npq}} = \frac{75 - 100(.73)}{\sqrt{100(.73)(.27)}} = \frac{75 - 73.00}{\sqrt{19.71}} = \frac{2.00}{4.44} = .45$$

When $z = .45$, column C in Table A $= .3264$, so there is about a 32.64% probability that at least 75 of any 100 college students agree that lying to avoid hurting another's feelings is sometimes acceptable.

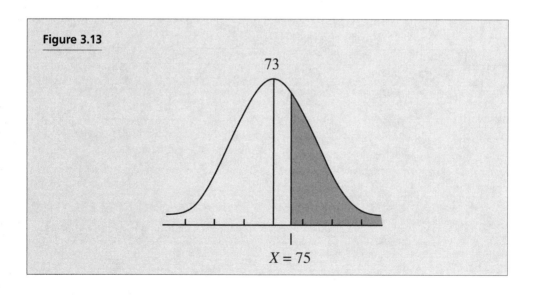

Figure 3.13

73

$X = 75$

This concludes our discussion of *continuous* binomial probabilities. Next, before turning to inferential statistics, we have one more topic: discrete binomial probabilities.

Probabilities for Discrete Binomial Outcomes

The previous section treats binomials as if they were continuous, but we may do so only under certain conditions (remember, if $np < 5$ or $nq < 5$, the binomial is discrete, not continuous). Binomial outcomes, as we know, are inherently discrete, so what if the binomial probability distribution does not mimic the normal curve? What if it looks like Figure 3.9(a), and the normal curve model does not work? Now what? For such discrete binomials, we must solve for the probability of each single value of X. There are no normal curve shortcuts, and we cannot simply identify areas of the curve. Now, we have to consider every single value of X and details, details, details.

When we calculate the likelihood of getting X successes out of n trials, we consider two things: (1) the probability that exactly X successes will actually occur out of all the X outcomes possible, and (2) all the different ways in which it might be possible to get *exactly* X successes over n trials.

The Probability of Exactly X Successes

First, we calculate the probability that a sequence of outcomes containing exactly X number of successes will actually happen. How likely is it that *exactly* X successes will occur? For example, we may consider the possibility of 2 heads in 3 coin flips, but we recognize that other outcomes are also possible: 0 heads, 1 head, or 3 heads. We are asking the probability of getting exactly 2 heads (or 2 successes). Not 0 heads, not 1 head, not 3 heads, but exactly 2 heads.

For the answer, we rely upon the multiplicative rule for **joint probabilities.** Joint probabilities refer to events that occur together or jointly. For example, what is the probability you will get B's on your next two statistics quizzes? This is the same as asking, What is the probability of a B *and* a B, the probability two B's will occur next, *with* each other, or jointly? Or consider, what is the probability two people selected at random do most of their shopping at large malls: P(Mall and Mall)? Whenever we have joint probabilities, the mathematical rule is to multiply their respective probabilities. For your statistics quizzes, you have P(B) $\cdot$ P(B). If you "guesstimate" your chances of B's, you can determine the probability of two high grades in a row. Similarly, knowing what percentage of people shop largely at major malls allows you to get the probability of two randomly chosen people both doing so.

When we apply this rule to binomial probabilities, we multiply the probabilities of success and failure on respective trials. For example, if your probability of a B on each quiz is 35% or .35, then $p = .35$, and q, or P(Not B), must be .65. For the probability of two B's, you have $p \cdot p$ or $(.35)(.35) = (.35)^2 = .1225$. Similarly, if the probability of someone shopping mainly in larger malls is .85 ($p = .85$), the probability that he or she does not is $q = .15$. The likelihood of two random shoppers preferring major malls becomes $p \cdot p$ or $(.85)(.85) = (.85)^2 = .7225$.

How do we actually use p and q and multiplication to find the $P(X$ successes over n trials)? The example of the mall shoppers provides a realistic situation. The example of your B's on statistics quizzes is less realistic because the quizzes may be few in number, if any are given at all, and your $P(B)$ may not remain constant from one exam to the next. But for shoppers we may extend n to cover any number of people and assume p is constant from one trial (or person) to the next. Let's now look at a problem that involves looking at $P(X$ successes over n trials).

8. *Nationally, 85% of U.S. shoppers tend to buy their general merchandise (other than food and cars) at major shopping malls. Of any 6 shoppers, what is the probability that 5 would tend to make their purchases at malls?*

Rewriting the question, we seek the probability that $X = 5$, given $n = 6$ and $p = .85$, or $P(X = 5 \mid n = 6, p = .85)$. To start, consider alternative ways in which 5 successes over 6 trials *could* happen. Table 3.3 illustrates just three of the different ways exactly $X = 5$ successes might occur (sequences A, B, and C). S denotes a success, which we are defining as shopping largely at malls, and F stands for a failure, which means avoiding malls. Next, consider the probabilities governing each sequence and the multiplicative rule. The probability of success (p) is substituted for each S, and the probability of failure (q) takes the place of each F.

Obviously, there is a pattern here. In each sequence, the $P(\text{Success})$ occurs five times. For Sequence A, that is $S \cdot S \cdot S \cdot S \cdot S \cdot F$. The first five favor malls, and the last one does not. In probability symbols, another way to express this is $p \cdot p \cdot p \cdot p \cdot p \cdot q$, which equals $p^5 q$. The overall probability of X successes over n

Table 3.3													
	Event/Outcome						Probability of Outcome						
Shopper:	1	2	3	4	5	6	1	2	3	4	5	6	
Sequence A:	S	S	S	S	S	F	p	p	p	p	p	q	$= p^5 q = p^X q^{n-X}$
Sequence B:	S	S	S	S	F	S	p	p	p	p	q	p	$= p^5 q = p^X q^{n-X}$
Sequence C:	S	S	S	F	S	S	p	p	p	q	p	p	$= p^5 q = p^X q^{n-X}$

trials now becomes a formula: It is the probability of success raised to the number of successes multiplied by the probability of failure raised to the number of failures. In general terms: $p^X q^{n-X}$; since $X = 5$ and $n = 6$, we get $p^5 q$,

$$(22) \qquad P(X \mid n, p) = p^X q^{n-X}$$

In numbers, for sequence A above, this becomes:

$$p^X q^{n-X} = p^5 q = (.85)(.85)(.85)(.85)(.85)(.15) = (.85)^5 (.15) = (.4437)(.15) = .0666$$

Thus, there is roughly a 7% chance (actually, a 6.66% chance) of that one sequence of outcomes (A) occurring. After that, sequences B and C each have identical probabilities of occurrence: .0666. So far, so good, but we also need to know the number of different ways X successes might happen to complete the problem.

Combinations: Different Ways to Get *X* Successes

To figure how many different ways an event might happen, we use a common mathematical concept and procedure known as combinations. **Combinations** *refer to the number of different ways we may select a set of items from a larger population of items.* In our example, in how many different ways or sequences could exactly 5 of 6 shoppers prefer malls? As another example, on a 12-member basketball team, in how many different ways could we pick 5 players to actually be on the court at one time? In other words, how many different combinations of 5 are possible if $n = 12$? Or, if a professor gives you 8 study questions and says she will randomly pick 4 for the test, how many different sets of 4 items could she possibly select? Do not be surprised by fairly large numbers of combinations being possible. Fully 792 different sets of 5 basketball players could be used if we have 12 from which to choose. The professor could select any of 70 possible combinations of 4 questions from her pool of 8.

For binomial probabilities, we ask: How many different ways could we possibly get X successes out of n trials? How many different sequences of outcomes will produce exactly X successes? Sometimes, if the number of trials (n) is fairly small, we may be able to do this in our heads, logically thinking it through. In how many different ways can we get 2 heads if we flip 3 coins? Coins 1 and 2 may come up heads while coin 3 is tails, or coins 1 and 3 could be heads, or coins 2 and 3. But that is it. There are only three different ways to get 2 heads out of 3 coins (i.e., 2 successes out of 3 trials), as shown in Table 3.4. This was quite easy. We logically thought through all the possible outcomes that would give us 2 heads. Larger numbers complicate matters, however.

Before considering situations with more trials, two cautions are relevant. First, it is important to recognize that other outcomes are obviously possible given $n =$

Table 3.4

	Coin 1	Coin 2	Coin 3
Sequence A	H	H	T
Sequence B	H	T	H
Sequence C	T	H	H

3 coins. All 3 coins might come up heads ($X = 3$), or we may get no heads at all ($X = 0$). Similarly, we may get just 1 head ($X = 1$). Regardless, the event or outcome of interest is $X = 2$, or 2 heads. This is the *only* outcome on which we focus. Given the way the question or problem was defined, we are not concerned with other outcomes, even though we recognize those possibilities exist. For the problem at hand, we ignore them. We wish to know in how many ways we may get *exactly 2 heads*, no more and no less. In the mall situation, we are interested 5 out of 6 favoring malls, no more and no less. When you read a question, focus on the outcome(s) or event(s) of interest only, not on other outcomes that could possibly occur.

Second, in our coin example, it is purely coincidental that the number of different ways to get 2 heads (3 ways) equals n (also 3 here). It just works out that way by chance. Do not assume it is that way all the time. In fact, recall the two problems mentioned earlier: from 12 basketball players ($n = 12$), we could select 792 sets of 5 people. From the professor's 8 questions ($n = 8$), 70 different sets of 4 items would be possible. Generally, the number of combinations with exactly X successes is considerably larger than n itself.

To continue the coin example, a logical solution works fine because $n = 3$ and $X = 2$, but suppose our numbers are larger. In how many different ways could we get 7 heads on 12 coins ($n = 12$, $X = 7$)? Or how did we get 792 possible combinations of basketball players when $n = 12$ and $X = 5$? Or 70 possible sets of test questions ($n = 8$, $X = 4$)? These examples are not so easy to see. Nor, in fact, are 5 mall shoppers out of 6 people. Given enough time, one could conceivably solve these questions by hand, but the procedures would be tedious. The combinations formula helps here. It tells us the number of different ways to get exactly X successes over n trials.

(23)
$$_nC_X = \frac{n!}{X!(n-X)!}$$

In equation 23, C is the symbol for combinations, and $_nC_X$ refers to the number of different combinations of X items possible from a pool of n items. Also, a number or symbol followed by "!" indicates a factorial. $n!$ is read as "n factorial," and $X!$ is "X factorial." **Factorials** *mean to multiply the given number by every* *number smaller than itself down to 1.* If $X = 5$, for instance, then $X! = 5 \cdot 4 \cdot 3 \cdot 2 \cdot 1 = 120$. If $n = 7$, $7!$ is $7 \cdot 6 \cdot 5 \cdot 4 \cdot 3 \cdot 2 \cdot 1 = 5040$. (Needless to say, factorials are a handy function on your calculator.)

We may see how this works with earlier examples. For the basketball team with $n = 12$ members, of which $X = 5$ may be in the game at any one time, how many different sets of 5 players could the coach select from the 12? Recall the answer was 792.

$$_{12}C_5 = \frac{12!}{5!(12-5)!} = \frac{12!}{5!\,7!} = \frac{479,001,600}{(120)(5040)} = \frac{479,001,600}{604,800} = 792$$

For the professor's 4-question exam, we found that 70 combinations were possible. She had 8 items from which to choose, so $n = 8$, and $X = 4$. By using equation 23, we get:

$$_8C_4 = \frac{8!}{4!\,4!} = \frac{40,320}{(24)(24)} = \frac{40,320}{576} = 70$$

A tip before moving on: Your answer when using the combinations formula must always be a whole number. You are calculating the *whole number* of ways in which one may get X successes over n trials. That might be 792. It might be 70. It might be 10 (which is $_5C_3$), or 2002 (which equals $_{14}C_9$), and so on. It is *never* 792.7 ways, or 70.5 ways, or 2002.8. Logically, there is always a finite and whole number of possible ways in which X successes can happen. No exceptions. If you get an answer other than a whole number, you have made a mistake; try it again.

Using these procedures with our shopping example, we may determine all the different ways 5 mall shoppers out of 6 people might occur. First, using simple logic, we soon exhaust all the possibilities. It turns out there are 6 sequences of outcomes (A through F in Table 3.5), resulting in 5 successes.

Table 3.5

			Event/Outcome			
	1	2	3	4	5	6
Sequence A:	S	S	S	S	S	F
Sequence B:	S	S	S	S	F	S
Sequence C:	S	S	S	F	S	S
Sequence D:	S	S	F	S	S	S
Sequence E:	S	F	S	S	S	S
Sequence F:	F	S	S	S	S	S

Alternatively, we can use the combinations formula (equation 23) to get the same answer much more easily:

$$_6C_5 = \frac{n!}{X!\,(n-X)!} = \frac{6!}{5!\,(6-5)!} = \frac{6!}{5!\,1!} = \frac{720}{(120)\,(1)} = \frac{720}{120} = 6$$

Finally, to get an answer to our shopping question, we multiply the probability of getting exactly X successes (recall it was .0666) by the number of different ways (6) in which that could happen: (.0666) (6) = .3996, or 39.96%. There is a .3996 probability that, of any 6 people, 5 will do most of their shopping at malls.

The two parts constitute the complete formula. We need the actual probability of getting X successes *and* the number of different ways this might happen. Therefore, the probability of X successes in n trials, where the probability of success is p is given by equation 24.

(24) $$P(X|n,p) = p^X q^{n-X} \left[\frac{n!}{X!\,(n-X)!} \right]$$

9. *What is the probability of getting 4 or more reds on 5 spins of a roulette wheel?*

Initial diagnoses help us to set up this problem. First, given the way the question is phrased, we let the roulette ball landing on a red number equal a success. Second, to determine the $P(\text{Red})$, a US roulette wheel has 38 slots or numbers, and 18 are red. Another 18 are black, and 2 are green.* $P(\text{Red})$, or p, equals 18/38, or .47 (rounded off). The probability of failure, or *not red* (q), includes the black and green numbers and equals 20/38 or .53. Note that q is also the complement of p: $q = 1 - p = 1 - .47 = .53$. The **complement** of an event is *the probability that the event will not happen*. Third, we realize that these probabilities, $p = .47$ and $q = .53$, will remain the same for all trials or spins of the wheel because each spin is independent of every other spin. Fourth, notice that the question asks about "4 or more" red numbers, or 4 or more successes in 5 spins. That means we have two successful outcomes or X values here, $P(X = 4 \text{ or } X = 5)$. If $n = 5$ spins, 5 successes is the maximum number possible. More than 5 successes cannot happen, so we stop at $X = 5$. We will therefore determine $P(X = 4)$ and then $P(X = 5)$. Finally, there is another consideration. What do we do with the two probabilities once we have them? Besides the multiplicative rule in probability, there is also the **additive rule,** which states that *if we want the probability of any of a set of outcomes,*

---

*The French or European roulette wheel differs from the US wheel. The green numbers on US wheels are 0 and 00. On the European wheel there is simply one 0.

we should add their respective probabilities. So, if we want $P(X = 4 \text{ or } X = 5)$, we should add $P(X = 4)$ and $P(X = 5)$.

Now we are ready to solve the problem:

$$P(X \geq 4 \text{ Reds}) = P(X = 4 \text{ or } 5) = P(X = 4) + P(X = 5)$$

For $P(X = 4)$, we have $n = 5$, $p = .47$, $q = .53$, and $n - X = 5 - 4 = 1$:

$$P(X = 4 \mid n = 5, p = .47) = p^X q^{n-X}\left(\frac{n!}{X!(n-X)!}\right) = (.47)^4(.53)\left(\frac{5!}{4!(1)!}\right)$$

$$= (.049)(.53)\left(\frac{120}{24}\right) = (.049)(.53)(5) = .1293$$

Next, we get $P(X = 5)$ in a similar fashion. These calculations require an explanation. Things are a little different whenever $X = 0$ or when n and X are the same, as is the case here: $n = 5$ and $X = 5$. First, we have to deal with a number raised to the 0 power and with 0 factorial (0!). Our formula includes the term $n - X$, and that will naturally be zero when n and X are the same. By mathematical convention, both 0! and any number raised to the 0 power each equal 1. Second, when n and X are the same or when $X = 0$, the combinations part of the formula always cancels down to 1. For $P(X = 5)$, we have $n = 5$, $p = .47$, $q = .53$, and $n - X = 5 - 5 = 0$.

$$P(X = 5 \mid n = 5, p = .47) = p^X q^{n-X}\left(\frac{n!}{X!(n-X)!}\right) = (.47)^5(.53)^0\left(\frac{5!}{5!(0)!}\right)$$

$$= (.023)(1)\frac{120}{120} = (.023)(1)(1) = .0229$$

For our final answer, we add $P(X = 4)$ and $P(X = 5)$:

$P(X \geq 4 \mid n = 5 \text{ and } p = .47) = P(X = 4) + P(X = 5) = .1293 + .0229 = .1522$, or 15.22%

There is slightly more than a 15% chance of the wheel showing four or more red numbers in six spins.

We use these procedures to derive the probability of getting X successes out of n trials. The previous pages have reviewed quite a few calculations and procedures. On occasion, however, other strategies may ease your work. First, it is sometimes quicker and easier to solve for the complement of X. Second, a binomial probabilities table may contain the actual probabilities you need.

Solving for Complements

Any event either will or will not occur. The probability of an event occurring *plus* the probability that it will *not* do so must therefore equal 1.00 or 100%. In binomial situations, if there is an 80% chance something will happen, there is, by definition, a 20% chance it will not. If certain outcomes have a 45% probability of occurring, there is a 55% chance they will not occur. Since the probability of an event and its complement must always equal 1.00, we get equation 25:

(25) $P(\text{Event}) = 1 - P(\text{Complement of the event})$

It makes sense to solve for the complement's probability when that strategy requires fewer steps and calculations than solving for the probability of the event itself. Fewer steps mean less chance for error. Suppose we have a situation involving 9 trials and are asked to solve for the probability of 4 or more successes. The complement and event become:

$$X = \underbrace{0 \quad 1 \quad 2 \quad 3}_{\text{Complement}} \qquad \underbrace{4 \quad 5 \quad 6 \quad 7 \quad 8 \quad 9}_{\text{Event}}$$

Solving for the probability of the event directly means running through equation 24 six times. We must solve for $P(X = 4)$, $P(X = 5)$, and so on up through $P(X = 9)$. Our final answer is the sum of these six probabilities:

$$P(X \geq 4 \mid n = 9) = \underbrace{P(X = 4) + P(X = 5) + P(X = 6) + P(X = 7) + P(X = 8) + P(X = 9)}_{\text{Event}}$$

In contrast, solving for the probability of the complement takes us through the formula only four times. We determine $P(X = 0)$, $P(X = 1)$, and up through $P(X = 3)$, and add these probabilities to get $P(\text{Complement})$. $P(\text{Complement})$ here means the probability we will *not* get 4 or more successes. We must then subtract that answer from 1.000 to get $P(\text{Event})$:

$$P(X \geq 4 \mid n = 9) = 1 - \underbrace{[\,P(X = 0) + P(X = 1) + P(X = 2) + P(X = 3)\,]}_{\text{Complement of the Event}}$$

Whenever faced with a discrete binomial situation, ask yourself whether getting the complement of an event is easier. That may be a more efficient way to solve the problem.

Similarly, a table of binomial probabilities may simplify the task. Some common binomial probabilities have been calculated already and appear in Table B. If the parameters of a binomial problem happen to match those found in the table, you may look up the relevant probabilities and not have to go through all the calculations. You still have to understand what you are doing and be able to correctly use the table's entries, columns, and rows, but it may substantially reduce the number of calculations needed.

A Handy Table of Binomial Probabilities

Table B lists binomial probabilities for common sets of conditions or circumstances. If a problem's exact values of n and p appear in the table, you can simply look up $P(X)$. Few, if any, actual calculations may be required. Sometimes, the numbers you are given will not appear in the table, and you will have to calculate the answer yourself by using equation 24.

Assuming Table B applies, we use n, X, and p to look up binomial probabilities. The values of n and X, respectively, appear in the two left-hand columns of the table. To the right, the column headings indicate probabilities of success, or p values ($p = .10, .15, .25$, and so on). Not all p values appear, of course, only the common ones; listing all would mean a prohibitively large table. Moreover, the sample sizes shown go from $n = 2$ to $n = 15$. These parameters cover most instances when we need such a table. With samples of more than 15 cases, we would most likely use the methods for continuous binomial outcomes and the normal curve. Table 3.6 reproduces a segment of Table B for $n = 5$ only. Notice

Table 3.6 Sample Entry from Table B: Selected Binomial Probabilities for $n = 5$

n	X	.10	.15	.20	.25	.30	.33	.35	.40	.45	.50
5	0	.590	.444	.328	.237	.168	.135	.116	.078	.050	.031
	1	.328	.392	.410	.396	.360	.332	.312	.259	.206	.156
	2	.073	.138	.205	.264	.309	.328	.336	.346	.337	.313
	3	.008	.024	.051	.088	.132	.161	.181	.230	.276	.313
	4		.002	.006	.015	.028	.040	.049	.077	.113	.156
	5				.001	.002	.004	.005	.010	.018	.031

Notes: Probabilities <.001 are not shown.
   To use the table, find the probability listed where X and p intersect: (1) locate n, (2) select the proper X row, and (3) locate the entry under the p value.

in Table 3.5 that 5 appears under *n* and entries are given for *X* values of 0 through 5, that is, for all the *X* values possible when *n* = 5.

10. *Assume that women convicted of felonies, once on parole, have a 20% recidivism rate; that is, these women have a 20% probability to return to jail for similar crimes. Of any five women felons, what is the probability three will violate parole and return to prison?*

We have *n* = 5 and *p* = 20% or .20. The question asks about three returning to prison. Therefore, *X* = 3, and the question translates into $P(X = 3 \mid n = 5$ and $p = .20)$. Before turning to equation 24, we check Table B. Luckily, all necessary parameters (*n* = 5, *p* = .20, and *X* = 3) are there. We first find the block of entries for *n* = 5. (Note that Table 3.6 reproduces these values.) Next, we locate the row corresponding to *X* = 3 given that *n* = 5. Finally, reading along that row, we look down the *p* = .20 column and we come to the entry of .051. So, $P(X = 3 \mid n = 5$ and $p = .20)$ is .051, or 5.1%. We conclude that, if *p* = .20, for any five women felons on parole, there is only about a 5% probability (or about 1 chance in 20) that exactly three will wind up remanded. Alternatively, by using equation 24 with *p* = .20, *q* = .80, *n* = 5, and *X* = 3, we get:

$$P(X = 3 \mid n = 5, p = .20) = p^X q^{n-X} \left(\frac{n!}{X!(n-X)!} \right) = (.20)^3 (.80)^2 \left(\frac{5!}{3!\,2!} \right)$$

$$= (.008)(.64)\frac{120}{12} = (.008)(.64)(10) = .051$$

11. *If one in three males at neighborhood yard sales and flea markets typically buys an unidentifiable gadget because it "looks cool and might someday be useful," what is the probability two or three of any seven males will buy such superfluous tools?*

To diagnose the problem: first, we are asked about seven males, so *n* = 7. At the same time, the question notes that "one in three" males buys a useless tool. Then *P*(Buys a useless tool) is 1/3, and *p* = .33. Finally, for *X*, we have the "probability two or three . . . buy," which, as we saw previously, equals *P*(*X* = 2) + *P*(*X* = 3). Consulting Table B for *n* = 7 and *p* = .33, we find the entries reproduced in Table 3.7.

Adding these probabilities, we get .562 or 56.2%. That is, when *p* = .33 and we have seven random males at yard sales, there is a better than even chance (56%) that two or three will buy spiffy-looking but useless gadgets.

Table 3.7

X:	2	3
P(X):	.309	.253

When necessary, we may also adapt our use of the binomial table to look up the probability of $n - X$ failures and use q rather than p. This method is illustrated in the following problem.

12. *Women employees are often described as working "double shifts," one at home and one at work. Despite claims of egalitarian ideals, a recent survey suggests household tasks are still gender-linked among young, two-income, and educated married couples. In fully 60% of such households, for instance, women do virtually all the grocery shopping and cooking. Of any 10 such couples, what is the probability the female partner will routinely perform these tasks in at least 6 households?*

Since n is small (10), the binomial probabilities table (Table B) will probably be useful. p, the probability of the female partner doing the grocery shopping, is a nice round number (.60), but it is unfortunately too large for the table. However, if we turn the problem around (or inside out, so to speak), and look at the number of "failures" rather than 6 or more "successes," we may use Table B. Having 6 or more successes is the same as getting 4 or fewer failures. They are actually the same event or outcome. Since the probability of failure $(1 - .60 = .40)$ *is* in the table, we may recast the problem using $n - X$:

When $X =$ 0 1 2 3 4 5 6 7 8 9 10 and $p = .60$

$$\underbrace{}_{\text{Event}}$$

Then $n - X =$ 10 9 8 7 6 5 4 3 2 1 0 when $q = .40$

$$\underbrace{}_{\text{Event}}$$

Now, we look up 4 or fewer failures and use $q = .40$ instead of $p = .60$. We simply use the q value as though it were p.

$$P(n - X \leq 4 \mid n = 10, q = .40) = P(X = 4) + P(X = 3) + P(X = 2) + P(X = 1) + P(X = 0)$$
$$= .251 + .215 + .121 + .040 + .006 = .633$$

The problem is solved by recognizing that X successes (given p) is the same event and has the same probability as $n - X$ failures (given q). There is about a 63% probability that at least 6 in 10 of the women in young and educated couples' marriages do just about all the food shopping and cooking.

Table B may save you quite a bit of time. Equation 24 for discrete binomial outcomes is obviously practical only with fairly small numbers. Calculating the necessary factorials or numbers to excessive powers when n and X are large is clearly unwieldy. We resolved this dilemma earlier in this chapter by using the alternative procedures of estimation with the normal curve or possibly by inverting the problem and looking up the probability of $n - X$ failures. For continuous variables, our use of the normal curve model is automatic.

Summary

Approach a probability situation by first determining whether you are dealing with a random or a binomial variable. If it is a binomial situation, follow the decision tree in Figure 3.14 through to a logical end: (1) Decide whether the binomial distribution is discrete or continuous (are np and $nq < 5$ or ≥ 5?). (2) If it is a discrete binomial, use equation 24 or Table B, inverting the problem if necessary. (3) If it is a continuous binomial and $n \geq 100$, use equation 23. (4) If n is less than 100, do you need the continuity correction? (5) If you do need that correction, determine whether to use $X - .5$ or $X + .5$. In other words, are you seeking an area above X or below X?

Now for the next question: How do we use this information to estimate population parameters? That is addressed in Chapter 4, which shows how to describe a population with information from a single sample.

Exercises

Continuous Random Probabilities

1. Surveys suggest that US teenagers watch TV (including videos and DVDs) an average of 26.58 hours per week, with a standard deviation of 7.83 hours. Assuming the distribution of TV viewing is approximately normal, what is the probability a randomly selected teenager watches television:

    a. 30 or more hours per week?
    b. 15 or fewer hours per week?
    c. within ±3 hours of the average viewing time per week?
    d. between 30 and 40 hours per week?

For these same data, answer the following questions:

Figure 3.14 Diagnosing a Binomial Probability Situation

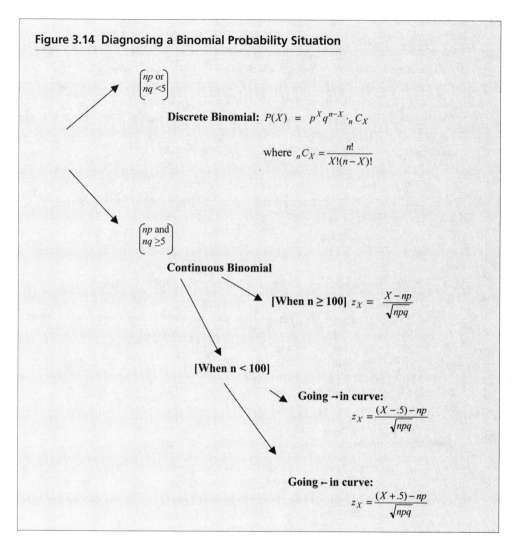

$\begin{bmatrix} np \text{ or} \\ nq < 5 \end{bmatrix}$

Discrete Binomial: $P(X) = p^X q^{n-X} \cdot {}_n C_X$

where ${}_n C_X = \dfrac{n!}{X!(n-X)!}$

$\begin{bmatrix} np \text{ and} \\ nq \geq 5 \end{bmatrix}$

Continuous Binomial

[When n ≥ 100] $z_X = \dfrac{X - np}{\sqrt{npq}}$

[When n < 100]

Going →in curve:

$z_X = \dfrac{(X - .5) - np}{\sqrt{npq}}$

Going ←in curve:

$z_X = \dfrac{(X + .5) - np}{\sqrt{npq}}$

e. How many hours per week does a teenager watch TV if he or she falls at the 40th percentile rank in this distribution?

f. The most extreme 15% of teenage viewers watch how many hours of TV per week?

g. What is the probability a teenager in this distribution watches TV exactly 35 hours per week? (*Hint:* This is a puzzler. What is your X value? What is the area of the curve being sought? Using the normal curve table, why could the answer be zero probability or .0000?)

2. College graduation requires 120 semester units (or the equivalent quarter units), but students often graduate with more. The actual numbers of semester units accumulated at graduation are normally distributed. About 8% of all students graduate with 139 semester units or more, and another 8% graduate with 129 or less. Given this information, what are the mean and standard deviation for the distribution of units accumulated at graduation? (*Hint:* Draw a model of the curve, and shade in the top and bottom 8%, respectively. Think how you would find the *z* scores that cut off those shaded areas in the tails.)

3. You plan to sell a piece of land. A real estate agent tells you recent sales prices of similar properties have been normally distributed. Some 5% of such properties have sold for $238,651 or more, and another 5% have sold for $198,349 or less. Given this information, what is the expected selling price for your property, and how much variation might you expect around this figure?

4. A chiropractor reported the weights of randomly selected college students' book bags and backpacks. She found the weights to be normally distributed around an average of 13.70 pounds, with a standard deviation of 3.59 pounds. In this distribution, what is the probability a student's bag or backpack weighs

   a. at least 20 pounds?
   b. between 10 and 15 pounds?
   c. no more than 5 pounds?
   d. between 18 and 20 pounds?

Answer the following two questions for this distribution:

   e. How much does a student's book bag weigh if it falls into the most extreme one-third (or 33.33%) of this distribution?
   f. How much does a student's book bag weigh if it is among the heaviest one-third of all bags?

5. In a local school district, the tenures or times teachers have spent on the job are normally distributed. The middle 40% of teachers in this distribution have been in the profession between 42 and 58 months. What are the mean and standard deviation of this distribution?

6. According to a national survey, the number of hours per week retired Americans spend online is normally distributed around an average of 14.80 hours, with a standard deviation of 4.76 hours. Given this distribution, what is the probability a retiree spends

    a. 12 or more hours per week online?

    b. 10 or fewer hours per week online?

    c. between 15 and 20 hours per week online?

    d. between 10 and 20 hours per week online?

Answer the following two questions for this distribution:

    e. How many hours per week does a retired person spend online if he or she is at the 20th percentile rank in this distribution?

    f. How many hours per week does a retiree spend online if he or she ranks in the top (or most online) 12% in this distribution?

    7. Gamblers' wins and losses (mostly losses) in US casinos are normally distributed around an average of $110 lost ($\mu = -\$110$), with a standard deviation of $76. Given this distribution, what is the probability a gambler could expect to

    a. win (i.e., leave the casino ahead or "in the black," so to speak)?

    b. lose $200 or more?

    c. win between $20 and $50?

Answer the following three questions for this distribution:

    d. How much does someone win if he or she is among the luckiest (i.e., most winning) 10% of all gamblers?

    e. How much does a player win or lose if he or she falls at the 65th percentile rank in this distribution?

    f. What is the probability a player breaks even, meaning he or she does not win but does not lose either?

    8. There are two parts to any model or drawing of the normal curve: (a) the horizontal (lower) axis, and (b) the arc or area below the curve and lying above the lower axis. What does each part of the curve represent? What does each tell us or show? (Please be as complete as you can; avoid one- or two-word answers.)

    a. The horizontal axis.

    b. The area of the curve.

    9. Why do we need z scores when we use the normal curve? Why can we not simply use raw scores or X values directly with the normal curve table?

    10. Why may z scores be positive or negative, whereas the standard deviation (on which z scores are based) must always be a positive number or value?

11. Why may we add or subtract areas of the curve but never add or subtract z scores?

12. How do the concepts of *relative frequency* and *probability* relate to the normal curve model?

13. Based upon your X score, suppose your z score was 2.03. Would your X score be fairly average or more of a rare event? Why so? (Try to answer this without consulting the normal curve table.)

14. We all have opinions about other drivers, right? Suppose you could assign z scores to them. In plain language, how would you describe drivers to whom you gave the following scores? (Try these without consulting the normal curve table.)

   a. $z = .13$
   b. $z = -1.89$
   c. $z = 2.05$
   d. $z = -.60$

15. Assume a set of test scores is normally distributed. Please describe a method by which letter grades (A through F) might be assigned by using students' z scores.

16. What effect does a small versus a large standard deviation (or σ) have on the shape of a normal distribution? (*Hint:* On the same scale or X axis, visualize normal curves with large versus small standard deviations.)

17. What effect would a different mean (μ) have on the shape of a normal distribution? What would change if the mean took on a different value?

18. Looking back and for review, could we assign z scores to respondents' answers in Tables 1.1 (Marital Status) and 1.3 ("Ever Told Minor Lies?")? Why or why not?

Binomial Probabilities

19. Criminologists tell us that about one-third of US embezzlers are female. Of any 6 embezzlers, what is the probability that less than half are female?

20. A survey found that 46% of American men believe women judge them by their looks, not their brains. Of any 50 US males, what is the probability at least half will believe they are judged mainly by their looks?

21. Describe the probability distribution for getting X number of heads when flipping a normal coin 8 times. This may be done (a) with a graph or drawing, (b) by showing the actual probabilities of X heads, or (c) by generally describing and explaining the distribution in prose.

22. As its academic year begins, a university must reconstitute a committee. Interested and eligible candidates include 7 students, 6 tenured faculty, 4 nontenured faculty, and 8 staff members. If 3 people from each category will be randomly selected to form the 12-person committee, in how many different ways may the final body be composed?

23. Death row lives up to its name in different ways. Media reports reveal that 20% of prisoners on death row are actually executed. Others die of natural causes (45%) or by suicide (35%).

> a. Of 10 prisoners on death row, what is the probability that 2 or fewer will be executed?
> b. The most populous state, California, has about 650 people on its death row. Given the data above, what is the probability that at least 290 of those men and women will die of old age or other natural causes?

24. Nationwide, about 33% of all telephone users have signed onto do-not-call lists to avoid telemarketers, appeals for donations, and similar unwanted calls.

> a. Of any 75 telephone subscribers, what is the probability that 20 or more have joined do-not-call lists?
> b. Of 12 telephone users, what is the probability that 5 or fewer have registered with do-not-call services?

25. You and a friend are in an eight-student seminar together. The instructor plans to randomly pair people for research projects. What is the probability that you and your friend end up working together?

26. In poker, the best possible hand is a royal flush: an ace, king, queen, jack, and ten all of the same suit. Considering all the poker hands one *could* get, what is the probability of getting a royal flush?

27. Every season, the Boston Red Sox play 81 baseball games at home in Fenway Park. However, there is a 9% chance of any Fenway game being rained out and postponed. Over the course of the season, what is the probability of 10 or fewer Red Sox games in Fenway being rained out?

28. Given the nature of the equipment used, garbage collecting is one of the most physically dangerous occupations in the United States. During any given year, about 16% of all workers report job-related injuries requiring time off. Over the course of a year, in a 12-person crew, what is the probability that at least 3 workers sustain injuries debilitating enough to keep them off the job?

29. In a survey, 595 college students responded to the statement, "I feel as though I belong in the campus community." Some 357 students answered in the affirmative or agreed. Based on this study, of any 15 students, what is the probability less than half (i.e., 7 or fewer) express similar sentiments?

30. Even in the best or most contentious of times, only about two of every three eligible voters go to the polls in the United States. In one small town, a particular precinct of 3507 registered voters is facing an especially divisive library board vote. (The issue: Should access to Internet pornography sites be denied on computers at the library's local branch?) What is the probability that more than the expected number of voters in this precinct turn out for the election?

31. Ho-hum college graduates? Nationally, of 100 graduates, 75 actually go through their graduation (cap and gown) ceremonies. Given these numbers,

   a. Of 10 graduates, what is the probability that 4 or more attend their graduation ceremonies?
   b. In a graduating class of 1570 students, what is the probability that 1200 or more show up for the graduation ceremonies?

32. US college students are mainly single and childless. However, surveys reveal that 16.5% do have children.

   a. Among a college population of 1000 students, how likely is it that 175 or more have children?
   b. Among a random sample of 5 college students, what is the probability all are childless?

33. Researchers tell us that young adult Japanese Americans are among the most likely to marry people of other ethnic backgrounds. Some studies report that as many as 70% do so. In an extended Japanese American family with 13 married cousins in their twenties and thirties, only 3 have married other Japanese Americans. Given the researchers' findings, what is the probability of this happening, that is, of 3 or fewer marrying other Japanese Americans?

34. University officials report that 65% of all graduates eventually land in careers related to their college majors. Of any 100 graduates, what is the probability that between 70 and 75 will have careers in fields similar to their majors?

35. A campus survey asked students about the main sources of stress in their lives, and 62% mentioned money or finances. Given this information:

    a. Of any 8 students, what is the probability that 6 or more find financial matters particularly stressful?

    b. Of any 15 students, what is the probability that less than half (i.e., 7 or fewer) are troubled by finances?

36. "If at first you don't succeed, try, try again." What does this statement tell you about the author's view of probability and repeated success-or-failure trials?

37. On each spin of a roulette wheel, the ball goes around and eventually falls on a particular number. You have been playing the same wheel for some time and losing, but you expect your luck to change. What is wrong, if anything, with assuming the long-run averages will now catch up for you and that you will start winning?

38. In your own words, *why* do we have two formulas for deriving binomial probabilities rather than using the easier normal curve procedure every time? (*Hint:* Do not merely say, "It is due to sample size." That is too superficial. And please do not describe steps in formulas. That misses the point. Explain *why* we use two different procedures.)

39. The products of *np* and *nq* comprise the criteria for continuity. Please explain in your own words what part each of the following factors plays in determining a normal approximation.

    a. *n* or sample size.
    b. The *p* and *q* values.

40. Why do we sometimes use the true limits of *X* rather than the *X* value itself when determining binomial probabilities?

41. In your own words, what is the complement of an event, and why is it sometimes useful in determining binomial probabilities?

Describing a Population: Estimation with a Single Sample

In this chapter, you will learn how to:

- Estimate unknown population averages and proportions
- Use the Student's *t* Distributions
- Determine whether *z* may be used with the sampling distribution of the proportion
- Apply the hypergeometric correction factor (fpc)
- Determine the optimum sample size for estimating a population average or proportion

THE CONCEPT OF THE SAMPLING DISTRIBUTION is the key to and the foundation of inferential statistics. This chapter introduces the two most useful sampling distributions and describes how they are used to estimate population parameters. These are: (1) the sampling distribution of the mean, and (2) the sampling distribution of the proportion, the latter being similar to the binomials of Chapter 3. We will also consider A Student's* *t* distributions as possible substitutes for the familiar z distribution when problems involve means or averages. We sometimes base our inferences and conclusions about population averages on small samples. Under these circumstances, the normal distribution or *z* curve is not accurate, so we substitute one of the *t* distributions. (And, yes, there is more than one *t* distribution. When we substitute *t* for *z,* selecting the correct one is

---

t distributions derive from the work of W.S. Gosset (1876–1937), who published under the appellation of "A Student." In Ireland, he systematically and meticulously compared measurements on small samples of beer (Guinness) to desired population parameters.

obviously important, but it is not difficult.) We start with the concept of sampling distributions and then turn to the two specific cases. In doing so, we also look closely at the *t* distributions and the hypergeometric correction factor for both *z* and *t*.

Following this introduction, the second part of the chapter applies these concepts to estimating unknown population means and unknown population proportions using data from single samples. We estimate a range or interval of values within which a parameter is expected to fall. We also introduce the concept of the **confidence level,** *the degree of confidence we have that the unknown population parameter actually does fall into our interval*. It may or it may not, but we naturally want this confidence level to be high, and we work only with confidence levels of 90% or more. There is still a chance, of course, that the population parameter does not fall into the interval we construct, but it should be a small chance.

Note the role of probability here. At the 90% confidence level, there is a .90 probability that the unknown population parameter lies within our interval and a 10% chance it does not. If we use the .95 confidence level, there is a 95% probability our interval contains the true population figure and a 5% chance it does not.

Finally, we determine (prior to gathering data) how large our sample must be to guarantee a confidence interval estimate with a certain exactness. **Confidence interval estimation** *is the construction of a range, based upon sample data, into which the unknown population is expected to fall with a certain level of confidence*. If we know what margin of error is tolerable for our estimate of μ or *p,* we may determine exactly how large a sample we need to give us this level of precision.

The Theory We Need

Curves of Many Sample Means

Up to now, we have considered the possibilities of data coming from either samples or, if we were very lucky, entire populations. We have seen that samples and populations have different symbols for their respective means and standard deviations. We have also seen that we may use the normal curve with data from either samples or populations, assuming those data are reasonably normally distributed. Now we use the normal curve in a more theoretical way. With inferential statistics, we apply the normal curve model to hypothetical or theoretical sampling distributions. They are hypothetical or theoretical distributions for the simple reason we do not often generate them empirically. We may easily generate sample and even population distributions; we collect data on samples and populations. Sampling distributions of any size would take so long to complete empirically that we traditionally rely on statistical theory and previous research to confirm their existence and describe their characteristics. We also may generate computer-simulated sampling distributions in a fraction of the time it takes to do it by hand. These computer-aided sampling distributions confirm all that statistical theory had suggested years ago.

What is a sampling distribution? *A* **sampling distribution** *is the hypothetical distribution of a statistic taken from a large number of random samples (of a given size) drawn from a particular population.* An example illustrates what this means. Imagine a university student body of 30,000 people. We draw a random sample of 100 students, and we ask each one how many times he or she attended any kind of organized religious service in the last month. We get the average for this sample; let's call it $\overline{X}_1$. We replace these students into the population and repeat the whole procedure. We get the mean number of services attended for our second random sample of 100 students and label it $\overline{X}_2$. We continue doing this ad infinitum, each time replacing the last sample, picking another 100 students at random, and recording the mean. We end up with a very large number of sample averages: $\overline{X}_1$, $\overline{X}_2$, $\overline{X}_3$, $\overline{X}_4$, $\overline{X}_5$, and so on.

Imagine now that we construct a graph or frequency polygon for all our sample means. In effect, we are treating each sample mean as a single observation. Our graph is the **sampling distribution of the mean** (Figure 4.1). The statistic recorded is the mean, and our graph shows *the distribution for all the possible*

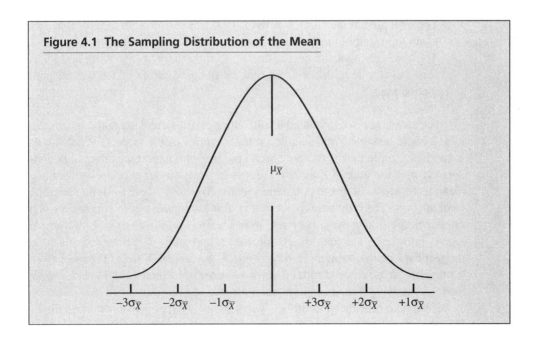

Figure 4.1 The Sampling Distribution of the Mean

$\mu_{\overline{X}}$

$-3\sigma_{\overline{X}}$ $-2\sigma_{\overline{X}}$ $-1\sigma_{\overline{X}}$ $+3\sigma_{\overline{X}}$ $+2\sigma_{\overline{X}}$ $+1\sigma_{\overline{X}}$

sample means, assuming random samples of *n* = 100. One can no doubt see why such sampling distributions are rarely attempted in practice—generating even one would be an enormous undertaking. But statistical theory, computer simulations, and the work of earlier statisticians actually reveal the characteristics of such sampling distributions. Those features, with slight variations in terminology, turn out to be the same as those highlighted earlier for normal distributions:

• The sampling distribution of the mean is normal in shape. It follows the normal curve, and this is true even if the original population is not normal, or is skewed. This is known formally as the central limit theorem. We are graphing sample means here, not individual *X* values. Given that sample means are the "centers" or balance points for our samples, they cluster around the true population mean, give or take a bit for expected sampling error. Our sampling distribution of the mean is normally distributed around the true population mean. Or, as it is sometimes put, the mean of all the means equals the true population mean.

• We have a standard deviation for this distribution, but with a sampling distribution it is known as the standard error. Here, we have the **standard error of the mean.**

• Virtually all the sample means will fall within ±3 standard errors of the central population mean. The sampling distribution curve is asymptotic, its tails theoretically extending into infinity, but practically all sample means will fall within this ±3 standard error range.

• The total area of the curve is set at 1.0000 and various segments of the area, again representing probabilities, are expressed as proportions.

• The same areas of the curve, that is, approximately 68%, 95%, and 99%, respectively, lie within ±1, ±2, and ±3 standard errors of the central population mean.

The standard error is related to **sampling error,** a fundamental concept. (In fact, it could be called "the standard sampling error of the mean.") What causes individual sample means to vary around the true population mean could be random error or the luck of the draw in sampling. We do not expect every sample mean to be identical to the population figure. We know there is some sampling error, always. The only way to avoid it is to measure the whole population—an unrealistic goal in most survey research. Due to sampling error, some sample means fall above the true population mean, and some fall below. So, just as the standard deviation measured how much the typical *X* value deviated from a single mean, now *the **standard error** measures how much the typical sample mean deviates from the true population mean.*

If we take this one step further, you will see why we say a sampling distribution is based upon means from many hypothetical samples of a certain size.

The larger the sample size, the closer our hypothetical sample means will cluster around the true population figure. Assuming random selection, larger samples are more accurate. With larger samples, we get less variation from one sample mean to the next. The larger the sample, the smaller the sampling error and the smaller the standard error. In fact, the standard error is inversely proportional to the square root of the sample size. As $\sqrt{n}$ goes up, the standard error goes down. This principle is plain to see in the formulas for the standard error (equations 27 and 28).

To return to the basic model, Figure 4.1 shows the sampling distribution of the mean. Apart from minor differences in labels, notice the obvious similarity to Figure 3.1, which shows the basic normal curve features. Note in particular the changes in symbols for the mean and standard deviation (standard error). The mean of this sampling distribution, to distinguish it (symbolically, not mathematically) from the population mean, is $\mu_{\bar{X}}$. The standard error of the mean also acquires a subscript; it becomes $\sigma_{\bar{X}}$.

We may also rewrite our original z formula to reflect the new symbols:

(26)
$$z_{\bar{X}} = \frac{\bar{X} - \mu_{\bar{X}}}{\sigma_{\bar{X}}}$$

where

(27)
$$\sigma_{\bar{X}} = \frac{\sigma}{\sqrt{n}}$$

or

(28)
$$s_{\bar{X}} = \frac{s}{\sqrt{n}}$$

If we know the population standard deviation (σ), equation 27 is appropriate and preferred. In many cases, however, we know only the sample standard deviations (s). We then use equation 28 for the standard error of the mean and substitute $s_{\bar{X}}$ in the denominator of equation 26.

Curves of Many Sample Proportions

A sampling distribution is the distribution of *any* statistic drawn from many samples. Conceivably, we could have a sampling distribution of the median, or a sampling distribution of the mode, or a sampling distribution of the standard deviation, and so on. We may have a sampling distribution for any measure possible on sample data. In fact, however, there are two sampling distributions that are used with any frequency. The first, as just discussed (Figure 4.1), is the sampling distribution of the mean. The second is the **sampling distribution of the proportion** (Figure 4.2).

Figure 4.2 The Sampling Distribution of the Proportion

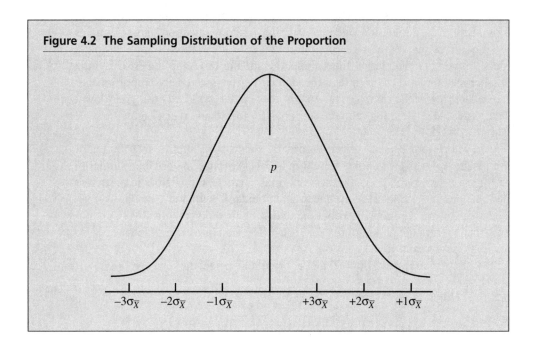

To review, recall that proportions are derived from variables (questions) framed in terms of Agree/Disagree or Yes/No. Researchers often structure their survey questions in such formats. We analyze data in terms of proportions or percentages answering "Agree" or "Yes." In Chapter 3, these sorts of variables were termed *binomials*. Our measurements consisted of two categories, success and failure. Now, however, instead of the whole number of successes, we record the *proportion* of cases falling into a designated category. There may be three or more response alternatives when the question is posed to the sample (e.g., Agree-Undecided-Disagree), but these are reduced to the proportion in agreement or the proportion saying "Yes" as the data are treated statistically.

For the theoretical sampling distribution of the proportion, imagine all the sample proportions we could derive from many random samples. Our symbol for the sample proportion is p', read as "p-prime," and we have p'_1, p'_2, p'_3, p'_4, and so on. Similar to the case for the mean, these sample proportions vary around the true population proportion as a function of sampling error. The **standard error of the proportion** *is the measure of variation in the sample proportions.*

Although we have Greek and Roman characters to distinguish between population and sample measures for averages (μ versus $\overline{X}$, respectively), we have no such symbols for proportions. The letter "p" represents both population and sample proportions. To distinguish between the two, we use p' to denote the sample proportion and p to symbolize the population proportion. q and q' represent $1 - p$ and $1 - p'$, respectively.

Our formulas use the frequency of agreement or positive responses over the population N, or a sample, n. For the population proportion:

$$(29) \qquad p = \frac{\text{Frequency Agree or Yes}}{N} = \frac{X}{N}$$

Or, for the sample proportion:

$$(30) \qquad p' = \frac{\text{Frequency Agree or Yes}}{n} = \frac{X}{n}$$

And for the standard error of the proportion:

$$(31) \qquad \sigma_p = \sqrt{\frac{pq}{n}}$$

or

$$(32) \qquad \sigma_p = \sqrt{\frac{p'q'}{n}}$$

As was the case with the standard error of the mean, if we know the population values of p and q, then equation 31 is appropriate and preferred. If we must, we may substitute our sample's p' and q' measures and use equation 32.

For any z score in the sampling distribution of the proportion, we have:

$$(33) \qquad z_p = \frac{p' - p}{\sigma_p}$$

In addition to the sampling distributions of the mean and proportion, we must also consider *t* **curves** or *t* **distributions.** These are important for sampling distributions of the mean when we cannot use the normal curve and z scores. (There are no *t* curves for the sampling distribution of the proportion.)

Working with Small Samples: *t* and Degrees of Freedom

When and Why Do We Use *t*?

Recall that, in any sampling distribution of the mean, due to sampling error, some hypothetical sample means fall above the true population mean and some fall below it. As sample size n increases, however, sampling error decreases, and the hypothetical sample means cluster nearer the central population mean. In general, the sampling distribution of the mean conforms to the normal curve. By implication, some curves are a bit more peaked than others (larger n's, less sampling error, and smaller standard errors), but these sampling distribution curves all follow the general shape of the normal curve.

What happens, however, if we have a very small *n*? What happens if the sample size is so small that there is considerable sampling error? We get a sampling distribution curve with a lot of spread and a comparatively large standard error. It does not peak as much in the middle as does the normal curve. It is still centered on the true population mean, but the individual sample means vary a lot around this central figure. In fact, this curve shows so much spread that it no longer conforms to even the general shape of the normal curve.

As we imagine sampling distributions based on smaller and smaller *n*'s, the distributions or curves become more variable. Once *n* falls below a certain threshold, the sampling distribution of the mean no longer conforms to the normal curve. Any *n* below that threshold produces a sampling distribution curve with more spread and less peak than the *z* curve. Below that threshold sample size, we get a slightly different curve for each individual *n*. The relationship between sample size and the shape of sampling distributions of the mean, illustrated in Figure 4.3, is as follows: With any *n* below a minimal threshold size, we must

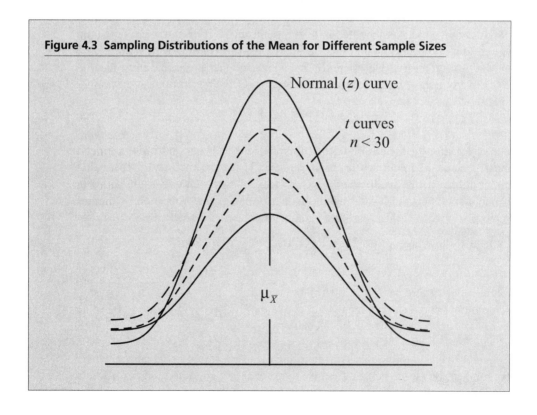

Figure 4.3 Sampling Distributions of the Mean for Different Sample Sizes

Normal (*z*) curve

t curves
n < 30

$\mu_{\bar{X}}$

use one of the *t* curves, and we must use the one *t* curve that corresponds to our particular sample size. As *n* approaches the threshold size, the sampling distributions of the mean gradually become more peaked until they reach the shape of the normal curve. With any *n* above that threshhold, we may use the normal curve and *z* scores.

The practical question becomes: What is that "magic threshold"? When is the sampling distribution of the mean peaked enough to approximate the normal curve? When do we use *z* and when must we use one of the *t* curves? *For the sampling distribution of the mean, the minimum sample size for using the normal curve and z scores is 30.* Whenever we have an *n* less than 30, we use one of the *t* curves. However, another factor does sometimes affect whether we use *z* or *t*.

The other consideration is whether we know the true population standard deviation, σ, and variance, σ^2, or we must use their sample counterparts, *s* and s^2. There is a difference of opinion on this issue. Some statisticians argue that the standard normal distribution or *z* curve may be used only when we have a large sample ($n \geq 30$) *and σ is known*. The argument is that we cannot be sure the sampling distribution curve is normal unless, besides being based upon a large enough sample, it is also defined by the population standard deviation. Using the sample standard deviation (*s*) suggests a more variable sampling distribution curve, and hence a *t* curve. The alternative argument is that when *n* is at least 30, any difference between σ and *s* is likely to be minor. Under those circumstances, the sample standard deviation is assumed to be an unbiased estimator for the true population figure. With larger samples, it becomes permissible to use the normal distribution even when σ is unknown.

Which view is correct, and which guideline do we follow? It depends. The "jury" of statisticians is still out on this issue. For this text, we adopt the more flexible approach, the second guideline: We assume that *s* may be substituted for σ when $n \geq 30$, and that we may use the *z* curve. We assume this is the case when we have random samples drawn such that all members of the population in question have had reasonably equal chances to be selected. Under those circumstances and with $n \geq 30$, we use *s* as an unbiased estimate for σ.

The *t* Table and Degrees of Freedom

t values are shown in Table C in Appendix I. Like *z*, *t* scores represent numbers of standard errors away from the mean, but the *t* table also differs from the normal curve and *z* table. It does not show specific *t* values associated with complete areas of the curve. It shows the *t* scores cutting off small areas in the tails of each *t* curve. There are two reasons for this format. First, each small sample size produces a differently shaped sampling distribution of the mean (Figure 4.3). With so many different *t* curves, it would be prohibitive to present each one in its entirety. Second, we do not need the complete curves. As you will see, we are interested only in *t* values that cut off relatively small areas in the tails of these curves. Although abbreviated in comparison to the normal curve table, the *t* table is quite sufficient for all we will require of it.

The columns and rows of the *t* table also differ from those of the *z* table. Down the left-hand side is a column labeled d.f. This stands for **degrees of freedom.** In brief, *this is the way we take sample size into account when using the* t *table.* By getting the correct row in this column we make sure we are using the proper *t* values (and *t* curve) for our sample size. In most circumstances, d.f. = *n* − 1, our sample size minus 1. Given *n* = 26, for example, we look along the row corresponding to 25 d.f. for a particular *t* value.

How do we get the term *degrees of freedom?* It is a bit complicated, but imagine that we have *n* number of *X* values and that they total to a certain sum. All but one of these *X* numbers (*n* − 1 of them) are free to vary, so to speak, or possibly be different. That final *X* value then must be a certain or predetermined number for the total to remain constant. It would *not* be free to vary. This is a somewhat esoteric concept, but think of it as merely a way of taking *n* into account when we use the *t* table.

There are three rows across the top of Table C. The topmost row is titled "Confidence Level." This row has a specific function and is used when we wish to estimate a population mean based on the data from a small sample (*n* < 30). The column headings (.90, .95, etc.) refer to broad central areas of *t* curves (e.g., the middle 90% or the middle 95%). The next row is labeled "Two-tailed Test." The various column headings here are .10, .05, and so on. These columns show the *t* values cutting off that much of a curve in *both tails combined.* For example, a *t* score in the .10 column cuts off a .05 (or 5%) area in each tail of the curve. Finally, the "One-Tailed Test" row also lists several areas of the curve as column headings: .05, .025, etc. The *t* values here mark off these areas in only one tail of the curve.

The concepts of confidence intervals and confidence levels are related to the two-tailed *t* values. Notice that the two-tailed values and the confidence level probability figures (for any given column) are complementary; they always add up to 1.00. The two-tailed entries refer to combined areas found in both the tails of the curve, while the confidence level or probability refers to the remaining area in the broad middle part of the curve. For instance, if there is 5% (.05) in the two tails, it means there is 95% (.95) of the curve remaining in the center. We may therefore look at the same *t* value in two ways: It either cuts off 5% in the tails or it circumscribes the middle 95%. Either way, it is the same *t* value.

As an example comparing the *z* and *t* tables, assume we need the specific *z* and *t* values that mark off 1% of the curve in one tail. The *z* score is the same for all sample sizes above the threshold *n* of 30. Find .0100 in column C of Table A and read off the *z* score of ±2.33. This means that for the *z* curve, we go out along the lower ($\overline{X}$) axis 2.33 standard errors from the mean to reach the final

1% of the curve. For t, assume we have a sample size of 19. With 18 d.f. we read across the appropriate row until we come to the column headed "One tail, .01," and we find a t value of ± 2.552. That means we must go fully 2.552 standard errors from the center before reaching the final 1% area of the t curve. Because t curves are more spread out and flatter than the z curves, we must go farther (i.e., more standard errors) from the mean than required to reach comparable points in the z curve. The descriptive term that means less peaked than the corresponding normal curve is *platykurtic*, so we may say that t curves are platykurtic.

There is one similarity between the z and t tables, however. Neither one includes negative values in the table proper. As with the z scores, t scores falling below the mean are always negative. Similarly, the upper and lower halves of t curves, just like the z curve, are symmetrical. It would be redundant to list the same t scores for the upper and lower halves separately. So only positive t values are shown. When working below the mean, as we did with z, one must remember to affix the minus sign (–) to any t score taken from the table. For example, for 18 d.f., a t score of ± 2.552 cuts off 1% in each tail of the curve. For the lower tail specifically, the t value is –2.552.

t distributions apply solely to the sampling distribution of the mean, however. The sampling distribution of the proportion is a different story, which we examine next.

Only *z*: The Unusual Case of Proportions

For the sampling distribution of the proportion, the question is not whether to use the z or t distribution, but rather, whether we may use z. There is no sampling distribution of the proportion that conforms to t curves. With small samples, such distributions lose any continuity altogether and resemble the discrete binomial distribution of Chapter 3. With proportions, the question is whether our sample size is sufficient for the appropriate sampling distribution of the proportion to conform to the z curve. Also, unlike the above situation for averages, there is no one set criterion or n by which to make the decision. There is a kind of sliding scale that depends upon the values of n and p and q. As we saw with binomials, with more extreme values of p and q (i.e., values close to either .00 or 1.00), we need larger sample sizes before it is appropriate to use the normal curve. Given extreme values of p and q (or p' and q'), it takes larger sample sizes for the sampling distribution curves to become less skewed and take on the shape of the symmetrical normal curve. With p and q scores close to .50, however, our sampling distribution curves follow the normal curve even with comparatively small sample sizes. For these middle-range values of p, the curves are naturally more symmetrical. All the hypothetical p' values cluster fairly near the middle of the p range anyway (i.e., around .50) and produce curves that peak at or near the center.

To determine whether a sampling distribution of the proportion approximates the continuous normal curve, we use the criteria shown in Table 4.1. Use the

Table 4.1 Minimum *n* Necessary for Using *z*, Given the Minimum Value of *p* or *q*

Minimum Value of p or q	Minimum n for Using the z Curve
.02	5,000
.05	2,000
.10	900
.20	300
.30	150
.40	60
.50	30

Source: From G. W. Summers, W. S. Peters, and C. P. Armstrong, *Basic Statistics in Business and Economics,* 3rd edition, page 232. ©1981. Reprinted with permission of Wadsworth, a division of Thomson Learning: www.Thomsonrights.com.

smaller value, p or q (or p' or q') to determine the minimum sample size necessary for using the z curve. If our sample is smaller than the minimum n listed, our sampling distribution of the proportion is too discrete or asymmetrical to assume it conforms to the normal curve. The proper use of Table 4.1 obviously requires interpolation in many instances to get the correct minimum n, but it does give us a workable guideline for using z with proportion problems.

To illustrate our use of Table 4.1, assume a problem includes a sample size of 600 and a p' value of .47 (thus, $q' = .53$). The smaller of our p' and q' figures is $p' = .47$. Locating where .47 falls in the left-hand column of Table 4.1, we read over to the right column. With this p' value, we need a minimum sample size between 30 and 60 in order to use the z curve. By interpolating, we find a minimum n of about 39 to produce a sampling distribution approximating the normal curve. With $n = 600$, it is safe to use the z curve in our analysis. That will not always be the case, however.

Imagine a problem for which $p' = .75$, $q' = .25$, and $n = 160$ cases. We use Table 4.1 with $q' = .25$ and find our minimum sample size to be halfway between the entries of $n = 150$ and $n = 300$, or $n = 225$. Our sample size of 160 is insufficient. Our sampling distribution is not symmetrical or continuous enough to justify using the normal curve. We would have to seek alternative ways to address our problem, but those are beyond the scope of this text. We simply note here that the z curve could not be used.

One additional feature of sampling distributions involves the occasional use of a helpful correction factor, which is discussed next.

A Correction for Small Populations

When we use the *z* and *t* distributions, we may be entitled to use a correction factor in our formulas, a correction factor that makes our work a little more accurate. *We may use this adjustment when (1) we are sampling without replacement, and (2) we are picking or sampling at least 10% of the population.* This is the **hypergeometric correction factor,** otherwise known as the **finite population correction,** or **fpc.**

Before we get to the correction factor itself, let's see why is it used only under certain circumstances. Recall the earlier discussion of sampling error. The standard error reflects the amount of sampling error. The larger our generally random sample, the smaller our sampling error (and standard error). That sample size inversely affects our standard error is clearly evident in equations 27, 28, 31, and 32. Sample size (*n*) occurs in the denominators of those formulas. The larger the *n*, the smaller the resulting standard error. Something else also affects our sampling error (and standard error), however. This is the variation in the population.

Variation in the population has a direct effect on sampling error and the standard error. It stands to reason that, for any given sample size, the more variation there is in our population, the more sampling error we could expect. Sampling a fairly homogeneous population is typically easier than sampling a comparatively heterogeneous one. With a heterogeneous population, a random sample of 400 people gives us a certain amount of sampling error and a certain standard error. In contrast, if we sample 400 people from a homogeneous population, we expect less error and more accuracy. Members of this second population are more alike, and therefore fewer random differences affect our sample's accuracy. So, generally speaking, the greater the population variation, the larger our sampling error and standard error.

But (and you knew there was a "but" coming), when we sample as much as 10% of the population *without replacement*, we argue that the population has continually and substantially changed over the course of our picks. Not only has the remaining population become progressively smaller, the variation in that remaining population has steadily shrunk as well. We expect a certain amount of variation in all populations, but if we reduce a population to 90% or less of what it once was (that is, take out 10% or more), then it may well be less varied than it was originally. This variation, of course, is reflected in our standard deviation. Although σ (or even *s*) may have been a true measure of the original population's variation, it may not be a true figure for the reduced population from which our later picks have been made. When we use that standard deviation to calculate the standard error of the mean, we may be using a slightly inflated figure. The hypergeometric correction factor (equation 34) adjusts that.

(34)   Hypergeometric correction factor (fpc) = $\sqrt{\dfrac{N-n}{N-1}}$

The hypergeometric correction factor becomes a multiplier for the standard error and slightly reduces our standard error. The correction factor itself is usually in the .85 to .99 range (and never more than 1.00). For instance, sampling 400 people from a population of 2,000 ($n = 400$, $N = 2,000$), would give us a multiplier of:

$$\sqrt{\frac{2000-400}{2000-1}} = \sqrt{\frac{1600}{1900}} = \sqrt{.800} = .895$$

Multiplying the standard error by a number less than 1.00 always gives us a smaller revised figure, and we want to get the most accurate (and usually smallest) possible standard error suitable for our data.

We will use this correction factor later in the chapter, but first, it usually goes by a shorter name. The term *hypergeometric correction factor* does not exactly roll off the tongue. Some people refer instead to the *sampling correction factor,* the argument being that its use is based upon how we sampled the population. A more common name, introduced earlier, is the *finite population correction*, or simply fpc. The rationale for this term is that we are, in all probability, dealing with a finite or small population when we use this correction factor. Many surveys are done on very large populations (e.g., the registered voters of South Dakota or of New York). In such situations we never come even remotely close to sampling 10% or more of the population; that is too unwieldy and expensive. With smaller (or finite) populations, however, sampling 10% or more is not only possible but quite realistic (e.g., a random sample of adults who serve as Girl Scout leaders in a particular county, high school principals in eastern Oregon, and so on). Since it is only with these smaller populations that this correction factor is ever used in practice, it is often called the finite population correction, the fpc, a term we adopt in this book.

Equation 35 actually applies this factor for the corrected standard error of the mean:

(35) $$\sigma_{\overline{X}} = \frac{\sigma}{\sqrt{n}}\sqrt{\frac{N-n}{N-1}}$$

Or, equation 36 uses the sample standard deviation:

(36) $$s_{\overline{X}} = \frac{s}{\sqrt{n}}\sqrt{\frac{N-n}{N-1}}$$

Similarly, the corrected standard error of the proportion becomes

(37)
$$\sigma_p = \sqrt{\frac{pq}{n}}\sqrt{\frac{N-n}{N-1}}$$

or,

(38)
$$\sigma_p = \sqrt{\frac{p'q'}{n}}\sqrt{\frac{N-n}{N-1}}$$

Up to this point, you should have a feel for several basic concepts related to inferental statistical analyses: (1) probability; (2) the sampling distribution of the mean; (3) the normal distribution and the use of z scores; (4) the t distributions and the use of t scores for some sampling distributions of the mean; (5) the sampling distribution of the proportion and its unusual shape and characteristics; and (6) the hypergeometric correction factor, or the fpc. The remainder of this chapter looks at a particular statistical procedure called estimation. We estimate the population average or the population proportion based upon sample data, and you will see how the concepts above are actually used.

Estimating Population Averages

Recall that the normal distribution is actually a probability distribution. The areas or segments of the curve give us probability figures. We determine areas of the curve (probabilities) falling above or below any X score. The same is true when we consider sampling distributions of the mean. Areas of the curve may be interpreted as probabilities, but now we determine areas of the curve (probabilities) falling above or below any hypothetical $\overline{X}$ value. Also, and of more importance for confidence intervals, we may determine the $\overline{X}$ values that cut off the very top and bottom 5% areas of the curve, or the 1% areas. We will turn to the mechanics and formulas for confidence intervals shortly, but first, how does the sampling distribution of the mean relate to the theory underlying confidence intervals?

There is a chain of logic or reasoning that explains the theory of confidence intervals. In making this argument, we rely on the concepts of random sampling, the sampling distribution of the mean, and areas of that sampling distribution being probabilities. First, a single sample mean, at least one based on a properly random sample, is an **unbiased estimator** for the unknown population mean. Many random and independent samples from a population yield that many unbiased estimates for the population mean. All the individual means would, when graphed, form a symmetrical curve around the true population figure. Given unavoidable sampling error, some sample means fall closer to the population figure than others, but all cluster around that center.

Second, the area of the curve lying over any interval on the horizontal axis (the $\overline{X}$ axis) tells us the probability of any single sample mean, or $\overline{X}$ value, falling into that interval. For instance, roughly 95% of the sampling distribution curve falls within ±2 standard errors of the center. Therefore, 95% of all possible sample means fall into this same interval. This means that 95% of the time, when we derive a sample mean to estimate the population mean, we get an estimate falling into this interval. If 95% of the unbiased estimates fall into a certain interval, we say there is a 95% probability that the unknown population mean itself falls into this same interval. The upper and lower limits of this range therefore delimit this .95 confidence interval for the population mean. The same principle holds, of course, when we construct the .90 or .99 confidence intervals. The area of the curve and the width of the interval along the $\overline{X}$ axis become smaller or larger, respectively.

Note that we are taking some liberties in this interpretation. The population mean either *does* or *does not* fall into this interval. It is very much a yes or no, does or does not, black or white proposition. Technically, there are no "shades of gray," such as a "95% yes it does" sort of conclusion. Nevertheless, even though we pay lip service to the statistical niceties or formalities, this *is* how we commonly interpret confidence intervals.

We establish confidence intervals using the broad area around the center of the sampling distribution curve. We want the middle 90%, 95%, or 99% of all possible $\overline{X}$ values, or the 90%, 95%, or 99% of all $\overline{X}$ values that are most likely to occur. We want a certain range for the *most probable* sample means. This is where the unknown population mean is *most likely* to fall. To make all this workable, there is one additional step or assumption.

The theory behind confidence intervals assumes our sampling distribution constitutes a graph or curve of many individual sample means. The population average is the central value of our theoretical curve (the mean of all the means). We do not know what that mean is, but we need a central figure around which to build our interval. In addition, the reality of the situation when we are actually working on a problem is that we have only *one* sample, not many. To blend theory and practicality, we assume that our one sample mean equals the unknown population mean. The sample mean becomes the assumed central value for our sampling distribution curve and the central value around which our confidence interval is established. It becomes the unbiased estimator for the unknown $\mu_{\overline{X}}$.

With all the components in place, we may begin to assemble and label the final confidence interval. Our sample mean ($\overline{X}$) becomes the **point estimate** for the confidence interval. It is called the point estimate because it occupies a single point or value along the $\overline{X}$ axis. It is the single value around which the confidence interval will be constructed. On either side of that point estimate we establish an

error term or **margin of error.** This is to be determined as we go outward from the central point estimate to the upper and lower limits of the interval in the tails of the curve. At the .95 confidence level, our interval spreads out to cover the middle 95% of the curve. How many standard errors do we go above and below the mean to get this middle 95%? We consult the normal curve table for this, and it happens to be 1.96 standard errors. We then have our point estimate, and we will construct our .95 confidence interval by going 1.96 standard errors on either side of it.

Earlier, in Chapter 3, we said the middle 95% of a normal distribution was circumscribed by "roughly ±2" standard errors or standard deviations. It actually is *exactly* ±1.96 standard errors, not 2.00. It is necessary to be exact when constructing specific confidence intervals. To get this figure, we look in column C of Table A for an entry of .0250. (Remember, the middle 95% of the curve leaves 2.50% or .0250 in each tail.) Reading column A and the *z* score to the left of .0250, we get 1.96. It becomes ±1.96, of course, because we are going this many standard errors both above *and* below the point estimate.

For our confidence interval (C.I.) estimate for the population average, we get equation 39. Equation 40 shows the same procedure with the fpc, or finite population correction, included.

(39)   C.I. Estimate for an Unknown Population $\mu = \overline{X} \pm z \dfrac{\sigma}{\sqrt{n}}$

(40)   C.I. Estimate for an Unknown Population $\mu = \overline{X} \pm z \dfrac{\sigma}{\sqrt{n}} \sqrt{\dfrac{N-n}{N-1}}$

What about using *t* distributions or the sample standard deviation, *s*? Substitute a *t* value for the *z* coefficient in equations 39 and 40 when our sample size falls below 30, and remember that *t* has $n - 1$ degrees of freedom. Similarly, if we do not know the population standard deviation (σ), we substitute our sample figure, *s*. The practice problems that follow illustrate how these estimates work.

1. A recent survey asked college students about their TV viewing habits. From a population of 30,000 students, 1303 were randomly selected. By their own estimation, sample members averaged 3.72 hours of viewing per weekend day, with a standard deviation of 2.64 hours per day. What is the .95 confidence interval estimate for the population's average number of viewing hours per weekend day?

To solve the problem, we first record the information and draw a model of the curve to illustrate what we are seeking. The sample average ($\overline{X}$) is 3.72 hours per day, with a standard deviation (*s*) of 2.64 hours per day. Since *n* = 1303, we use *z* rather than *t* and substitute *s* for σ. For the .95 confidence level, *z* = ±1.96. *N* = 30,000, and since *n* is < 10% of that, we may not use the fpc.

For the actual calculations:

$$.95 \text{ C.I. for } \mu = \overline{X} \pm z\frac{s}{\sqrt{n}} = 3.72 \pm 1.96\frac{2.64}{\sqrt{1303}} = 3.72 \pm 1.96(.07) = 3.72 \pm .14$$

For the C.I.'s lower limit: 3.72 − .14 = 3.58 hours per day, and for the C.I's upper limit: 3.72 + .14 = 3.86 hours per day. Figure 4.4 shows the distribution curve for this problem.

Our conclusion may be stated in either of two ways. First, we may make reference to the overall interval: We are 95% confident that the average student in this population watches between 3.58 and 3.86 hours per day of television on weekends. Alternatively, we may cite the point estimate and margin of error. At the .95 confidence level, our error term, or margin of error, is ±.14 hours: We estimate the average student watches 3.72 hours of TV per weekend, and we are 95% confident that the true figure is within ±.14 hours of this estimate, that is, 3.72 hours ±.14 hours. There is only a 5% chance that the true (unknown)

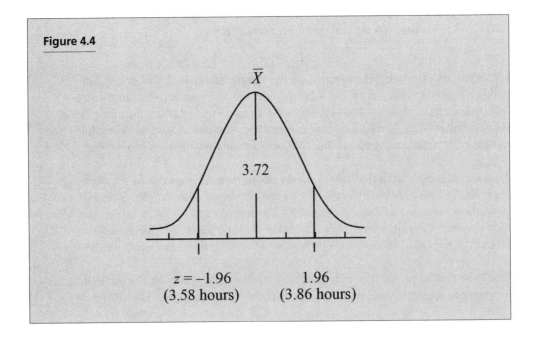

Figure 4.4

$\overline{X}$

3.72

$z = -1.96$
(3.58 hours)

1.96
(3.86 hours)

average would fall outside the 3.58- to 3.86-hour interval, that is, that it would be greater or lesser than our estimated range.

2. Many students from other countries study at US colleges. A particular campus has a 139-member International Students' Engineering Club. One member took a brief poll to see how long, on the average, his fellow members had been in the United States. Nineteen students took his survey and averaged 5.42 years in the country with a standard deviation of 2.19 years. What is the .90 confidence interval estimate for the club members' average time in the United States?

First, we transcribe the information given and diagnose the problem. The sample average ($\overline{X}$) is 5.42 years. The standard deviation (s) is 2.19 years. Since $n = 19$, which is less than 30, we will use t with $n - 1$ degrees of freedom. With 18 d.f. at the .90 confidence level (see Table C), $t = \pm 1.734$. Since n is more than 10% of N (139), we may use the fpc.

Then, for the .90 confidence interval estimate for the population mean:

$$.90 \text{ C.I. Est. for } \mu = \overline{X} \pm t \frac{s}{\sqrt{n}} \sqrt{\frac{N-n}{N-1}} = 5.42 \pm 1.734 \frac{2.19}{\sqrt{19}} \sqrt{\frac{139-19}{139-1}}$$

$$= 5.42 \pm 1.734[(.50)(.93)] = 5.42 \pm 1.734(.47) = 5.42 \pm .82$$

For the C.I.'s lower limit: $5.42 - .82 = 4.60$ years, and for the C.I.'s upper limit: $5.42 + .82 = 6.24$ years. Figure 4.5 shows the distribution curve.

The student may be 90% confident that the average club member has been in the United States between 4.60 and 6.24 years. Alternatively, he could say that, with 90% confidence, the average length of residence is 5.42 years with a margin of error of ±.82 years.

You may have noticed something in these examples that confirms an earlier point. In the first problem, with TV viewing, we had a comparatively large sample of $n = 1303$, and we had a standard error term of .07 hours. In the second situation, with a much smaller sample of $n = 19$, we got a standard error of .47 years, even with the fpc. The standard deviations in the two problems, however, were similar: 2.64 and 2.19, respectively. The standard error reflects the sample size as much as any other single factor; these examples underscore the fact presented in earlier sections: *the greater the sample size (n), the smaller the sampling error and the standard error.* Again, this is an important principle in inferential statistics. A third example illustrates a different confidence level, .99.

3. A large city operates a series of community gardens in which residents may lease plots. There is also a lengthy waiting list for spaces. The garden managers

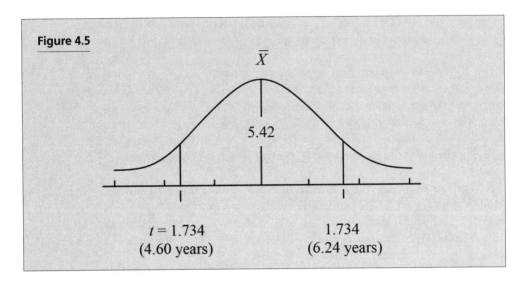

Figure 4.5

are planning a public education campaign regarding new policies and practices to be introduced, and they suspect the receptivitiy of prospective gardeners to the changes may be related to their educational levels. From the 210 people on the waiting list, managers sampled 24 and discovered an average of 15.29 years of formal education, with a standard deviation of 2.46 years. At the .99 confidence level, what is the managers' estimate for the average years of education completed by those waiting for garden plots?

Here, $\overline{X}$ = 15.29 years, and s = 2.46 years. With n = 24, t is appropriate, and given $n - 1$ or 23 degrees of freedom at the .99 confidence level, Table C shows $t = \pm2.807$. Finally, if N is 210 and n is 24, we may use the fpc.

To derive the .99 confidence interval estimate for the average years of education, we have:

$$.99 \text{ C.I. Est. for } \mu = \overline{X} \pm t\frac{s}{\sqrt{n}}\sqrt{\frac{N-n}{N-1}} = 15.29 \pm 2.807\frac{2.46}{\sqrt{24}}\sqrt{\frac{210-14}{210-1}}$$

$$= 15.29 \pm 2.807[(.50)(.94)] = 15.29 \pm 2.807(.47) = 15.29 \pm 1.32$$

For the C.I.'s lower limit: 15.29 − 1.32 = 13.97 years of education, and for the C.I.'s upper limit: 15.29 + 1.32 = 16.61 years of education. Figure 4.6 shows the distribution curve.

The garden managers may be 99% confident that the average formal education for those on the waiting list falls between 13.97 and 16.61 years. Alternatively,

Figure 4.6

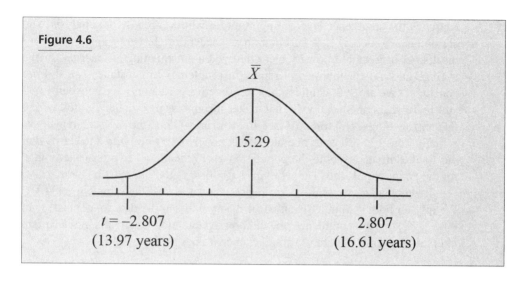

they can say that they are 99% confident that the average years of education completed is 15.29 with a margin of error of ±1.32 years.

Before turning to confidence intervals for population proportions, there are two more important points to be taken from the examples above. First, in addition to a larger sample size producing a smaller standard error, we can see that the confidence level has a *direct* effect on the overall margin of error. The confidence level determines our *z* or *t* value. The higher that confidence level, the larger *z* or *t* will be. The *z* and *t* scores tell us how many standard errors we must go from the mean to encompass our confidence interval. A higher confidence level and a wider interval, the more standard errors we must travel from the central mean. Other things being equal, then, the higher our confidence level, the larger our final margin of error. This is one reason we may not always opt for the .99 confidence level. We may be willing to settle for a slightly lower confidence level provided our margin of error is smaller and our interval more precise.

Why do we not use the 100% confidence level to be sure our interval contains the unknown parameter? Unfortunately, that is impossible. We may never be 100% sure of anything in inferential statistics. The quirkiness of sampling and sampling error are ever present. Also, our confidence level corresponds to an area of the normal curve (or a *t* curve). Since the tails of these curves are open, we can never contain the total or 100% area of a curve in any interval. We could go to the 99.99% confidence level, but there would still be *some* small chance for error, *some* small part of the curve outside our interval.

Second, notice how the conclusions for problems 1 to 3 are presented. They each mention the confidence level and also make it clear an *average* is estimated.

Knowing the confidence level is important for whoever reads your analysis. You are obliged to include that information so readers might better appreciate your results. Also, it should be very clear that you are estimating an interval for the average and not for the overall range. In problem 3, for instance, we did *not* conclude: "We are 99% confident that prospective gardeners have between 13.97 and 16.16 years of education." That makes it sound as though they *all* fell within that narrow range. Not true. (In fact, in the original data, the *sample's* range was actually from 12 to 20 years of education completed.) It is important to clarify that our final estimate refers to the population *average* and not to a projected range for the entire population. The word *average*, or a synonym, must be there.

To summarize, we build a confidence interval around the sample mean $(\overline{X})$. We rely on both (grounded) statistical theory and sample data to say that, with 90%, 95%, or 99% confidence, the unknown population mean falls into a certain interval; conversely, there is a small chance it does not.

Estimating Population Proportions

What about proportions? Do they work the same way? Confidence intervals for the unknown population proportion (p) follow the same statistical principles and theory as do those for the mean. A difference, however, as noted previously, is that we do not use t curves with proportions. It is z or nothing. With or without the fpc, we use only a z coefficient, and we construct a confidence interval around our sample proportion, p'. Equation 41 gives the estimate without the fpc, and equation 42 gives the estimate with the fpc.

(41)   C.I. Estimate for an Unknown Population $p = p' \pm z\sqrt{\dfrac{p'q'}{n}}$

(42)   C.I. Estimate for an Unknown Population $p = p' \pm z\sqrt{\dfrac{p'q'}{n}}\sqrt{\dfrac{N-n}{N-1}}$

Again, examples will illustrate how these two formulas work in practice.

4. *Various national security measures followed the events of September 11, 2001: airport searches, tightened immigration regulations, and more extensive use of cameras to monitor public places, among others. Having lived for a period with these new routines and as part of a larger survey, a sample of college students was asked to respond to the statement, "National security is more important than personal privacy." The 888 respondents were randomly drawn from a population of about 30,000, and 320 answered in the affirmative. At the .90 confidence level, what percentage of the population might be estimated to agree that national security should take precedence over personal privacy?*

To diagnose the question, we first retrieve the relevant information: We need the p' and q' values. Given the way the question is phrased, we let p equal the proportion agreeing with the statement. Using equation 30, $p' = X/n$. $X = 320$ people agreeing, and $n = 888$ responded altogether. Therefore, $p' = 320/888 = .36$, and $q' = 1 - p' = .64$.

Is our sampling distribution curve for $p' = .36$ and $n = 888$ sufficiently continuous and symmetrical to conform to the normal curve? May we use z? Consulting Table 4.1 and interpolating, we realize that when $p' = .36$, a sample of about 96 cases or more allows the use of z. Our sample size of 888 obviously meets this requirement, and at the .90 confidence level, $z = \pm1.645$.

Finally, our population size is about 30,000 and our sample size is 888. Since n is less than 10% of N (or 3000), we may not use the fpc. Equation 41 is appropriate.

$$90 \text{ C.I. Est. for } p = p' \pm z\sqrt{\frac{p'q'}{n}} = .36 \pm 1.645\sqrt{\frac{.36(.64)}{888}}$$

$$= .36 \pm 1.645(.016) = .36 \pm .026 = .36 \pm .03$$

For the C.I.'s lower limit: $.36 - .03 = .33$ or 33%, and for the C.I.'s upper limit: $.36 + .03 = .39$ or 39%. Figure 4.7 shows the distribution curve.

Based on our sample of 888 cases, we are .90 confident that between 33% and 39% of this college population agrees that national security is more important than personal privacy. Put another way, we are 90% confident that 36% support such a proposition, with a margin of error of ±3%.

Before considering another example, notice that the z score in this problem (±1.645) is taken to three decimal places, not the customary two. In fact, this is one of the two instances in which we take a z score to three decimal places. For the 90% confidence level, we need the z score that cuts off *exactly* .0500 in each tail. Examining column C of Table A, we find entries of .0505 and .0495, respectively, and z values of 1.64 and 1.65 in column A. Since our .0500 falls exactly in the middle of the two adjacent areas listed, we take the midpoint of their z scores: 1.645.

The other time we do take a z score to three decimal places is with the 99% confidence level. In this case, we want the z score cutting off .0050 (or .5%) in each tail. We again have to split the difference between the two closest entries in column C of Table A: .0051 and .0049. Our .0050 is midway between these two. We take the midpoint of the two z scores shown in column A, 2.57 and 2.58, respectively, and we use $z = 2.575$.

We simplified things in Chapter 3's introduction to the normal curve by merely using the nearest z score when necessary. Now, however, we want to be as precise

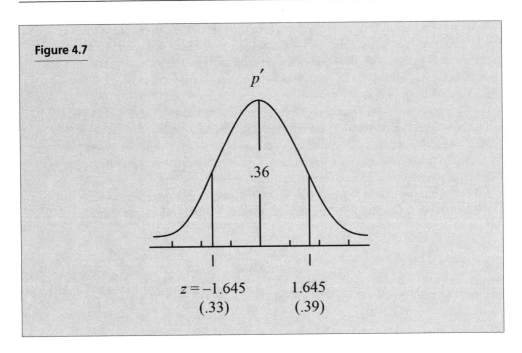

Figure 4.7

and accurate as possible with our confidence interval estimates. When it is more accurate to take the midpoint between two listed *z* scores, we go to that third decimal place. (In contrast, as you have no doubt noticed, the entire *t* table goes to three decimal places, and we always use all three.)

The next example problem shows how the .99 confidence level and the fpc work with proportions.

5. *A random sample of 520 was selected from a population of 4300 graduate students. Sample members responded to the statement, "I prefer to socialize with people of my own ethnic background." Some 121 respondents agreed with the statement. At the .99 confidence level, what is the estimate for the percentage in this graduate student population agreeing with this point of view?*

As usual, we start by recording the pertinent information. Since we are asked about agreement with the statement, we let p' equal the proportion agreeing, and $p' = X/n$. $X = 121$ people in agreement, and $n = 520$ people sampled. Therefore, $p' = 121/520 = .23$, and $q' = 1 - p' = .77$.

May we use the *z* curve? Taking the smaller of our p' and q' values, we consult Table 4.1 with $p' = .23$. By interpolating, we see we would need a minimum 255 respondents when $p' = .23$. With $n = 520$, our sampling distribution curve clearly approximates the normal distribution. At the .99 confidence level, z always equals ±2.575.

What about the fpc? $N = 4300$, and $n = 520$. Clearly, n is greater than 10% of N (i.e., 430), so we may use the correction factor with the standard error. That means equation 42 is best. To determine the actual C.I. estimate, we have:

$$99 \text{ C.I. Est. for } p = p' \pm z\sqrt{\frac{p'q'}{n}}\sqrt{\frac{N-n}{N-1}} = .23 \pm 2.575\sqrt{\frac{.23(.77)}{520}}\sqrt{\frac{4300-520}{4300-1}}$$

$$= .23 \pm 2.575 \,[(.018)\,(.938)] = .23 \pm 2.575\,(.017) = .23 \pm .043 = .23 \pm .04$$

For the C.I.'s lower limit: $.23 - .04 = .19$, and for the C.I.'s upper limit: $.23 + .04 = .27$. Figure 4.8 shows the distribution curve.

Based on our sample, we may be 99% confident that between 19 and 27 percent of this graduate student population prefers to socialize with ethnically similar people. Rephrased: At the .99 confidence level, we estimate that 23% of our population feels this way, with a margin of error of ±4%.

Up to this point, with proportions, we have been able to use the normal curve to derive our confidence interval estimates. Table 4.1 suggested our p' and q' values and our sample sizes produced sampling distributions that did, in fact,

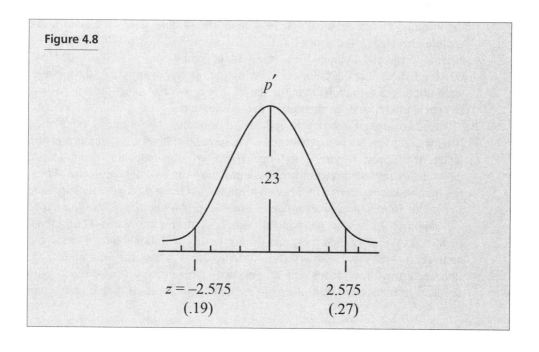

Figure 4.8

p'

.23

$z = -2.575$
(.19)

2.575
(.27)

conform to the normal curve. This may not always be the case, however, as the next example illustrates. It involves a different item from the survey noted in problem 4: reactions to national security measures introduced since the events of September 11, 2001.

6. College students were asked whether the new security measures introduced following September 11, 2001, had actually affected their personal lifestyles. In another random sample, this time, of 883 respondents, taken from a population of 30,000, 87 agreed that the new circumstances had had a personal impact. In this population, what is the .95 confidence interval estimate for the percentage personally affected by recent security measures?

To analyze the problem, we let p' equal the proportion actually reporting a personal effect. That makes sense, considering how the question is worded. We also note that $p' = X/n$. That becomes 87 (personally affected) out of 883 (responding to the question): $87/883 = .099 = .10$. Thereafter, q' becomes $1 - .10$, or .90.

Consulting Table 4.1 and looking for the minimum n necessary when $p' = .10$, we find a required sample size of 900. Uh-oh. Our $n = 883$ respondents to the question falls short.

Our sampling distribution "curve" is not quite a curve at all. With $p' = .10$, 833 cases are not enough to produce a sufficiently continuous and symmetrical sampling distribution that mimics the normal curve. Whether we may use the fpc is moot. Given our result for this item and sample size, we cannot make an inference using the procedures we have covered. We would have to use alternative methods, which do exist but are beyond the scope of this text. Here, we simply note that the theory underlying the normal distribution, the z curve, and sampling distributions in particular cannot be used as a justification for any statistical inference or extrapolation in this situation.

We have looked at varied procedures for estimating two unknown population parameters: averages and proportions. We have seen how z, t, and a correction factor may apply. Before switching to another topic relevant to confidence intervals, we present three tips to consider as you work on actual problems. These tips involve keeping a watchful eye on formulas, a reference for z coefficients, and a checklist for diagnosing a confidence interval problem or research situation.

The first tip is very mechanical: Watch what you are doing. Think about each step in your calculations, and do not hastily and thoughtlessly crank out numbers that may make little sense. In particular, it is easy to overlook that the standard error formulas for both $\sigma_{\overline{X}}$ and σ_p involve taking square roots. With averages or means, we want the *square root* of n, not n itself (and certainly not

N). With proportions, the standard error (σ_p) will almost certainly be in the .01 to .03 range. If you record a number such as .00034 for σ_p, you probably have forgotten to take the square root. (In problem 5, for example, our σ_p was .018, the square root of .00034.) Imagine trying to explain an utterly nonsensical answer based on using .00034 instead of .018. People sometimes do such things because they are not thinking about what their answers should *logically* be. As you work the problems, also *think* about them. Practice putting your mind a step ahead of your pencil. Developing this habit as you work statistical problems may help you catch errors resulting from haste, fatigue, distractions, wrong keypad entries, simple disinterest, and so on.

The second tip involves *z* coefficients. For all *t* values, you will have to consult Table C, but *z* coefficients are constant and unchanging. Unlike the *t* curves, there is only *one z* curve. When we use *z* in our confidence interval equations, it is handy to have a quick reference for *z* and the three commonly used confidence levels. Table 4.2 provides just such a reference.

For the third tip, at this point in the chapter, it may be useful to have a checklist for diagnosing estimation problems, especially word problems. Such a list or procedure has been implicit in the way solutions have been presented, but it may be useful to have it spelled out for reference. Ask yourself the following questions when you estimate a population parameter. Your answers should lead you to one specific procedure and formula. Thereafter, if all the numbers are in place, it is just a matter of doing the math and interpreting the results.

1. Does the problem involve means/averages or proportions/percentages?
2. Is a *z* or *t* distribution appropriate to solve the problem?
   a. Averages: *z* or *t*? When $n < 30$, use *t* with $(n - 1)$ d.f.; otherwise, use *z*.
   b. Proportions: May the *z* distribution be used? Consult Table 4.1.

Table 4.2	
Confidence Level	*z* Coefficient
.90	± 1.645 standard errors
.95	± 1.960 standard errors
.99	± 2.575 standard errors

3. May the finite population correction (the fpc) be used? Consider the ratio *n*:*N*.

This list of three steps should bring home the point that, as a student of statistics, your most important tasks are diagnosing data problems, determining the appropriate procedures in any situation, and then objectively interpreting the results. The actual number-crunching is an intermediate and somewhat secondary exercise.

At this point, we switch to a related but somewhat different procedure. It may appear from the above that our final results, especially our margin of error, are at the mercy of our sample size, or *n*. That is true, of course. As you know, the larger our sample size, the smaller our error term will be. However, when doing planned or consigned research and rather than simply waiting to see how large our eventual sample is going to be, we can actually determine beforehand how large a sample is needed to *give us a desired margin of error at a given confidence level*. The procedure is not difficult and merely consists of re-arranging earlier formulas.

Sample Sizes: How Many Cases Do We Need?

Estimating a Population Average

When we are calculating a confidence interval for a population mean, the basic procedure involves a point estimate and error term (see equation 39), repeated here:

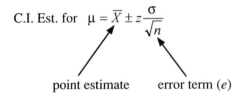

$$\text{C.I. Est. for } \mu = \overline{X} \pm z\frac{\sigma}{\sqrt{n}}$$

point estimate   error term (*e*)

Essentially, we build a confidence interval around the point estimate. The error term, our plus-and-minus margin of error, is the half-width of the overall interval. It is this error term that we wish to keep at a certain size. We start by giving the error term a symbol, *e*. Thus,

(43) $$e = z\frac{\sigma}{\sqrt{n}}$$

Once we have decided upon a certain value for *e*, that is, how large or small we wish our margin of error to be, we know all the values in equation 43, except for the n we wish to determine. The z coefficient is decided by our confidence level, so once we have picked a confidence level, z is determined for us. Similarly, even though σ may not be a constant, it is determined at the time of our study, and we cannot change its value. That leaves n as the one factor in the equation that we *can* manipulate. If we wish to control our error term (e), the only way to do it is by manipulating n. We rewrite equation 43 so that n appears by itself on one side of the equation:

(44)
$$n = \frac{z^2 \sigma^2}{e^2}$$

7. *A campus administrator wishes to estimate the average amount of money full-time students spend on books per quarter. How many students should she sample if she wishes to use the 95% confidence level, if she has an idea (from previous research adjusted for inflation) that the population standard deviation may be about $70, and if she wishes her estimate to have a margin of error of no more than ±$10?*

The administrator's desired margin of error is $10, so $e = 10$. Taking earlier results, she assumes the standard deviation to be about $70, σ = 70. Finally, at the .95 confidence level, $z = ±1.96$. Then, by using equation 44,

$$n = \frac{z^2 \sigma^2}{e^2} = \frac{(1.96)^2 (70)^2}{(10)^2} = \frac{3.84(4900)}{100} = \frac{18,816.00}{100} = 188.16 = 189$$

Her initial answer is 188.16 students, but she cannot take 16/100 of a student and, technically, 188 students are not enough, so she rounds up to a sample size of 189. Note that we *always* round upwards when estimating a sample size needed. Note here also that $n = 189$ is probably well less than 10% of her college's population. She therefore does not have to take it to a next step and get a revised answer using an fpc version of the formula (discussed below).

To see how the administrator's $n = 189$ works in practice, we plug her numbers into equation 43:

$$e = z \frac{\sigma}{\sqrt{n}} = 1.96 \frac{70}{\sqrt{189}} = 1.96 \frac{70}{13.75} = 1.96(5.09) = 9.98$$

Allowing for rounding off and the fact that she increased her sample size from 188 to 189, the administrator may actually anticipate a margin of error of ±$9.98, within her criterion of "no more than ±$10," as expected.

So far, so good, but there is a potential problem with this situation. In reality, if we do not know and are eventually hoping to estimate the population mean, μ, how likely is it we would know the population standard deviation, σ? The answer, of course, is "not very likely." What do we do if σ is unknown? There are three possibilities.

First, we may rely on a pilot study to get an estimate for σ. A pilot study is a dry run, or a rehearsal prior to an actual survey or experiment. Are our planned methods feasible? Will they work? In a survey, for instance, we contact a small number of potential respondents to see whether our planned measurements are possible. What should the administrator in problem 7 ask students? What time during the quarter is best for her survey? Are students able and willing to answer the questions she plans to use? From such pilot studies we may be able to get enough information to calculate a tentative sample standard deviation, or *s*. This may serve as at least a general indicator for the unknown σ. We then substitute *s* for σ in equation 44, to get:

$$(45) \qquad n = \frac{z^2 s^2}{e^2}$$

Second, as described in problem 7, we may have previous data available. We may be able to use or possibly estimate a value for σ based upon earlier research (remember to adjust for inflation, if necessary).

Finally, in lieu of anything else, we can use the range to generate a figure for σ. If we are able to reasonably guess or estimate a range for our *X* values, the standard deviation is sometimes taken as a certain fraction of that range:

$$(46) \qquad \text{Estimated } \sigma = \frac{\text{Range}}{4}$$

In problem 7, if the administrator conducted a pilot study and/or consulted with faculty in various departments and with her campus's bookstore employees, she may get a rough estimate of the high and low dollar amounts full-time students could expect to spend for books. Assuming she reasoned a high figure of around $450 per quarter and a low of about $130, the range is $450 − $130, or $320. One-fourth of $320 is $80, the value she can then use for σ when she derives her required sample size.

Under some circumstances, equation 44 or equation 45 is only our first step. A problem may require an additional step and formula for a final answer. Suppose the estimated sample size using equation 44 or 45 turns out to be 10% or more of the population. Under those circumstances before, we used the finite population correction, or fpc (equation 34).

How do we employ the fpc when estimating a sample size? It is a two-step process. First, we estimate the needed *n* by using equation 44 or 45. Then, if that answer is 10% or more of the population and if we plan to sample without replacement (which may be taken for granted in most studies), we get a revised figure by using a formula that includes the fpc. This revised number using the fpc is always smaller than the original answer, which represents real savings in both time and money for the actual field work. As discussed earlier, when taking such a comparatively large segment of the population into the sample ($\geq 10\%$), we gradually reduce the variation in that population. That reduced variation entitles us to use a corrected standard error, modified by the fpc. That fpc adjustment also finds its way into our formulas to determine sample sizes. The method is similar to that used to find n in equation 44 without the fpc.

When constructing a confidence interval for the mean using the fpc, we had equation 40, repeated here:

$$\text{C.I. Estimate for } \mu = \overline{X} \pm z \frac{\sigma}{\sqrt{n}} \sqrt{\frac{N-n}{N-1}}$$

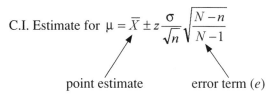

point estimate   error term (e)

We now rewrite the error term as:

(47)
$$e = z \frac{\sigma}{\sqrt{n}} \sqrt{\frac{N-n}{N-1}}$$

Rewriting equation 47 so that n appears by itself on one side of the equation, the revised formula for estimating a sample size now becomes:

(48)
$$n = \frac{Nz^2\sigma^2}{(N-1)e^2 + z^2\sigma^2}$$

8. *A local senior center was recently given 10 new computers. Going online has become a popular pastime for some 300 senior visitors, with most spending between 5 and 45 minutes at a time "surfing the 'Net." If the staff wishes to be 90% confident of estimating the average time spent online to within ±2 minutes, how many senior users should be sampled?*

Pulling out the pertinent information, the error term is ±2 minutes, so $e = 2.00$. The population size, N, is 300. At the .90 confidence level, the z coefficient equals ±1.645. No standard deviation is given, but the range is from 5 to 45 minutes, so we can estimate σ by using equation 46:

$$\text{Estimated } \sigma = \frac{Range}{4} = \frac{45-5}{4} = \frac{40}{4} = 10$$

As a caution, always start the process of determining a necessary sample size with one of the basic formulas, that is, equation 44 or 45. Here, we start with equation 44:

$$n = \frac{z^2\sigma^2}{e^2} = \frac{(1.645)^2(10)^2}{2^2} = \frac{2.706(100)}{4} = \frac{270.603}{4} = 67.65 = 68$$

Rounding to the next whole number (because we can't have 6/100 of a senior), we have a tentative answer of 68 seniors to be randomly sampled. But that is more than 10% of the $n = 300$ population. So the original estimate for n must now be revised downward by using equation 48. (The calculation of $z^2\sigma^2$ from our previous work may be used in this revised step; no sense repeating the work.)

$$n = \frac{Nz^2\sigma^2}{(N-1)e^2 + z^2\sigma^2} = \frac{300(270.603)}{(300-1)4 + 270.603} = \frac{81,180.90}{1196 + 270.603}$$

$$= \frac{81,180.90}{1466.60} = 55.35 = 56 \text{ seniors}$$

Since they will be sampling such a sizable segment of the population, staff members are entitled to reduce their n with no loss in accuracy. If they wish to be 90% confident of estimating the true average time online to within ±2 minutes, they will need data from 56 randomly selected seniors.

Note two additional points here. First, as before, it may be necessary to substitute s, a pilot sample standard deviation, instead of σ, into equation 48. That is acceptable. Second, do not immediately opt for the fpc formula on the assumption you may need it eventually. It is only knowing that we must sample at least 10% of the population that justifies using the fpc and equation 48, and that knowledge comes from either equation 44 or equation 45.

These procedures allow us to determine the sample sizes necessary to produce sampling distribution curves having certain degrees of peakedness or spread over the horizontal axis. In problem 8, we have a sampling distribution curve for which the middle 90% occupies an interval of ±2 minutes along the horizontal axis. If staff members opted for a larger error term, they would require a smaller sample. They would need only a large enough n so that the middle 90% of the sampling distribution falls over a wider interval along the lower axis. The same principle operates when we determine the sample size necessary to estimate a population proportion.

Estimating a Population Proportion

How large an n do we need to produce a sampling distribution curve in which the middle 90% or 95% falls over an interval of ±3% (.03) or ±5% (.05) along the horizontal axis? As we did for the population average, when estimating a population proportion we first isolate the error term in the confidence interval formula. To derive a confidence interval for the population proportion we use equation 41, repeated here:

$$\text{C.I. Estimate for } p = p' \pm z\sqrt{\frac{p'q'}{n}}$$

point estimate   error term (e)

The error term (e) becomes:

(49)
$$e = z\sqrt{\frac{p'q'}{n}}$$

Rewriting to solve for n, we get:

(50)
$$n = \frac{z^2 pq}{e^2}$$

    Two of the variables in equation 50 are easily quantified; the others present some options. As before, z is determined by our confidence level, and e is either given in the question or becomes whatever error factor we decide is tolerable for our estimate of p. However, we must also decide upon values for p and q. After all, we have not done any actual sampling yet, so we cannot use the p' and q' values—notice the absence of the prime for the p and q symbols in equation 50. We have no actual p' and q' values with which to work at this point. This is similar to our lacking σ or s in the estimation of the population average.

    As with averages, we have three ways of determining values for p and q. First, a pilot study may provide data upon which to base reasonable estimates for p and q. Second, previous research and data may provide clues as to probable p and q figures. Third, it is common to let p and q each equal .50. In equation 50, p and q are multiplied together. Values of .50 and .50 give us the largest possible product for this multiplication, and hence the largest possible numerator. This, in turn, results in the largest possible n we might ever need, given our z and e values. In a way, this strategy anticipates the worst possible scenario. How many cases must we sample if we assume our population is as varied as possible—split right down the middle, 50/50? In reality, of course, p may actually equal .20, .40, or anything else. It is possible that, by letting p and q equal .50, we are estimating a larger sample size than is actually needed. It is obviously better to have a conservatively large sample, however, than one that may turn out to be too small.

9. *In a community of approximately 25,000 households, a market research firm wishes to estimate the proportion of all households that include at least one pet. If researchers wish to be 95% confident of estimating the true figure to within ±3%, how many households should they sample?*

The error margin is given as ±3%, so $e = .03$. At the .95 confidence level, $z = ±1.96$, or 1.96. Since we have no p and q values and are given no clues, we let each equal .5. By using equation 50, we get:

$$n = \frac{z^2 pq}{e^2} = \frac{(1.96)^2(.5)(.5)}{(.03)^2} = \frac{3.84(.25)}{.0009} = \frac{.96}{.0009} = 1066.67 = 1067$$

The company should randomly sample 1067 households to meet its criteria. In doing so, it is 95% confident of estimating the actual proportion owning pets to within ±3%. The projected sample size is obviously less than 10% of n (25,000), so the researchers need not turn to any fpc formula.

The parameters for problem 9 are very common. The .95 confidence level seems to be the one most frequently used, possibly because it represents a middle ground between .90 and .99. It is equally true that a margin of error of ±3 percentage points is widely cited. It seems to be an acceptable margin of error for many population estimates. In fact, along with the .95 confidence level, ±3 percentage points has become a somewhat conventional margin of error in surveys of public opinion. Many polls are now conducted at a confidence level of .95 and with about 1000–1200 respondents—the researchers wanting a high level of confidence and a fairly small (acceptable) margin of error.

As was the case earlier, there are occasions when we sample comparatively large segments of our populations. When estimating population proportions and determining optimum sample sizes, we may also include the fpc, as shown in equation 42, repeated here:

$$\text{C.I. Estimate for } p = p' \pm z\sqrt{\frac{p'q'}{n}}\sqrt{\frac{N-n}{N-1}}$$

point estimate   error term (e)

To reproduce the error term (again, without the primes for p and q):

(51)
$$e = z\sqrt{\frac{pq}{n}}\sqrt{\frac{N-n}{N-1}}$$

We rearrange this equation to isolate n on the left side:

(52)
$$n = \frac{Nz^2 pq}{(N-1)e^2 + z^2 pq}$$

10. *A journalism professor wishes to estimate the percentage of liberal arts majors on her campus who regularly read a daily newspaper (other than just the sports section or comics). She knows that about 35% of all college students do so. Is this figure true of the 2330 local liberal arts students? How many should she sample if she wishes to be 90% confident of estimating the actual percentage to within ±4%?*

First, if p equals the proportion reading a daily paper, we have an actual figure: $p = .35$. Then q equals $1 - .35$, or $.65$. The z coefficient is ± 1.645 (1.645) at the .90 confidence level, and $n = 2330$. We start with equation 50, which yields:

$$n = \frac{z^2 pq}{e^2} = \frac{(1.645)^2 (.35)(.65)}{(.04)^2} = \frac{.616}{.0016} = 384.76 = 385$$

This is clearly more than 10% of her population, so we revise her answer by using the fpc and equation 52. (Again, we may take advantage of some calculations already done.)

$$n = \frac{Nz^2 pq}{(N-1)e^2 + z^2 pq} = \frac{2330(.616)}{(2330-1)(.0016)+.616} = \frac{1435.28}{3.726+.616}$$

$$= \frac{1435.28}{4.342} = 330.56 = 331$$

The final answer is 331 liberal arts majors. The journalism professor has to sample that many students to be 90% confident of estimating the population figure to within ±4 percentage points. Even though it is still more than 10% of her population, the fpc allows her to reduce the required sample size while maintaining the same criteria for her eventual confidence interval estimate. This is the value of the fpc. It makes our work somewhat more precise and accurate or, as in the present context, less expensive and time-consuming with no loss of accuracy.

Summary

To summarize, we have looked at confidence intervals for estimating unknown population means and population proportions. The statistical procedures here are based on the concept of sampling distributions. We estimate the 90%, 95%, or 99% of all sample means or proportions that would *most likely occur* and conclude there is that level of certainty the unknown parameter actually falls into our interval. We may also determine the sample size necessary to give us a sampling distribution curve of a desired shape. This shape (or spread), in turn, determines the margin of error of our estimate.

Earlier in this chapter, we considered a possible checklist for diagnosing an estimation problem. Now we add one more step to that list. Our initial diagnostic question now concerns whether we are establishing a confidence interval or determining the sample size for such an interval. Our revised checklist asks four basic questions:

1. Are we asked to establish an actual confidence interval estimate or to determine a sample size for such an estimate?
2. Does the problem involve means/averages or proportions/percents?
3. Is a z or t distribution appropriate to solve the problem?
   a. Averages: z or t? When $n < 30$, use t with $n - 1$ d.f.; otherwise use z.
   b. Proportions: May the z distribution be used? Consult Table 4.1.
4. May the finite population correction (the fpc) be used? Consider the ratio $n:N$.

The next chapter turns to hypothesis testing. We will continue to rely heavily upon the concepts of sampling distributions and the standard error. With confidence intervals, however, we have focused upon the broad middle (or most probable) area of the curve. Just the opposite occurs when we test hypotheses: Our focus is upon the tails—very improbable or rare-event areas of sampling distribution curves.

Exercises

For each exercise below, include your conclusions in a sentence or two. It is important that you know how to clearly express what your results mean.

1. Do today's older workers expect to have to work for pay after they retire from full-time employment? With the cooperation of a large regional company, 750 of its 2900 employees aged 50 and older were sampled, and 315 said they expected to have to work to supplement their pensions and Social Security. Given

these results, what is the .90 confidence interval estimate for the percentage of older employees expecting to have to continue to work after formally retiring?

2. The cost effectiveness of sobriety roadblocks has been questioned. As part of assessing that effectiveness, a police consultant examined county records. She reviewed records for 20 randomly selected roadblocks from the 142 held in the county in the last five years. She found an average of 6.94 legally impaired drivers arrested per roadblock, with a standard deviation of 2.31 drivers. At the .99 confidence level, what is her estimate for the average number of impaired drivers arrested per roadblock?

3. LubeNow advertises "Twenty minutes or it's free! If your oil change or lube is not started in 20 minutes or less, it's free." Randomly sampling records from 30 of the 314 LubeNow outlets in the Midwest and eastern United States, you find an average time of 18.05 minutes between the time a driver came in and service on his or her vehicle is started, with a standard deviation of 2.87 minutes. At the .95 confidence level, what would you conclude? Is LubeNow living up to its advertising?

4. A political science professor has been asked to estimate the percentage of students in his department registered to vote. Of the 376 political science majors at his university, how many should he sample if he wishes to be 90% confident of estimating the actual percentage to within ±5%?

5. Assume the Sportswriters' Association of America has approximately 5000 members. A researcher randomly sampled 600 members, and 168 agreed that "A professional athlete convicted of knowingly using illegal performance-enhancing drugs should be banned for life from his or her sport." At the .99 confidence level, what is the estimate for the percentage of all Association members who support such a position?

6. A campus placement office wishes to estimate the average number of positions or jobs held by graduates in the 10-year period following graduation. Census information and Bureau of Labor Statistics data suggest the national average may be about 4.60 jobs, with a standard deviation of 1.37 jobs. The university in question produced 2147 BA/BS graduates a decade ago. How many graduates should be sampled and contacted if researchers wish to be 95% confident of estimating the average number of jobs to within ±.25 jobs?

7. As part of a year-end review, the East-West Gateway Council of St. Louis looked at how many of last June's area graduates had enrolled in post-high

school courses (technical school or college, public or private). The Council staff sampled 400 of 4208 graduates and found that 136 had enrolled in post-graduation courses. What is the .95 confidence interval estimate for the percentage of all St. Louis area graduates enrolling in additional courses?

8. In a troubled fuel market, Via Con Dios Airlines is considering offering a discount to all passengers whose luggage (checked plus carry-on) weighs at least 20% less than the average load per passenger. To get its own baseline figures, Via Con Dios must sample its passengers' baggage weight. National figures are estimated to average 54.20 pounds per flyer, with a standard deviation of 4.38 pounds. Airline researchers have one week to establish a baseline and wish to use the .99 confidence level with a ±.5 pound margin of error. Via Con Dios, moreover, expects about 19,000 passengers next week. The checked plus carry-on bags of how many passengers should be randomly sampled for weight?

9. An automobile insurance company is planning to survey its policyholders regarding various driving habits: Ever used a mobile phone while driving? Ever used a mobile phone to report a road emergency? Ever driven in the carpool lane while alone? And so on. It wishes to be 95% confident of estimating the true percentages to within ±2 percentage points. How many of its 35,000 policyholders should it sample for its survey?

10. An economist wondered how much families had in savings. Working with a local company's list of 800 employees, she randomly selected 28 for personal interviews. She found the average worker's family had $10,430 in savings, with a standard deviation of $3276. Given these results, what is her .90 confidence interval estimate for the average amount saved?

11. A university research team recently completed a campus survey regarding students' risks of exposure to HIV and other sexually transmitted diseases. It randomly sampled 500 of the university's 5400 unmarried seniors under age 30. In the sample, 155 students met its definition of "high risk" behavior (unprotected sex with two or more different partners in the last 60 days). At the .99 confidence level, what is the team's estimate for the percentage of all unmarried seniors under age 30 engaging in high-risk sexual behaviors?

12. A professional association of commercial realtors wishes to estimate the average number of square feet of workspace per employee among businesses in a particular county. Five years ago, a statewide survey put the average at 49.73 square feet per employee, with a standard deviation of 9.44 square feet. For its

own purposes, the association must estimate the overall average to within ±1 square foot at the .95 confidence level. How many of the county's 4187 businesses should it select for a survey?

13. A particular university's College of Business has 4700 students. A recent campus-wide survey included 206 business majors, and 125 of them agreed with the statement, "Television has too much sexual content." That resulted in a sample proportion of .61 (125/206), or 61% of business majors agreeing there was too much sex on TV. The .99 confidence interval estimate was .61 ±.09. Given this information, how many business majors should have been sampled to reduce the margin of error to ±.04 or 4%?

14. A particular retailer has 65 department stores covering the western United States. The chain relies on Christmas revenues for a considerable part of its profit and usually hires some 210 Santas for the holidays. The retailer knows that the more young children visiting Santa, the more adults and potential shoppers in the stores. Wishing to find out how many children a typical Santa sees, chain managers randomly select 25 Santas and record an average of 31.49 children seen per hour, with a standard deviation of 3.96 per hour. What is the .95 confidence interval estimate for the average number of children visiting a Santa per hour?

15. A public health nurse wishes to estimate how much the elderly in a certain community must spend on medications per month. He knows a previous study put the range at $30 to $370 per month, but he wishes to be more accurate. From the membership or client lists of community health agencies, clinics, churches, volunteer organizations, and so on, he estimates he has a population of about 7600 from which to pick a sample. If he wishes to estimate the average amount spent per month to within ±$5 at the .90 confidence level, how many senior citizens should he sample?

16. A high school social studies class conducted a small poll. The class members randomly sampled 114 of the 1720 students at their school and found that 41 responded "Yes," meaning they and their families celebrated particular ethnic cultural events and holidays. What is the class's .95 confidence level estimate for the percentage of all students celebrating ethnically related events and holidays?

17. According to the county Bar Association, a particular city has about 197 criminal defense attorneys experienced in handling DUI (driving under the influence) cases. A friend of yours now needs such a lawyer, so you volunteer to find

out how much they typically charge for handling a DUI case. From calls to a random sample of 24 attorneys' offices, you calculate an average fee of $2987, with a standard deviation of $182. At the .90 confidence level, what is the estimate for the average charge among all the city's DUI attorneys?

18. In a university student body, 4424 students are married. How many students in this population should be sampled if we wished to be 90% confident of estimating the percentage employed to the following margins of error?

    a. To the ± .02 margin of error?
    b. To the ± .03 margin of error?
    c. To the ± .04 margin of error?
    d. To the ± .05 margin of error?

19. In a small ($n = 32$) pilot study of a team's season ticket holders, the average age was 41.56 years, with a standard deviation of 6.27 years. How many of the 18,500 ticket holders should we sample if we wish to have a margin of error of ±1 year when estimating the population's average age at the:

    a. .90 confidence level?
    b. .95 confidence level?
    c. .99 confidence level?

20. Students from a university population of 30,000 responded to the statement: "Women in the military should serve in front-line combat roles alongside their male counterparts." Of the 657 students in the sample, 350 agreed. What is the .90 confidence level estimate for the percentage of all students who would agree with the statement?

21. Yosemite National Park officials are concerned about the number of vehicles entering environmentally fragile Yosemite Valley. Some visitors bypass that crowded area altogether, and others spend more time in the Valley than expected. Given the need to restrict vehicular traffic there, a Department of the Interior statistician recorded the exact number of visitor vehicles in the Valley for 20 days, randomly selected from the park season of 254 visitor days last year. In the 20 days sampled, the average number of vehicles in the Valley per day was 5076, with a standard deviation of 792 vehicles. At the .95 confidence level, what is the estimate for the daily average number of vehicles in Yosemite Valley during last year's park season?

22. A certain neighborhood is known as a bellwether precinct for national elections. The 4072-voter precinct has been an infallible predictor of presidential election results for the last 36 years. In a sample of 500 voters, 115 believed the

president and vice president should be elected separately rather than as a pair or team. At the .95 confidence level, what is the estimate for the percentage of all precinct voters who favor such an election format?

23. Demographers project that a particular city's population includes roughly 8000 households with same-sex couples. A researcher wishes to estimate the percentage of such homes that include children. If she wishes to be 95% confident of estimating the actual figure to within ±4%, how many same-sex households should she sample?

24. From a population of 30,000 college students and a sample of $n = 661$ answering the question, the average number of hours worked per week during the school year is 19.86, with a standard deviation of 8.48 hours per week. At the .99 confidence level, what is the estimate for the average number of hours employed per week for all students?

25. What are sampling distributions and how are they used in confidence interval estimation?

26. What are "unbiased estimators"? Please give examples and explain why we consider them unbiased.

27. In general, what is the theory we use, the sampling assumption we make, and the procedures we follow to get a confidence interval estimate?

28. Sampling error is related to estimation.

    a. What is sampling error?
    b. Why is sampling error important in confidence interval estimation?
    c. What factors affect sampling error?

29. How is the concept of probability related to sampling distributions and specifically to confidence interval estimation?

30. In confidence interval estimation, why do we never use the 100% confidence level?

31. What are t curves, and how are they related to both sample size and the concept of sampling error?

32. How do the z and t distributions differ? In what ways are they the same?

33. Why is there more than one *t* curve?

34. Why are *t* values larger than comparable *z* values?

35. Why can we never use *t* distributions with proportions?

36. With inferential statistical analyses, why do we use the standard error rather than simply using the standard deviation?

37. Other things being equal (or staying the same), how would each of the following changes affect the ± margin of error (the ± error term) and why?

    a. a larger population size
    b. a smaller sample size
    c. a higher or larger point estimate
    d. using *t* rather than *z*
    e. a lower confidence level
    f. greater variation in the population
    g. less variation in the sample
    h. using the finite population correction

38. *Why* (not how) do we use degrees of freedom with the *t* distributions but not with the *z* distribution?

39. Some people question the effectiveness of female police officers. A large metropolitan police department recently released data pertaining to this question. Among 50 female and 50 male officers matched for duty assignments and using the .95 confidence level, it reported the average number of felony arrests per ten working days among female officers to be 7.37, with a margin or error of ±1.34 arrests. For male officers, the figures were 9.08 felony arrests, with a margin or error of ±1.67 arrests. How would you interpret these confidence interval estimates? Would they support or contradict the claim of female officers averaging significantly fewer felony arrests, and why?

40. When determining a sample size needed to estimate a population proportion or percent, why do we sometimes let *p* and *q* each equal .5? What effect does this have on the results we get?

Testing a Hypothesis: Is Your Sample a Rare Case?

In this chapter, you will learn how to:

- Test null hypotheses for sample means and sample proportions
- Correctly reject or fail to reject null hypotheses
- Apply the finite population correction
- Determine the power of a hypothesis test

CHAPTER 5 LOOKS FIRST AT THE theory and logic underlying hypothesis testing, and then at the mechanics of how it is actually done. It focuses on the single-sample situation. It is also possible to test hypotheses involving two (or even more) samples. For instance, we may wish to see whether there are statistically significant differences between an experimental group and a control group (two samples). It is easier to introduce the theory and mechanics of hypothesis testing, however, by considering the single-sample research situation, as presented in this chapter. The theory and logic introduced here also apply to the two-sample case in Chapter 6, alternative hypothesis testing procedures for ordinal and categorical data in Chapter 7, and testing for significant differences among three or more samples in Chapter 8.

The Theory Behind the Test

In the single-sample case of hypothesis testing, we compare a sample mean (or proportion) to a known population mean (or proportion) to see whether the two differ significantly. Does our sample statistic differ significantly from the known population parameter? In contrast to the confidence interval from which we estimate an unknown population figure, as discussed in the previous chapter, the population parameter is now known or is assumed to be known. This chapter's hypothesis testing procedures will not work unless we have a value for the population parameter. There are other ways of testing hypotheses that do not require knowing the population mean or proportion or, indeed, make any assumptions about the population at all. These are known, quite logically, as nonparametric tests. We will consider these tests in Chapter 7. For now, we are comparing our sample statistic to a known population parameter.

Also, note the term "differ significantly." It is not enough that the sample statistic and the population parameter simply differ. They must differ by a certain minimum, or *statistically significant* amount. Recall that, because of normal sampling error, we do not necessarily expect a sample statistic (the mean or proportion) to equal the population figure exactly. We expect some difference. In hypothesis testing, however, we look for a difference judged to be unusually high or large. We look, in other words, for a difference that is statistically significant. We often have a vested interest in showing that our sample average or percentage, after the experimental treatment or due to certain sample characteristics, differs from the population figure. However, we must be fair, reasonable, and consistent with other researchers as to what constitutes a "significant" difference from that population parameter. Within certain guidelines, we decide what criterion or standard to apply in deciding whether a difference is significant. In a way, the situation is similar to that in the last chapter. In working situations, we decide what confidence level or margin of error is most appropriate to estimate an unknown population parameter. In testing a hypothesis, we decide what magnitude of difference between the population parameter and sample statistic is enough to be judged *significant*.

Many Samples and the Rare Case

From Chapter 4, we know that sampling distributions consist of many independent sample statistics that, when graphed, form a symmetrical, bell-shaped (and possibly normal) curve around the respective population parameter. In practice, of course, we generally use two such sampling distributions: the sampling distribution of the mean, centered around μ, and the sampling distribution of the proportion, centered around p. If know a single sample mean or proportion, we can place it at the correct location along the horizontal axis on one of these curves. Using z or t scores, we can determine whether this sample figure falls near the tails of the curve or near the central peak.

When testing hypotheses we have a particular interest in sample statistics falling near the tails of the curve. Those sample means or proportions are very unusual and improbable. We judge them to differ significantly from their respective population parameters in the middle of the curve. Recall that the normal curve and the *t* curves are probability distributions. Areas of the curves are probabilities. If a sample mean or proportion falls near the tail of such a curve, we say that it is very unlikely given the central parameter for this curve. Under a curve with our known central value, a random sample having such an extreme mean or proportion is very improbable. When a sample statistic differs from the population figure by a substantial amount, we regard that difference as statistically significant.

The question now, of course, is: Exactly what constitutes a substantial or significant difference? Well, that answer varies a bit. What constituted a high confidence level for the confidence intervals in Chapter 4? It varied somewhat, but there was general agreement that the 90%, 95%, and 99% levels were high. Which level was used depended upon the precision desired and how much confidence in the estimate was needed. It was the researcher's decision.

The same is now true as to what constitutes a substantial or statistically significant difference. There is agreement as to the general guidelines, but each researcher decides upon his or her specific criterion in a given situation (and, as before, there is a trade-off involved). The 90%, 95%, and 99% levels, respectively, relate to areas in the tails of sampling distribution curves of 10%, 5%, and 1%. We are especially concerned with 5% and 1% when determining statistical significance. *Specifically, if a sample mean or proportion falls into the most extreme 5% or 1% (and occasionally the most extreme 10%), we are going to assume it differs significantly from the known population figure.* After all, there is only a 5% or even a 1% chance of a sample statistic falling there by random chance alone. Only a very small percentage of all sample means or proportions fall way out in the tails just by random occurrence. When a sample statistic *does* fall there, we conclude that it is not an accident and not due to random chance, but that it differs significantly from the known population parameter. There is something actually going on that is causing the sample to be different. That something should be the **independent or predictor variable(s)** or the experimental treatment. The **dependent variable** is influenced by the independent variable(s).

For example, assume we have a sample mean that falls into the top 5%, the very upper tail of a sampling distribution curve. By falling here, it is well above the known population mean at the center of the curve. Also assume we have adopted the top 5% as our criterion for deciding whether the sample and population means differ. According to our test guidelines, then, they do. Our decision is based on probability. There is only a 5% chance of our sample mean landing in the upper tail just by accident or chance alone, so we conclude it has done so because it genuinely differs from the population mean. It constitutes a *rare event* and so differs significantly from the population average. Figure 5.1 shows the sampling distribution curve for such a hypothesis test.

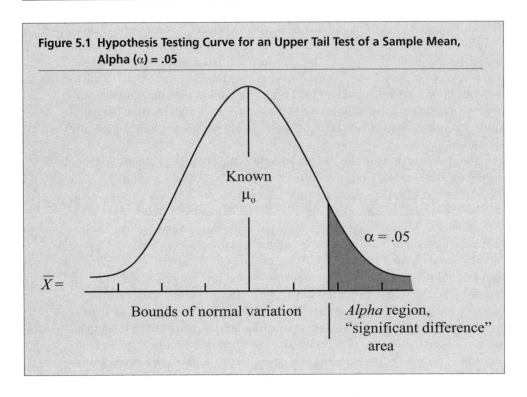

Figure 5.1 Hypothesis Testing Curve for an Upper Tail Test of a Sample Mean, Alpha (α) = .05

Known
μ_o

$\alpha = .05$

$\overline{X} =$

Bounds of normal variation

Alpha region, "significant difference" area

We use such a curve when testing to see if a sample average is significantly greater than the known population average. Figure 5.1 depicts a hypothesis testing curve with the top 5% as the significance area, where μ_o is the population mean assumed for the hypothesis test or "under the null hypothesis." A common name for that area is the **alpha region,** and 5% or .05 is known as our **alpha level.** We determine where our sample average, $\overline{X}$, falls along the horizontal axis. If it falls into the shaded upper tail, we conclude it is significantly greater than the population figure, or μ. Alternatively, if $\overline{X}$ falls below or to the left of the alpha region, it is within the bounds of normal variation. In that case, $\overline{X}$ is *not* significantly greater than μ when allowing for normal sampling error. $\overline{X}$ may differ from μ and may even be larger than μ, but it falls within the normal or expected range for sample averages, or it is within the range of expected sampling error.

Two potential problems may occur to you at this point. First, what if we are wrong about a sample being different? What if it comes from that top 5% alpha region because of random chance? What if our result *is* due to an inevitable quirk of random sampling? What if our independent variable(s) or experimental treatment(s) have nothing to do with this unusually high mean and our sample average belongs in the upper tail regardless?

This is a valid concern. There is always a risk of error in hypothesis testing. When we adopt that extreme 5% as a criterion for assuming there is a difference, we realize that 5% of the time we will make errors. We assume that *any* sample average landing there differs legitimately from the population figure when, in fact, 5% of all samples land there no matter what. For that 5% of cases, we do reach incorrect conclusions. We claim they differ from the population due to the independent variable or our treatment's effect when they really do not. It is simply a matter of random chance. Another way of saying this is that, for every 100 times we conduct our test this way, we reach erroneous conclusions 5 times. We can reduce this possibility of error to 1% (by adopting the top 1% of the curve as our alpha region), but we cannot eliminate the possibility of error altogether. An alpha region, no matter how small, is necessary to conduct the test. We refer to this error as an alpha error, and its possibility is an inevitable part of hypothesis testing.

The second problem involves this question: What if our test sample lands in the *lower* 5% of the curve? After all, samples in the lower tail represent extreme cases, too. Do we still conclude the sample and population differ significantly? Well, it depends. It depends upon the sort of difference for which we are looking, specifically, the *direction* of the difference. We are sometimes interested in sample statistics that are greater than comparable parameters. Figure 5.1 shows this kind of **upper-tailed test.** At other times, we looking for sample statistics that are lower or less than the population figures. Here, we conduct **lower-tailed tests,** and our alpha regions appear only in the lower tails of the curves. These two **one-tailed tests** are directional hypotheses tests. We sometimes look for samples that differ in *either* direction from the population parameter. In these **two-tailed tests,** we have alpha regions in each tail of the curves, and our sample statistics are judged significantly different if they fall into either tail. These are called non-directional hypotheses tests.

Note that we conduct one of these tests, not all three at the same time. If we set up our test to look specifically for a sample figure that is greater than the population parameter, as in Figure 5.1, we ignore (not completely overlook, just ignore statistically) anything else. This is an aspect of hypothesis testing that we will soon consider in more detail.

For now, to conclude this introduction, we look for sample averages or proportions that would occur very rarely by random chance. We focus on the tails of our hypothetical sampling distribution curves (just one tail or both at the same time), and we look for sample statistics falling into these regions. We do have a little discretion as to the size of our alpha regions in the tails (5% or just 1%), but when sample statistics land there we conclude they differ significantly from population figures at the curves' centers. If sample statistics lie closer to the centers of the curves, we consider them to be within the bounds of normal variation and conclude they do not differ significantly from their population parameters, allowing for normal or expected sampling error.

Null and Alternative Hypotheses

When we systematically compare a sample statistic and population parameter, we formally state hypotheses. In fact, we state two contradictory hypotheses. If one of them is true, the other is false. Moreover, we specifically test only one of these hypotheses. Our test leads us to one conclusion: we either accept that one hypothesis or we reject it in favor of the other. These two propositions are known as the null hypothesis (symbolized by H_O) and the alternative hypothesis (or H_A). The null hypothesis always goes by the same name and symbol (H_O), but different names are sometimes used for what we are calling the alternative hypothesis. Some texts refer to this as the working, experimental, or research hypothesis and often use the symbol H_E.

The null hypothesis (H_O). The **null hypothesis** always states that *there is no significant difference between our sample statistic and the population figure.* It states, for instance, that the test sample mean and the known population mean are essentially the same, allowing for sampling error. The sample and population values may not be *exactly* the same, but if the sample statistic falls within our area of normal variation, the difference falls within the range of normal and expected sampling error. We then conclude that the test sample does not differ significantly from the population, and we accept (fail to reject) the null hypothesis.

The null hypothesis may not seem too interesting or important at this point. After all, we generally do hope to find differences between our sample and population values and to reject this hypothesis. The null hypothesis is critical for constructing our hypothesis curve, however. Figure 5.1 shows such a null hypothesis curve (an H_O curve), a known sampling distribution curve. It is always H_O that we actually test and subsequently reject or fail to reject. Figure 5.2 slightly revises Figure 5.1 to indicate the *rejection region* for the null hypothesis. In the overall picture, the alternative hypothesis plays an important role as well.

The alternative hypothesis (H_A). The **alternative hypothesis** states that *there is a difference between the sample and population figures.* Our test sample and our population differ if our sample statistic falls in the alpha region of our hypothesis curve (our H_O curve). We conclude that the difference between the sample and population is statistically significant, and we reject H_O in favor of H_A. Allowing always for the small chance of an alpha error, if the sample statistic lies well away from the known population value at the center of the curve, we have evidence to support H_A.

The alternative hypothesis also has an importance of its own besides merely being a complement to the null hypothesis. H_A expresses the specific nature of the

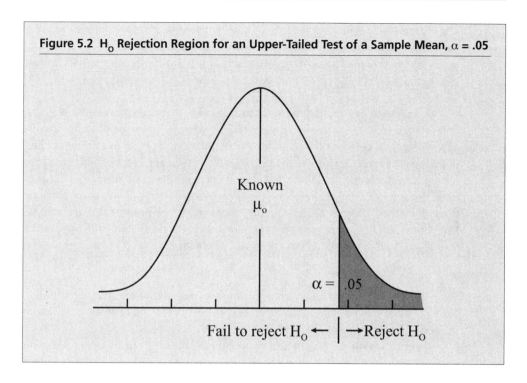

Figure 5.2 H_O Rejection Region for an Upper-Tailed Test of a Sample Mean, $\alpha = .05$

Known
μ_o

$\alpha = .05$

Fail to reject $H_O \leftarrow$ | $\rightarrow$ Reject H_O

H_O test itself. Although we always test and reject or fail to reject H_O, H_A determines how the test will be conducted. It is H_A that determines where our critical area or alpha region will be for rejecting H_O.

Alternative Hypotheses and One- or Two-Tailed Tests

Consider again the situation illustrated in Figure 5.1. We want to see if our sample statistic is significantly greater than the population value. This presumes we have some theoretical argument or rationale for expecting the sample mean to be greater than the known population figure. In symbolic form, using the example of averages, H_A becomes $\overline{X} > \mu_o$, dictating an upper-tailed test. We focus on the upper 5% of our curve and test whether our sample mean falls into this remote area. Our null hypothesis, the complement of H_A, becomes $\overline{X} \leq \mu_o$. Since H_A dictates a focus on the upper tail here, we fail to reject H_O if our sample statistic lands anywhere else, that is, anywhere below the upper tail. The nature of the alternative hypothesis determines how we phrase our null hypothesis and how we conduct our H_O test.

Other situations follow the same logic as the upper-tailed test. If we suspect the sample mean or proportion falls below the population figure, we conduct a lower-tailed test. For averages, we express H_A as $\overline{X} < \mu_o$, and H_O becomes $\overline{X} \geq \mu_o$.

The H_0 rejection region is at the lower tail. Alternatively, if we are not sure whether the sample figure will fall above or below the population value and if we wish to test for either eventuality, a two-tailed test is appropriate. The alternative hypothesis, again for averages, is written as $\overline{X} \neq \mu_0$, the null hypothesis is $\overline{X} = \mu_0$, and we have alpha regions in both tails of the H_0 curve.

No matter what the nature of the test, it is H_A that determines how we conduct the test and the specific form of H_0. H_0 always states there is no significant difference between the sample and population, at least in the anticipated direction(s). Figure 5.3 shows the possible H_0 curves and sets of hypotheses for testing sample averages. Figure 5.4 shows the same for tests of sample proportions. Notice in Figures 5.3 and 5.4 that H_0 always contains an equal sign of one kind or another. It expresses an equality: Allowing for sampling error, the sample and population figures do not differ. H_A alerts us to the direction of the hypothesis test, if any, and to the specific form of H_0 to be tested.

Always Test the Null

There is a final point to consider regarding the null and alternative hypotheses: Why do we always test the null hypothesis in particular? If our primary research interest is in verifying H_A, why do we put so much emphasis on H_0?

In concert with the population parameter, H_0 locates our hypothesis testing curve along the $\overline{X}$ or p' axis (the horizontal axis). We know where the H_0 curve falls along the lower axis. It is centered around the known population figure. We have no choice but to test our sample figure against this known parameter. Even

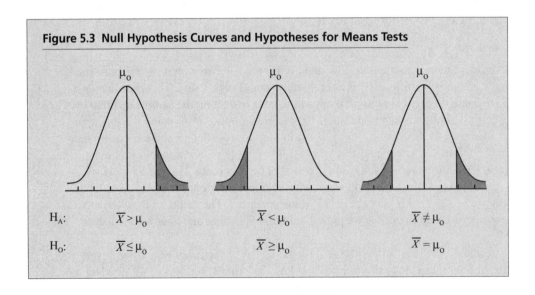

Figure 5.3 Null Hypothesis Curves and Hypotheses for Means Tests

H_A:	$\overline{X} > \mu_0$	$\overline{X} < \mu_0$	$\overline{X} \neq \mu_0$
H_0:	$\overline{X} \leq \mu_0$	$\overline{X} \geq \mu_0$	$\overline{X} = \mu_0$

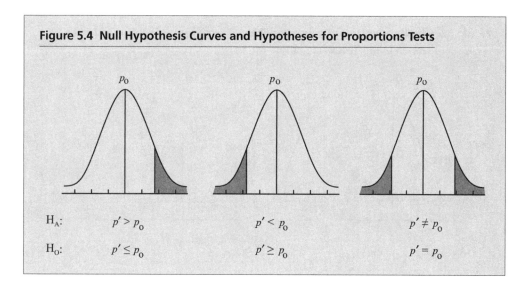

Figure 5.4 Null Hypothesis Curves and Hypotheses for Proportions Tests

H_A:	$p' > p_O$	$p' < p_O$	$p' \neq p_O$
H_O:	$p' \leq p_O$	$p' \geq p_O$	$p' = p_O$

if we do not include its actual numerical value, the population figure is incorporated into our H_O; it is the (understood) value for μ_O or p_O. In fact, we express these values as μ_O and p_O precisely because they are the assumed population values under (according to) the null hypothesis. And the null hypotheses states that our sample figures equal or at least do not significantly differ from these known quantities. In stating such equalities, the null hypothesis expresses something specific, concrete, and testable. Do our sample figures equal these known population parameters or do they not?

By contrast, the alternative hypothesis does not express anything nearly as specific; it merely states inequalities. It states, respectively, that the sample $\overline{X}$ or p' is greater than, less than, or different from the population figure. It does not state what $\overline{X}$ or p' *should* equal, simply that they do not equal μ_O or p_O. In the case of averages, for example, H_A does not state that $\overline{X}$ should equal the mean of a specific alternative curve: μ_A. We do not know and could only assume a value for a possible μ_A. What if our sample, following some experimental treatment or due to having a particular characteristic, *does* belong to another population? What if $\overline{X}$ belongs to a population centered around μ_A, and what if μ_A falls well above (or below) the known μ_O on the horizontal axis? We have no way of knowing. We might imagine 10 or 100 alternative populations whose means might fall at very precise and quite different spots on the horizontal axis. Each would have a sampling distribution curve associated with it, and these curves might naturally overlap (see Figure 5.5). Under which of these curves might our $\overline{X}$ belong? We do not know, and H_A itself says nothing to specify a possible H_A or μ_A curve.

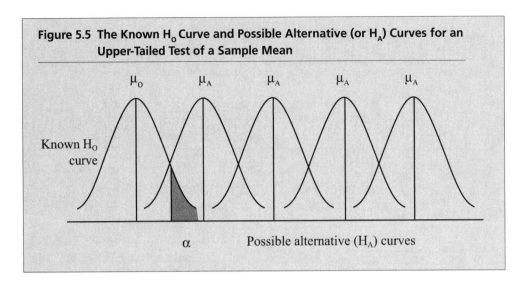

Figure 5.5 The Known H$_O$ Curve and Possible Alternative (or H$_A$) Curves for an Upper-Tailed Test of a Sample Mean

In the final analysis, these alternative curves are, of course, supposition. They are simply assumed. They *may* be there; they *may* exist. They may not. We do not know. The H$_O$ curve is the only one known to exist and the only one whose location (along the horizontal axis) is clearly marked by μ_O or p_O. We must therefore see if our sample statistic belongs to this known curve. It is the only option, the only choice we have. It is the null hypothesis, H$_O$, that expresses this identity, this equality between the sample and the known curve, so that is the one we must test, not the more vague and less specific H$_A$. Does $\overline{X}$ or p' fall within the bounds of normal variation of this known H$_O$ curve, or does it fall into an alpha or rejection area at the tail?

Wrong Decisions?

Inferential statistics are based upon probability, and hypothesis testing is no exception. We have seen that certain areas of the H$_O$ curve become either *rejection regions* or *areas of normal variation* based upon the probability of $\overline{X}$ or p' values landing there. We have also seen that if we reject a null hypothesis while using a 5% alpha region, we run a corresponding 5% chance of reaching a wrong conclusion. Perhaps our sample statistic landed in the tail of the curve by pure chance (5% of them would) and not because there was something unusual about the sample itself. We assume the latter, of course, but we could be wrong. It is appropriate to consider this ever-present error (called Type I or alpha error) factor in more detail and also its counterpart, the Type II or beta error.

As you read on, keep the overall procedure in mind. In terms of hard data, we have one sample statistic, the average or the proportion, and its standard deviation. We also know the population parameters, and we are trying to locate our sample average or proportion in a sampling distribution. To locate the $\overline{X}$ or p' in the H_O curve, we convert that statistic into a z or t score (a standard score) and determine where it falls along the horizontal axis. How likely is it to occur? Is it a common value and within the bounds of normal variation, or is it a very unusual and extreme case? As we do this, however, by following the rules and proper procedures, we may unavoidably make errors.

A true null is rejected: Type I alpha errors. No matter what null hypothesis decision we make, we risk making an error. We may make a **Type I** or **alpha error** only when we reject our null hypothesis. We do this when we reject a true H_O. We reject the H_O precisely because our sample has landed in our designated rejection region. We therefore claim that our sample statistic differs significantly from the population parameter and that our independent variable or treatment is apparently having some effect. Even though we say this, we also know there is a small chance (5% or 1%) that our test statistic landed in the rejection region just by chance alone. There is a small possibility that our results are merely a quirk of random sampling, that any independent variable or experimental treatment only *appears* to be affecting the sample. In truth, our sample $\overline{X}$ or p' may differ from μ_O or p_O just by chance alone, and our independent variable or experimental treatment may actually be irrelevant and have no effect. The sample just happens to be one of those odd cases.

What might we do? Even though we followed all the rules of hypothesis testing, as we must, we unknowingly reject an H_O that is really true. We are unaware of committing an alpha error when it happens. We might repeat the study with another sample, of course, but this is not always practical. We typically have to live with the results from one sample.

Although we may never eliminate the possibility of an alpha error, however, we can minimize it. We cannot eliminate it altogether because we must have a rejection or alpha region to conduct the test. Whatever the size of that region, we run that corresponding risk of making a Type I error. With a 5% rejection region, we run a 5% risk of committing an alpha error. If we reduce our rejection region to 1%, the risk of an alpha error is correspondingly reduced to 1%. We may even have an alpha or rejection region as small as .1% if we wish. But we must have such an area, however small, and that means there is always the prospect of an alpha or Type I error if we reject our H_O.

It may seem that the smallest possible rejection or alpha region is the best, but this is not necessarily so. Although it reduces the possibility of an alpha error there are two drawbacks to reducing our rejection region (again, the trade-off). First, we make it more difficult to reject our H_O. A sample statistic must now land even farther out in the tail before we may reject our H_O. We would require even more difference between the sample and population figures before we could judge it to be statistically significant. That raises the question of standards or rigor.

How demanding or rigorous do we have to be? Given the circumstances and the variables of our study, what is a *reasonable* standard or criterion? Perhaps a very small rejection region (a very low alpha level) *should* be required in medical research. Researchers want to be especially sure, for instance, of their claim that an experimental drug is significantly more effective than a known remedy. But what about our studies in the social, behavioral, or management sciences? Instead of alpha levels around .001 or even .01, is .05 or even a rather high .10 acceptable? There is no one answer, of course. It is a judgment call and a decision made by every researcher in every hypothesis-testing situation. The point is that we may not want to use the lowest alpha level or smallest rejection region statistically possible. There are other considerations.

Second, if we lower the alpha level, we may increase our chances of a Type II or beta error. P(Beta error), or $P(\beta)$, is just the opposite of an alpha error.

A false null is not rejected: Type II beta errors. Beta errors are a little more complicated than alpha errors. A **Type II error (beta error)** *happens when we do not reject an H_O that is really false, or accepting a false H_O.* We conclude there is no significant difference between the sample and population figures (again, following all the rules), but unbeknownst to us, a difference actually does exist. We run the risk of a Type II error whenever we assume our H_O is true.

As a researcher, you determine the size of your alpha or rejection region (usually 5% or 1%), and know exactly what risk you run of an alpha error. In contrast, even though a beta or Type II error is easy to define, it is not quite as easy to understand nor is its probability quite as easy to calculate. Naturally, there is a procedure for determining that probability, but one has to use imagination and make an assumption or two to understand what is going on and to make the calculations possible.

To illustrate this by an upper-tailed example, assume first of all that H_O really is false. Assume that, in reality, the sample and population figures *do* differ significantly. Assume that the sample figure *is* greater than the population parameter. H_A is actually the true hypothesis. However, even though H_O is genuinely false, our sample statistic ($\overline{X}$ or p') falls into the acceptance region, that is, within the bounds of normal variation of our H_O curve. Naturally, we follow all the rules and do just as we should: we accept H_O. Our test statistic lands in the acceptance region only through a quirk of random sampling. Our sample $\overline{X}$ or p' does not exhibit the significant difference a statistic normally would. By random chance, it appears normal. This is not a result of anything we have done or an indication of how other samples may differ from the population; it is merely random chance. Again, our sample happens to be one of those odd cases.

This possibility is illustrated in the H_O curve of Figure 5.6. To appreciate that figure, it is necessary to understand one more bit of the overall picture. We must assume a value for μ_A or p_A, that is, a location for the H_A curve along the horizontal axis. The argument goes as follows: H_O really is false; in other words, our test sample really does belong to a population other than the known H_O population. To calculate the P(Beta error), we must assume a mean value for this alternative population curve, that is, a value for μ_A or p_A. Once we have a value for the center of this alternative curve, we can locate or draw it at the correct spot along the horizontal axis. Calculating the P(Beta error) involves finding the area of this second curve that overlaps the H_O acceptance region in our original H_O curve. Figure 5.6 illustrates this situation with an upper-tailed test for a sample proportion. If our p' value happens to fall into the beta area it is under the H_A curve and part of the alternative population. However, since the alternative and null curves overlap, p' also falls within the bounds of normal variation for the H_O curve. Since we must make our decision in terms of that known H_O curve, we fail to reject our null hypothesis and fail to recognize that p' really does belong to a different H_A population. We have done everything we should and made the correct decision given the information available. Unbeknownst to us, our p' value is an unusual case of the H_A population.

There is one other important point about calculating the probability of a Type II or beta error. As above, we assume a value for the second curve's center, a

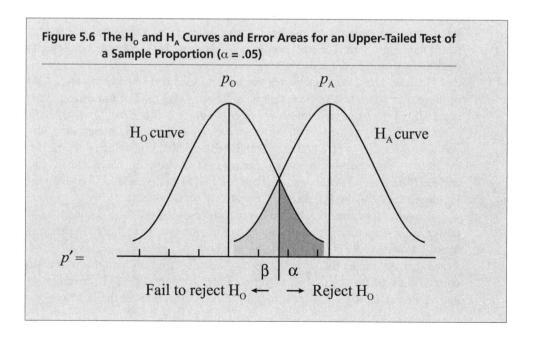

Figure 5.6 The H_O and H_A Curves and Error Areas for an Upper-Tailed Test of a Sample Proportion ($\alpha = .05$)

figure for either μ_A or p_A. This means we could select any one of many possible values. What specific value we use determines where we locate that alternative curve (along the horizontal axis), and this, in turn, determines how much the H_A curve overlaps the original H_O curve. There are many possible alternative curves, each overlapping the H_O curve to a slightly different degree. It follows, then, that there are many different probabilities for a beta error. Everything depends upon just where we put that second curve (see Figure 5.5).

At this point, you may be thinking, "If that is the case, why bother? I can calculate the probability of committing a beta error, but someone else may use a different mean value for the H_A curve and come up with a completely different figure. There is no one right answer." True, but that is okay. It is what we expect. In fact, for any hypothesis-testing situation, we would generally calculate the P(Beta error) for various possible locations of that second curve, or for numerous different values of μ_A or p_A. In a case involving minutes and averages, for instance, we might ask ourselves, "What is the probability of committing a Type II error if we assume H_O is false and that the test sample really belongs to a population with a mean of 28.5 minutes (if $\mu_A = 28.5$ minutes)? A mean of 28 minutes (if $\mu_A = 28.0$ minutes)? A mean of 27.5 minutes? A mean of 27 minutes?" We can, in other words, say to ourselves, "If our independent variable really has an effect and if our test sample *does* belong to a population that averages 28.5 minutes, what is the probability our test will fail to reveal that fact and that we will mistakenly accept our H_O? What is the probability of this happening if our test sample is part of a population averaging 28 minutes, 27.5 minutes, or 27 minutes?" We might rephrase the question and ask, "Under different circumstances and for different locations of an H_A curve, how good is our test? How often will it correctly detect a false null hypothesis?"

Two complementary measures are used to evaluate hypotheses tests. One of these refers to a test's operating characteristics (O.C.). The **operating characteristic** *of a test is the probability of committing a beta error, given specific values (or locations) for the alternative curve.* We calculate the P(Beta error) for various values of μ_A or p_A, and may graph the respective probabilities. As the distance (or $\mu_O - \mu_A$ **shift**) between the means of the two curves increases, the overlap and the P(Beta error) decreases. At some point, the H_A curve ceases to overlap the H_O curve at all, and the P(Beta error) effectively becomes zero.

The other measure by which a hypothesis tests is evaluated is known as its power. This is merely the complement or flip-side of a test's operating characteristic. **Power** *is the ability of a hypothesis test to correctly reject a false null hypothesis.* Assuming the H_O is false, how likely is it that our hypothesis test will correctly reveal that fact? The power of a test is simply $1 - P$(Beta error). As above, we calculate the P(Beta error) for different locations of the H_A curve. We

then subtract our beta probabilities from 1, that is, $1 - P$(Beta error). If we graph
the $1 - P$(Beta error) figures, we have a curve showing us the power of our test
for various possible locations of the alternative curve. In lieu of more detailed
graphs, Figure 5.7 shows the O.C. and power probabilities using the previous
curves (Figure 5.6). Determining the operating characteristics or the power of a
hypothesis test is not especially difficult, but it is obviously based upon calculating
P(Beta error). We will return to this procedure when we look at the specific formulas and mechanics of hypothesis testing. For now, Figure 5.7 shows the O.C. and
power areas of the H_A curve for an upper tail test of a null hypothesis.

Up to this point we have considered the theory and logic of hypothesis testing. We test to see whether a sample statistic differs *significantly* from a known
population parameter. Our null hypothesis says it does not, and the alternative
hypothesis states that it does. We actually test the H_O and either reject it in favor
of the H_A or we fail to do so. We are never completely certain of our decision,
however. No matter what conclusion we reach, there is always the possibility of
error. At a cost, we may to some extent control the respective probabilities of
committing these errors. With one other note about terminology, we consider hypothesis testing's nuts and bolts. How do we actually conduct a hypothesis test?

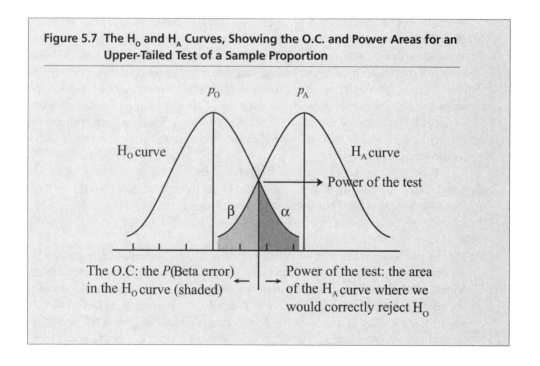

**Figure 5.7 The H_O and H_A Curves, Showing the O.C. and Power Areas for an
Upper-Tailed Test of a Sample Proportion**

Please recall the terminology and phrasing introduced so far in this chapter. We test H_0, and we have sufficient evidence to reject it, or we fail to reject it. We tend to avoid saying we "accept the null hypothesis." That has the ring of certainty. Instead, our language hedges somewhat. We conclude that, based upon evidence from our one sample and given our alpha level, we are unable to reject H_0—or conversely, that we do have sufficient evidence to reject it. Our conclusion is firm but suitably modest and tentative. Given our evidence, our decision is correct, but there is a small chance we are wrong. Our conclusion should reflect that slight degree of uncertainty.

Nuts and Bolts: Conducting the Test

We will now consider different forms of single-sample hypotheses tests. We look first at one-tailed tests and then at a two-tailed test. The same basic procedures apply to problems involving means and those involving proportions, and we will consider examples of each. We will also review examples of z tests and t test alternatives for procedures involving averages.

Some aspects of hypothesis testing are similar to those governing confidence interval estimation. Means or proportions? z or t? The criteria for deciding whether to use z or t are the same as before. When testing a sample mean, is n less than 30? For proportions, we again consult Table 4.1 to see whether our sampling distribution approximates the z curve. Now, however, we are using an H_0 curve centered around a known p_0 value. Therefore, we use p_0 or q_0 for Table 4.1. We also use the fpc in hypothesis testing and under the same conditions: when n is 10% or more of N and sampling is done without replacement. When used, the fpc appears in the denominators of our formulas. It is a multiplier for the standard error, and that standard error is now our denominator. Finally, as something new, we will take a closer look at Type II errors and how the P(Beta error) is influenced by our sample size.

There are two ways in which hypotheses tests using the z and t curves are conducted. The one involving z or t tests is far more common than the one involving critical values, but both work equally well.

Options: *z, t,* or a Critical Value

As we go through the mechanics of hypothesis testing, let us initially consider a lower-tailed test of a sample mean using the .05 alpha level. Is our sample mean significantly less than the known population mean? In order to be considered *significantly* lower, $\overline{X}$ must fall into the lowest 5% of the curve, or, to be technically correct, it must fall at a spot on the $\overline{X}$ axis that is under this lowest 5% area. The gist of our hypothesis test is to determine whether it does or does not, and there are two ways to do this.

z and t tests. We conduct our hypothesis test in terms of z scores—or, in the case of means, possibly t scores. This is the way most hypotheses tests are done. In our H_O curve, we find the z or t value that cuts off the lowest 5% of the curve. This point in the curve is critical. It separates our alpha region from the area of normal variation. If our sample average falls on one side of this point, we reject our H_O. If it falls on the other side, we fail to do so. The z or t scores cutting off this critical 5% of the curve are readily available in the z and t tables, respectively, and they are known as **z-critical** and **t-critical** (z_C and t_C) values. For our lower-tailed test at the .05 alpha level, $z_C = -1.645$, a familiar z score marking off the lower 5% in any normal distribution. The t value depends upon our degrees of freedom (d.f. $= n - 1$). The actual test itself compares our $\overline{X}$, once we convert it to a z score (Chapter 3)—to our z_C value. According to its z score, where does our sample mean fall: in the rejection region or within the bounds of normal variation and sampling error? If it is right on the line or falls in the alpha region, we reject our H_O. If not, we fail to reject our H_O. We refer to our sample statistic's z score as **z-observed** (z_O). It is based upon what we observe in our sample. Our test consists of comparing z_O to z_C (or t_O to t_C) and making the appropriate decision regarding our null hypothesis. Figure 5.8 illustrates a lower-tailed test of a sample mean at the .05 alpha level.

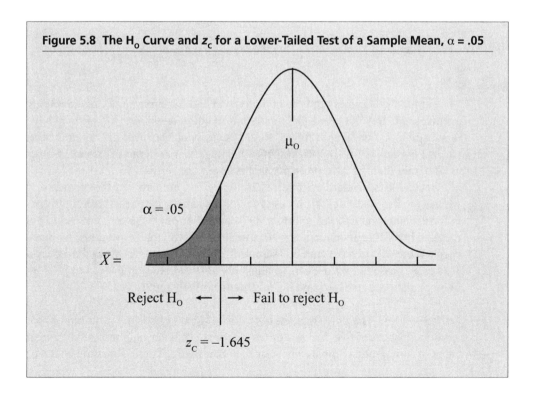

Figure 5.8 The H_O Curve and z_C for a Lower-Tailed Test of a Sample Mean, $\alpha = .05$

To actually conduct the test, we convert our sample statistic into a z score (z_0). For averages, equation 26 is slightly changed and becomes equation 53. If our problem requires the use of t, we calculate t_0 instead, as in equation 54. For proportions, equation 33 translates a sample proportion to a z score. This is now slightly revised as equation 55. These are our *observed z* and *t* scores. The final step compares our observed and critical values of z (or t) to make the appropriate decision regarding our H_0.

$$(53) \qquad z_0 = \frac{p' - p_0}{\sigma_p} = \frac{p' - p_0}{\sqrt{\dfrac{p_0 \, q_0}{n}}}$$

$$(54) \qquad t_0 = \frac{\overline{X} - \mu_0}{s_{\overline{X}}} = \frac{\overline{X} - \mu_0}{\dfrac{s}{\sqrt{n}}}$$

$$(55) \qquad z_0 = \frac{p' - p_0}{\sigma_p} = \frac{p' - p_0}{\sqrt{\dfrac{p_0 \, q_0}{n}}}$$

Familiar guidelines apply to equations 53 and 54. First, for a case involving averages, we may not know the population standard deviation (σ). We may have to substitute s, our sample figure, to get the standard error in the denominator: $(s/\sqrt{n})$ instead of $(\sigma/\sqrt{n})$. In doing so, we may wish to write the standard error as $s_{\overline{X}}$ rather than $\sigma_{\overline{X}}$. No problem; that is common.

Second, the standard error terms are in the denominators of the formulas. If we use the fpc (the rules for its use remain the same), it appears as a multiplier in those denominators. We multiply the standard error by the fpc, just as we did previously with confidence interval estimates. Finally, for proportions, we must check Table 4.1 to make sure our sampling distribution conforms to the z curve. This time, however, we use our smallest *population* value, p_0 or q_0, not p' or q'. The population figures, p_0 and q_0, now define our H_0 curve.

Critical values. The other method of testing a hypothesis is not as common, but it works. Instead of two z scores or two t scores, we determine the actual critical value of $\overline{X}$ or p' and compare our $\overline{X}$ or p' to that figure. There is no z-observed or *t*-observed. In a lower-tailed test, if our sample statistic falls below or to the left of the critical point, $\overline{X}_{\text{C}}$ or p'_{C}, we will reject our H$_\text{O}$ (Figure 5.9). If it falls above or to the right of this point, we fail to reject H$_\text{O}$. $\overline{X}_{\text{C}}$ or p'_{C} is a critical spot on the lower axis, as shown in Figure 5.9.

We calculate $\overline{X}_{\text{C}}$ or p'_{C} using variations of earlier formulas. In Chapter 3, we solved for an *X* value falling at a particular point under the normal curve. We now use the same basic procedure to solve for a similar $\overline{X}$ value, specifically $\overline{X}_{\text{C}}$ (equation 56), or similarly for p'_{C} (equation 57). As before, if we do not know the true population standard deviation (σ), we may substitute *s* instead. And of course, with small samples and averages, we will substitute *t* (with *n* − 1 d.f.) for *z*.

(56)
$$\overline{X}_{\text{C}} = \mu_{\text{O}} + z_{\text{C}} \frac{\sigma}{\sqrt{n}}$$

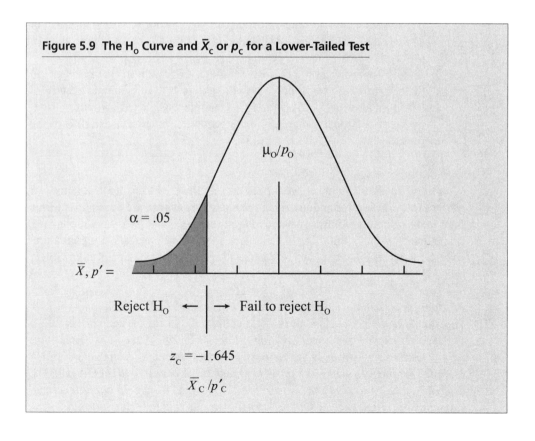

Figure 5.9 The H$_\text{O}$ Curve and $\overline{X}_{\text{C}}$ or p_C for a Lower-Tailed Test

In Chapter 4, we converted p' values into z scores. Now we revise the formula to determine p'_c.

(57)
$$p'_c = p_o + z_c \sqrt{\frac{p_o \, q_o}{n}}$$

The final step compares our actual $\overline{X}$ to $\overline{X}_c$ (or p' to p'_c), and we make the appropriate decision regarding H_o.

Before examples and actual hypotheses tests are conducted by looking at several examples, a checklist may be helpful. These are the questions to ask yourself in diagnosing a hypothesis testing situation:

1. Does the situation involve means or proportions? Am I testing a sample mean or a sample proportion?
2. What kind of test is called for? Do I wish to know whether the sample statistic is significantly greater than, significantly less than, or merely differs from the population parameter? Should I be doing a one-tailed test or a two-tailed test? If it is a one-tailed test, should it be an upper-tailed or lower-tailed test? (Think about your H_A here.)
3. Should I be using z or t? In the case of means, what is my sample size? In the case of proportions, according to Table 4.1, may I conduct a z test?
4. May I include the fpc? Assuming the sample was picked without replacement, as is typically the case, is the sample size (n) at least 10% of the population size (N)?

One-Tailed Tests: Directional Differences?

With these questions in mind, we turn to actual hypothesis testing problems. For now, to make this introduction easier, the examples are all one-tailed tests. Even though the second checklist item above therefore becomes somewhat moot here, it is a question that should be asked in any general hypothesis testing situation. Even here, we still have to decide whether the test should be upper tail or lower tail.

1. *Does alcohol influence reaction times? Based upon tests using a timing device attached to a car's brake pedal, it was determined that the average driver had a .88-second reaction time, with a standard deviation of .15 seconds. Twenty-five people were selected from 200 volunteers attending a weekend traffic school. Each of the 25 people was then given three 12-ounce beers over a period of 90 minutes. Thereafter, their reaction times averaged .93 seconds. At the .05 alpha level, did the sample average show a significantly slower reaction time than normal?*

To diagnose the question of "average reaction time" clearly indicates a means or averages problem here, not proportions. We are asked to determine whether the test sample's time is "significantly slower." Another way of reading this is: was the sample's time longer? Did it take significantly more time to hit the brake pedal? This dictates an upper-tailed test. The population average is .88 seconds. Our sample average is .93 seconds. Exactly where does .93 fall along a continuum and curve centered on .88? In the alpha region or not?

To actually solve the problem, we first write down all the information we are given, present our hypotheses, and sketch a model of our H_O curve. As above, μ_O = .88 seconds, σ = .15 seconds, and $\overline{X}$ = .93 seconds. Since n = 25, we use a t test with 24 degrees of freedom, and from from Table C that t_C = 1.711 for a one-tailed, upper-tailed test at α = .05. Finally, N = 200 and n = 25, justifying using the fpc. Then our hypotheses are $H_A: \overline{X} > \mu_O$ and $H_O: \leq \mu_O$, and the distribution curve is shown in Figure 5.10.

We calculate t_O by using equation 54 (with $s = \sigma$) and the fpc:

$$t_O = \frac{\overline{X} - \mu_O}{\dfrac{\sigma}{\sqrt{n}}\sqrt{\dfrac{N-n}{N-1}}} = \frac{.93 - .88}{\dfrac{.15}{\sqrt{25}}\sqrt{\dfrac{200-25}{200-1}}} = \frac{.05}{(.030)(.938)} = \frac{.05}{.028} = 1.786$$

We have sufficient evidence to reject the H_O at the .05 alpha level because t_C = 1.711 and t_O = 1.786. At the .05 alpha level, alcohol *has* significantly slowed reaction times. On the average, the sample took significantly longer to hit the brake pedal than did a population of nondrinkers.

Figure 5.10

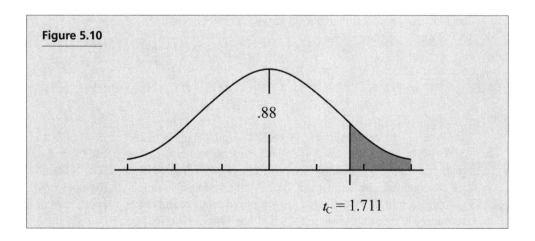

Before looking at another example, consider how different circumstances might have affected our results here. First, if the fpc had not been used, t_O would have been $.05/.030 = 1.667$ (instead of 1.786). In that case, we would have failed to reject H_O. When the observed and critical values are close, using the fpc (or being unable to do so) may make the difference between being able to reject H_O or failing to do so.

Second, if we had used the $.01$ alpha level, we would not have rejected H_O. t_C would then have been 2.492. We would have failed to reject our H_O with $t_O = 1.786$ and we would have concluded that the sample's average reaction time was not significantly greater. In effect, we have enough evidence to be 95% confident alcohol has slowed reaction times but not enough to be 99% confident.

2. *Based upon previous studies, it is known that 45% of all college students tend to daydream in class. In a random sample of 50 statistics students (drawn from a population of 612 statistics students campus-wide), 17 said they tended to daydream in these particular classes. At the .05 alpha level, is the percentage of students who daydream in statistics classes significantly less than in classes as a whole?*

This is a proportions problem (proportions and percentages are essentially the same). We are asked whether the sample proportion is "significantly less" than the known population proportion. Therefore, a lower-tailed test is appropriate. We consult Table 4.1 to see whether we may conduct a z test: $n = 50$ and $p_O = .45$ (so $q_O = 1 - .45 = .55$). For our p_O value, the table reveals a minimum n of 45 for meeting the z curve's requirements. Therefore, a z test is justified, and with $\alpha = .05$, $z_C = -1.645$. Given that $n = 50$ is less than 10% of $N = 612$, we may not use the fpc. Finally, $p' = X/n = 17/50 = .34$. Our hypotheses are $H_A: p' < p_O$ and $H_O: p' \geq p_O$. The distribution curve is shown in Figure 5.11.

From equation 55, we get:

$$z_O = \frac{p' - p_O}{\sqrt{\dfrac{p_O \, q_O}{n}}} = \frac{.34 - .45}{\sqrt{\dfrac{.45(.55)}{50}}} = \frac{-.11}{.07} = -1.57$$

We fail to reject our H_O at the $.05$ alpha level because $z_C = -1.645$ and $z_O = -1.57$. When compared to college students and classes in general, there is no evidence that a significantly lower percentage daydreams in statistics classes.

Problems 1 and 2 illustrate the one-tailed, single-sample hypothesis test. We compare a sample statistic to a known population parameter, and specifically determine whether that statistic falls into one particular tail of our H_O curve. Based

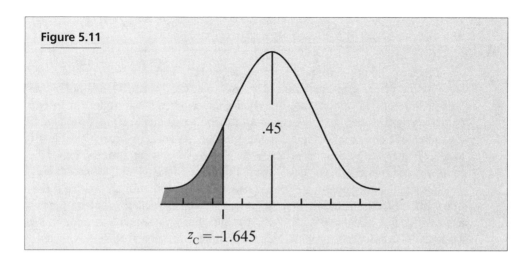

Figure 5.11

.45

$z_C = -1.645$

upon the direction of the sample/population difference in which we are interested, we conduct either an upper-tailed test or a lower-tailed test.

Two-Tailed Tests: Any Differences?

If we wish to actually test for a difference in either direction from a central population value or if we suspect a sample/population difference but are not sure in which direction it might lie, we should use a two-tailed test. In the two-tailed test, we consider a sample/population difference of *any* sort. It does not have to be only in a particular direction; we look for any difference at all. As the name implies, in the two-tailed test we have an H_O rejection region in each tail of the curve. For example, if our alpha level is .05, we have a .025 or 2.5% rejection region in each tail. Our total alpha area is always split evenly between the two tails. We reject the H_O if our test statistic falls into either tail.

Problems 3 and 4 obviously involve two-tailed tests, but it is important to keep in mind the four checklist questions offered earlier in this chapter. We must still diagnose each situation as one involving means or proportions, as a z or t test for means, as one permitting a z test for proportions, or as one justifying the fpc.

3. *You are in casino and have just had a winning streak at a roulette wheel. US roulette wheels alternate 18 red numbers, 18 black numbers, and 2 green numbers. You have just bet on "any red number" for 40 straight spins and won 28 times. You are elated, but figure that if there is something unusual about this wheel, you could just as easily have lost as much as you have won. At the .01 alpha level, does your success rate differ at all from the range of successes one*

might expect purely by random chance? In other words, at the .01 alpha level, does there appear to be something wrong with this wheel?

To set up the problem, the term "success rate" suggests a proportions problem. On the wheel itself, the choices are red, black, or green. This is not a success-or-failure, two-option, true binomial choice, but we may reduce it to two outcomes by considering red versus non-red. We let p equal the proportion coming up red, the outcome of interest in the question. As usual, we consult Table 4.1 to see whether a z test is appropriate. We are also asked to determine whether the sample success rate "differs at all" from a random chance "range of successes." The random chance percentage for red becomes the population figure. No directional difference is specified, and the phrase "range of successes" also suggests a two-tailed test. Finally, we know $n = 40$ spins, but we have no figure for N. We assume N is infinity. The probabilities we assume here, such as $P(\text{Red})$, are based upon a hypothetical infinite number of spins. Therefore, if N theoretically equals infinity, we may not use the fpc.

We get: $p_O = P(\text{Red}) = 18/38 = .474$, so $q_O = P(\text{Non-red}) = 1 - .474 = .526$. According to Table 4.1, if $p_O = .47$, we need $n \geq 39$ to use the z test. Since $n = 40$, we may conduct a z test. Our sample size is 40 spins, and $p' = 28/40 = .70$. The hypotheses are $H_A: p' \neq p_O$ and $H_O: p' = p_O$. Figure 5.12 shows the distribution curve.

Then, using equation 54, we get:

$$z_O = \frac{p' - p_O}{\sqrt{\dfrac{p_O\, q_O}{n}}} = \frac{.70 - .474}{\sqrt{\dfrac{.474(.526)}{40}}} = \frac{.226}{.079} = 2.861$$

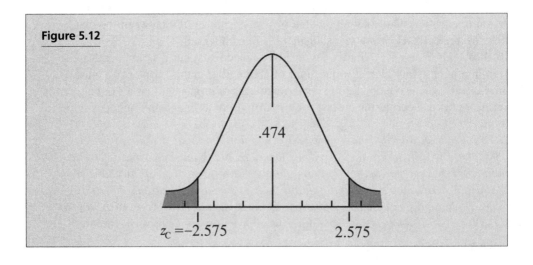

Figure 5.12

.474

$z_C = -2.575$ 2.575

We have sufficient evidence to reject our null hypothesis at the .01 alpha level because $z_C = \pm2.575$ and $z_O = 2.861$. There does appear to be something wrong with this wheel. It deviates significantly from random chance at the .01 alpha level. You got lucky this time. You beat the house, but it would be best try a different wheel next time.

Note one additional point before we look at another practice problem: Having rejected our H_O at the .01 alpha level, is it clear that we would (automatically) have done so at the .05 alpha level also? After all, the .01 level is the more demanding of the two. In fact, any difference significant at the .01 level is automatically significant at the higher alpha level (.05). In problem 3, z_C at the .05 level is ±1.96, and with $z_O = 2.861$, we would have had clear and overwhelming evidence to reject H_O.

4. *Youth service organizations receiving public funds must follow federal guidelines in integrating physically disabled youngsters into all activities. To be eligible for funding, these organizations must periodically submit data showing that the participation of children with disabilities does not differ significantly from that of participants overall. Last year, in Los Angeles County's 410 service groups, the average time boys and girls spent at summer day camps was 11.31 days, with a standard deviation of 4.16 days. A random sample of 30 organizations revealed that the average for disabled youngsters was 9.62 days at summer camps, with a standard deviation of 2.85 days. At the .05 alpha level, did the average number of camp days for disabled children differ at all from that of youngsters in general? In other words, are Los Angeles County's organizations complying with federal guidelines?*

To sort out the question, the problem involves averages, not proportions. We are given population and sample averages. It also asks whether the camp days of disabled children "differ at all," not whether they it do so in a particular direction. This suggests a two-tailed test. The sample size is 30, so a z test is appropriate. Since $N = 410$ organizations and $n = 30$, we may not use the fpc. In addition, $\mu_O = 11.31$ days, $\sigma = 4.16$ days, $\overline{X} = 9.62$ days, and $s = 2.85$ days. Our hypotheses are $H_A: \overline{X} \neq \mu_O$ and $H_O: \overline{X} = \mu_O$, and the distribution curve is shown in Figure 5.13.

We calculate z_O by using equation 53:

$$z_O = \frac{\overline{X} - \mu_O}{\dfrac{\sigma}{\sqrt{n}}} = \frac{9.62 - 11.31}{\dfrac{4.16}{\sqrt{30}}} = \frac{-1.69}{.76} = -2.22$$

At the .05 alpha level, there is sufficient evidence to reject the null hypothesis because $z_C = \pm1.96$, and $z_O = -2.22$. For the 30 groups sampled, at the .05

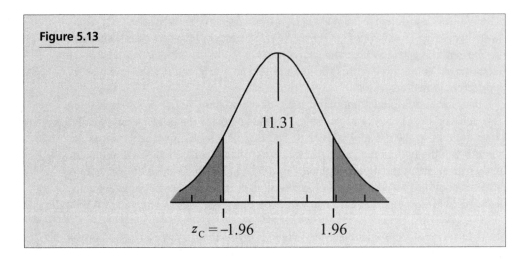

Figure 5.13

11.31

$z_c = -1.96$ 1.96

alpha level, the average camp days of disabled youngsters differs significantly from that of children in general. The Los Angeles County organizations appear not to be in compliance with federal guidelines.

Note two further points about this question: First, the alpha level makes a difference. At the .01 alpha level, there is insufficient evidence to reject H_O. At that level, $z_c = \pm2.575$, so our −2.22 is not statistically significant, and the associations meet federal guidelines. Second, notice how the conclusion is stated despite the fact that z_O clearly falls into the *lower* alpha region. We merely conclude that the average for youngsters with disabilities *differs,* not that it is significantly less than the norm. The conclusion is always consistent with the test. Since we set up a two-tailed, nondirectional test, all we may claim is a difference. To legitimately claim the sample average is *less than* the population average, we must rewrite our hypotheses and design a strictly lower-tailed test.

With the two-tailed test of a sample mean or proportion, in contrast to the one-tailed test, we are prepared to reject the null hypothesis should our sample statistic fall in *either* tail of the H_O curve. We do not require a sample/population difference in a particular direction. Should the sample statistic be sufficiently larger or smaller than the population parameter, we judge it to differ significantly.

There are two final points to note in reference to a two-tailed test. First, as with its one-tailed counterpart, we may conduct a test by finding the critical values of the test statistic itself (by using equations 56 and 57) rather than in terms of z_O or t_O values. For instance, we may find the $\overline{X}_c$ values for the upper and lower tails and compare our observed sample mean to these figures. The range between the lower and upper $\overline{X}_c$ values constitutes the broad central area of the curve and the bounds of normal variation. If our sample mean falls into this central range, we fail to reject the null hypothesis. Conversely, if our sample mean falls outside this range and into one of the tails, we have evidence to reject H_O. We may do the same sort of test with proportions, of course. We find the lower and upper values for p'_c, locate our observed sample proportion (p') in relation to this range, and accept or reject our null hypothesis accordingly.

Second, it is sometimes argued that a two-tailed test is generally to be preferred over the one-tailed procedure, and that when in doubt, one should opt for the two-tailed test. The argument is that, for any given alpha level, a two-tailed test is more rigorous and therefore it is more difficult to reject H_O. In contrast, the one-tailed test consolidates all the alpha area at one tail, and, at least for that tail, makes it easier to reject H_O. This point is debatable, however. The two-tailed test does make it more difficult to reject H_O, but only for a constant alpha level. Because the alpha level is a matter of the researcher's discretion, it may be changed. With a one-tailed test we may simply lower our alpha level from .05 to .025 or to .01, from .01 to .005, and so on. It appears easier to control the stringency of our test through the alpha level than by necessarily opting for the two-tailed format. There are times when we very definitely expect differences in a certain direction and are (statistically) disinterested in anything else. A one-tailed test with a suitable alpha level is as valid as a two-tailed test and, indeed, may well be more congruent with the general sense of a research situation.

We have thus far considered one- and two-tailed hypotheses tests. Which one is appropriate depends upon the research situation and specifically upon our alternative hypothesis. No matter which sort of test we conduct, however, there is always a chance for error simply due to random chance. The probability of a Type I or alpha error is under the researcher's direct control because he or she sets the alpha level. The probability of a Type II, or beta, error is a more nebulous figure, however, and its management requires a little more effort and investment than simply setting a suitable significance level, as we will see next.

Coping with Type II Beta Errors: More Power

A Type II, or beta, error occurs when we fail to reject a false null hypothesis. Our sample belongs to a second or alternative population's curve, which overlaps our original and known H_O curve. The two curves overlap to such an extent that, along the horizontal axis, our test sample falls at a spot within the acceptance region for the null hypothesis (recall Figure 5.7). We are conducting our test in terms of this known H_O curve, as we must, so we follow the rules and do not reject our H_O. In doing so, we have committed a Type II error. We have accepted a null hypothesis that is actually false.

We may not be able to eliminate the possibility of a beta error altogether. Provided we make some assumptions, however, we can calculate the probability of committing a beta error for any given hypothesis test. (Recall the earlier discussion of a test's operating characteristics and power.) Also, if we are able to increase our sample size (n), we may be able to minimize the P(Beta error).

There are three steps to calculating *P*(Beta error). Before the specifics, however, imagine a hypothetical test situation. Assume we have an upper-tailed test of a sample mean. We know the population mean (μ_O) and have assumed a value for the mean of the alternative curve (μ_A). We know the sample mean ($\overline{X}$) and standard deviation (*s*). Given *s* and *n*, we can calculate the standard error, and assume we may conduct the H_O test by using *z*.

We must consider two other points: First, the standard error of the H_A curve is equal to the standard error we calculate for the known H_O curve. The hypothetical H_A sampling distribution curve is based upon the same sample size and standard deviation as the H_O curve. In other words, this second curve assumes the same shape (spread) as the original H_O curve. The only difference is the value around which it is centered (μ_A). This means that it falls at a different location or spot along the lower axis, but its shape is the same.

Second, we must use the *z* curve when calculating *P*(Beta error). We cannot use a *t* curve. We cannot use the method described here unless our situation meets the appropriate criterion for using the normal curve. This is because *P*(Beta error) is equivalent to an area of the alternative curve, and only the *z* table gives us a complete set of scores and total areas of the curve. The *t* table shows only selected *t* scores along with very small areas in the tails of the curves, and *P*(Beta error) may well be equivalent to a considerably larger area.

The first step for calculating *P*(Beta error) uses the H_O curve; the last two steps involve only the H_A curve. For the case involving averages or means:

• Step 1 (*in the H_O curve*):
We calculate the critical value of the test statistic ($\overline{X}_C$) using equation 56.

• Step 2 (*in the H_A curve*):
We calculate the *z* score for $\overline{X}_C$ using equation 58. Depending upon the information available, we use either σ or *s*, with σ being preferred.

$$(58) \qquad z_{\overline{X}_C} = \frac{\overline{X}_C - \mu_A}{\dfrac{\sigma}{\sqrt{n}}} = \frac{\overline{X}_C - \mu_A}{\dfrac{s}{\sqrt{n}}}$$

• Step 3 (*in the H_A curve*):
Using the *z* score from step 2 and the normal curve table, we look up the area of the H_A curve corresponding to *P*(Beta error). This is the area in the tail of the H_A curve (column C in Table A).

To illustrate this procedure, Figure 5.14 depicts the alpha and beta error areas of the curves for an upper-tailed test of a sample mean and also numbers

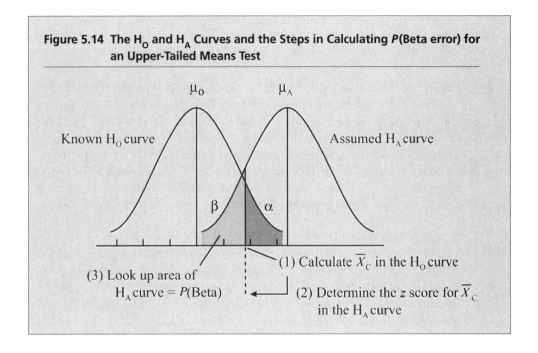

Figure 5.14 The H$_O$ and H$_A$ Curves and the Steps in Calculating *P*(Beta error) for an Upper-Tailed Means Test

the steps involved in calculating *P*(Beta error). To summarize the steps, we first determine the critical value ($\overline{X}_C$) of our statistic along the lower axis. This spot marks the critical acceptance/rejection point for H$_O$. For an upper-tailed test, if a sample mean falls to the right of this point, we reject the null hypothesis. If it falls to the left, we fail to reject H$_O$. We now focus on the H$_A$ curve. To the left of $\overline{X}_C$, besides the H$_O$ acceptance region, is a certain overlapping portion of the H$_A$ curve. To find this latter area, we must get the *z* score for $\overline{X}_C$ in terms of the H$_A$ curve. Once we have this *z* score, we look up the overlapping lower-tailed area of the alternative curve (column C in Table A). This area tells us the probability of a sample statistic from the alternative population falling into this lower tail. Even though it is from a different population, any such sample results in our accepting H$_O$. We are conducting our hypothesis test in terms of the H$_O$ curve. That is all we consider in deciding whether to accept or reject H$_O$. It is a clear-cut proposition: In the H$_O$ curve, does our sample fall to the left or to the right of the critical acceptance/rejection point? In making our all-important H$_O$ decision, we do not even consider a second curve.

The general procedure is the same for proportions. We first concentrate on the H$_O$ curve and establish a value for p'_C (equation 57). Once we have this figure, we look to the H$_A$ curve. The next step is to calculate a *z* score for p'_C in terms of the H$_A$ curve (equation 59).

$$(59) \qquad z_{p'_C} = \frac{p'_C - p_A}{\sqrt{\dfrac{p_A q_A}{n}}}$$

Finally, we look up an area of this latter curve corresponding to P(Beta error).

To show different situations, problem 5 involves means and problem 6 involves proportions. One problem includes calculations of P(Beta error) for a lower-tailed H_O test, and the other shows the calculations for an upper-tailed test. Examples could also employ the fpc to correct the standard errors. It would be added as a multiplier for the standard error term, just as before. To simplify things, however, the fpc is omitted in the following problems to more clearly convey the main points regarding beta errors.

5. *A large residential treatment program for alcoholism operates several hospitals around the country. The average patient stays 43 days, with a standard deviation of 10 days. It is believed that a newly available medication might significantly shorten the treatment period. To test the new drug, we plan to take a random sample of 35 patients from a pool of some 700 volunteers. We will conduct a lower-tailed test and use the .01 alpha level. If the new drug does indeed shorten the treatment period by 6 days, on the average, what is the probability our test will mistakenly suggest it has no effect? That is, what is P(Beta error) given these test conditions?*

Obviously, the problem involves an average, not a proportion. We have the mean and standard deviation for our population ($\mu_O = 43$ days, and $\sigma = 10$ days). A lower-tailed test is appropriate to see whether the sample average is significantly shorter or less than the population figure, the .01 alpha level has been chosen, and we have a value of 37 days for μ_A (43 days minus 6 days). If the new drug really is effective, what is the probability that our way of testing it might fail to reveal this fact? Note that with $n = 35$ and $N = 700$, we may not reduce our standard error by using the fpc.

Notice that the information given omits a mean ($\overline{X}$) for the test sample. We would need that to conduct an actual H_O test, of course, but that is not our purpose here. We have the following values: $\mu_O = 43$ days, $\sigma = 10$ days, our assumed value for $\mu_A = 37$ days, and $n = 35$. Our hypotheses are H_A: $\overline{X} < \mu_O$ and H_O: $\overline{X} \geq \mu_O$, and from Table A we see that at the .01 alpha level, $z_C = -2.33$. Figure 5.15 shows the H_O and H_A distributions.

To calculate P(Beta error), we follow the three steps:

• Step 1 (using equation 56):

$$\overline{X}_C = \mu_O + z_C \frac{\sigma}{\sqrt{n}} = 43 + (-2.33)\frac{10}{\sqrt{35}} = 43 + (-2.33)(1.69) = 43 - 3.94 = 39.06 \text{ days}$$

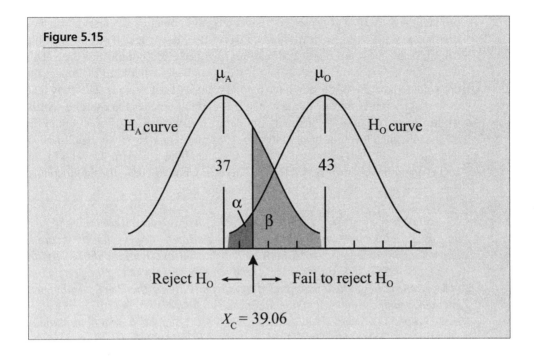

Figure 5.15

- Step 2 (using equation 57):

$$z_{\overline{X}_C} = \frac{\overline{X}_C - \mu_A}{\frac{\sigma}{\sqrt{n}}} = \frac{39.06 - 37}{1.69} = \frac{2.06}{1.69} = 1.22$$

- Step 3 (using Table A):
  P(Beta error) = area to the right of $z_{\overline{X}_C}$ in the H_A curve = .1112, or 11.12%

Assuming our test sample, after being treated with the new drug, now belongs to a population that averages only 37 days at a hospital, there is still an 11.12% chance that the sample (due to normal sampling error) would average a stay of 39.06 days or longer. Put another way, if the new medication does reduce the mean stay by 6 days and if we conduct our test with 35 people, there is an 11.12% chance of a sample average still falling into the H_O acceptance range of 39.06 days or more. We run an 11.12% risk of committing a Type II error and mistakenly concluding the new drug did *not* shorten the average stay at a facility. Conversely, we have a $(1 - .1112)$ or a .8888 chance of correctly rejecting the H_O because our test is 88.88% powerful.

We could go on to develop an operating characteristics curve or a power curve. We could also calculate P(Beta error) for different μ_A values. What is the probability of a Type II error (or, conversely, the power of the test) if the new drug reduces the average stay by only 3 days? By 7 days? By 10 days? We could determine the power of our test for various assumed effects of the new drug. This would not tell us the *true* effect, but it would indicate what kind of impact the drug had to have for our test to have a good chance of detecting it. This may lead us to reassess and improve our test, and that certainly makes the exercise worthwhile. Just as we do not construct confidence interval estimates at less than the 90% confidence level, or have alpha levels of more than 10%, we may not wish to conduct an H_O test suspected of being less than 90% powerful. Developing an actual operating characteristics (O.C.) curve or a power curve for the hospitals' drug test is quite possible but beyond our scope here.

6. *Assume that about 36% of adult Americans favor a renewed campaign to pass an Equal Rights Amendment (ERA). We have reason to suspect that the figure is higher among college students and, moreover, that the figure might be slightly over half, or 55%. We are planning a study to determine whether the percentage of favorable college students is indeed significantly higher than the favorable percentage among the general public. If the student figure is around 55% and we use the .05 alpha level, and if we plan to interview 100 students from a population of about 30,000, what is the probability we might commit a Type II or beta error?*

Problem 6 involves proportions using an upper-tail test. For the general public, $p_O = .36$, and for students, $p_A = .55$. Given an alpha level of .05 and $n = 100$, what is the risk of our test mistakenly failing to show a significant difference between the general public's and college students' opinions? Finally, because $n = 100$ and $N = 30,000$, we may not use the fpc, but $n = 100$ justifies assuming both the H_O and H_A curves conform to the normal distribution (see Table 4.1). The hypotheses are $H_A: p' > p_O$ and $H_O: p' \leq p_O$. We are given $p_O = .36$ and $q_O = .64$, and our assumed values are $p_A = .55$ and $q_A = .45$. From Table A, at the .05 alpha level, $z_C = 1.645$. Figure 5.16 shows the distributions.

- Step 1 (using equation 57):

$$p'_C = p_O + z_C \sqrt{\frac{p_O \, q_O}{n}} = .36 + 1.645 \sqrt{\frac{.36(.64)}{100}}$$

$$= .36 + 1.645(.048) = .36 + .08 = .44$$

- Step 2 (using equation 59):
  Find the z score for p'_C in the H_A curve:

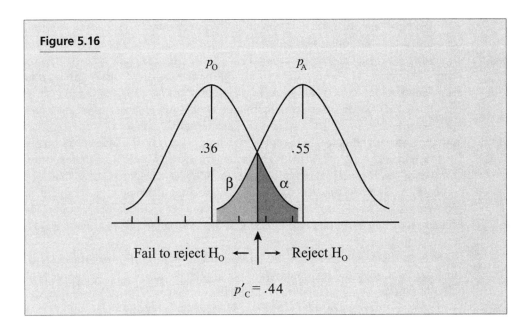

Figure 5.16

$$z_{p'c} = \frac{p'_c - p_A}{\sqrt{\dfrac{p_A \, q_A}{n}}} = \frac{.44 - .55}{\sqrt{\dfrac{.55(.45)}{100}}} = \frac{-.11}{.05} = -2.20$$

Note that the denominator, the standard error, now uses the p and q values for the H_A curve. The two curves should have nearly identical standard errors, however. The value of n is (or is assumed to be) the same in each case, and the respective p and q values should not be too dissimilar if the two curves are close enough to overlap. For example, to three decimal places, the standard error for the H_O curve is .048, and for the H_A curve it is .050.

• Step 3 (using Table A):

P(Beta error) = area below $z_{p'c}$ in the H_A curve = .0139 = 1.39%

Assuming an alpha level of .05, an n of 100, and a p_A of .55, we would run only a 1.39% chance of committing a beta error. Only 1.39% of the samples from such a student population have p values of less than .44. Only 1.39% of all the sample proportions are low enough to result in us mistakenly accepting the null hypothesis. Conversely, our test is 98.61% powerful. With the conditions above, we have a 98.61% chance of correctly rejecting H_O. Quite good.

We have seen that for the probability of committing a Type II error with a one-tailed test, we look to an alternative curve lying either above or below our H_O curve. But what about a two-tailed test? In this case, the procedures are generally the same, but we consider the possibility of alternative curves falling both above and below the central H_O curve (see Figure 5.17).

In such a case, there are two alternative means and two beta areas. To establish the alternative means (or, of course, the alternative p_A values), we go an equal distance above and below our central μ_O (or p_O). Each of the alternative curves overlaps the H_O curve by an identical amount or area. We calculate P(Beta error) for one of the alternative curves and simply double that figure for the total beta probability. Since we do not know whether the true alternative curve might lie above or below our H_O curve, we have to take both possibilities into account. The actual procedure and calculations are the same as with the one-tailed test and are not repeated here.

Besides merely calculating the probability of a Type II error, however, we may also control it by manipulating our sample size. Generally speaking, with all other things being equal, we reduce P(Beta error) as we increase our sample size. As the sample size increases, the hypothetical sampling distribution curve becomes more peaked. It clusters more around the central population mean or proportion. The spread of the curve, as reflected in the standard error, decreases.

To visualize this principle, just as we had previously with the H_O and H_A curves, imagine two curves side by side. Moreover, imagine the centers of these curves a set distance apart, one having a mean of 20 and the other a mean of 30 (see Figure 5.18). If these two curves are based upon a small sample size, they will

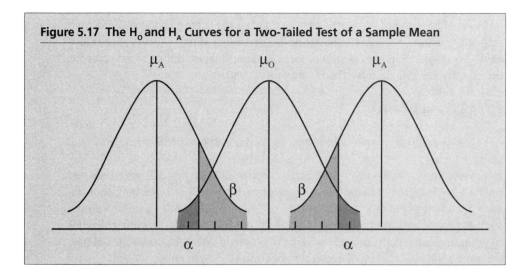

Figure 5.17 The H_O and H_A Curves for a Two-Tailed Test of a Sample Mean

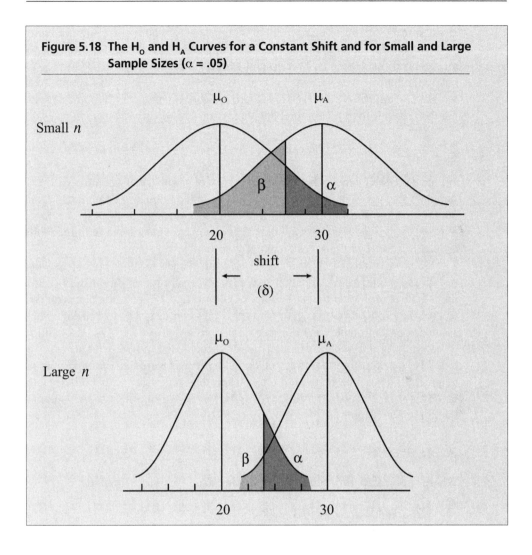

Figure 5.18 The H$_o$ and H$_A$ Curves for a Constant Shift and for Small and Large Sample Sizes (α = .05)

each have considerable spread and will probably overlap quite a bit. If we imagine two curves based upon a large sample size, however, each appears taller and more peaked and has little spread compared to the small *n* case. There is little overlap, if any. *P*(Beta error), as we saw, is a function of this overlap: The more the two curves overlap, the greater the probability of committing a beta error. For any constant or set shift (δ) between the means of the two curves, *P*(Beta error) varies inversely with the sample size: as one goes up, the other goes down. Another way of saying this is that, for any assumed value of μ_A or p_A. (i.e., for any assumed shift from μ_0 or p_0), our sample size is going to be a major factor in the amount of spread, the degree of overlap, and P(Beta). It follows that we may control the degree of overlap and hence the probability of a beta error by manipulating our sample size. Procedures exist for determining the sample sizes required to minimize Type II errors and to increase the power of hypothesis tests, but they are left to more comprehensive texts.

Summary

We have covered a lot in this chapter. By way of summary, we compare a single sample's statistic to its known distribution around a population parameter. Does our sample statistic fall into a normal or expected range or does it fall into a relatively rare range? Is there evidence for assuming our sample is significantly different or for regarding it as essentially normal?

We never make a definite decision, of course, because all of our calculations rest on probabilities. We are always alert to the errors we might make. We have immediate control over the probability of committing a Type I or alpha error. We control the probability of rejecting a true H_O. In contrast, we must make certain assumptions and manipulate our sample size in order to control the probability of committing a Type II or beta error. How much risk do we wish to take for possibly *not* detecting the effect of our independent variables or experimental treatments? These are the conditions, indeed the exigencies, of hypothesis testing. This is only the single-sample case, however. We turn next to the situation involving two samples.

Exercises

Where appropriate for each of the following exercises, please include:

- Your null and alternative hypotheses
- Shaded and labeled models of your H_O (and possibly H_A) curves
- Your formula(s), a clear indication of whether you are using z or t, and the z_C and t_C values
- Your decision regarding your H_O, P(Beta error), or a sample size and a sentence or two explaining your answer in plain language.

1. In one particular Midwestern state, the average homeless person is without a formal residence for an average of 27.46 months, with a standard deviation of 9.35 months. A fairly prosperous county in the state, known for its tourist attractions, has about 3000 homeless children and adults. In a sample of 200

homeless people, the county found an average of 26.07 months without a formal residence. At the .05 and .01 alpha levels, is the county average significantly less than the statewide average?

2. The last census revealed that the percentages of black and white Californians, respectively, were either holding constant or dropping. Nevertheless, blacks and whites combined still accounted for 81% of all California voters. A just-completed survey examined a random sample of 800 drawn from the state's 7800 elected officeholders. Some 664 were either black or white. At the .05 and .01 alpha levels, is the total percentage of black or white officeholders significantly higher than their combined percentage among California voters?

3. National surveys suggest that, among young adults, 64% believe in love at first sight. A sample of 150 males was drawn from a regional fraternity's 1530 members, and 108 professed to believe in love at first sight. At the .05 alpha level, does the fraternity's percentage differ at all from the national figure?

4. Are students' cars older than those of the average adult driver? Insurance companies tell us that the typical adult drives a vehicle 4.72 years old, with a standard deviation of 2.06 years. For a class project, a research team asks 25 fellow high school students, drawn from a population of about 200 with cars, how old *their* vehicles are. The average age is 5.34 years, with a standard deviation of 1.43 years. On the average and at the .05 and .01 alpha levels, are the students' vehicles significantly older than the national norm?

5. Surveys suggest that only about 23% of all employed adults have ever used public transportation to get to work. A particular city has some 1400 public employees. In a sample of 300, 60 say they regularly or at least occasionally use such means to get to their jobs. At the .01 alpha level, does the percentage of public employees using public transportation for work differ from that of adults in general? What about at the .05 alpha level?

6. A college counselor suspects that students' grade point averages (GPAs) tend to fall when they move in mid-semester. She knows the overall GPA for juniors and seniors to be 2.74, with a standard deviation of .42. To test her concerns, she searches her university's records and finds 489 upper-division students who reported changes of address last semester. Randomly selecting 26 cases from this population, she calculates a combined sample GPA of 2.53 with a standard deviation of .51 for that term. At the .05 alpha level, is the counselor correct? Does the evidence suggest the GPAs of moving students, at least for that

one semester, are significantly below the junior-senior average? What about at the .01 alpha level?

7. A local politician running for re-election hires new consultants to run his present campaign. In his last campaign, the average contribution received was $37.20, with a standard deviation of $10.70. He now wonders whether that figure differs at all (allowing for inflation) with new people in charge of his campaign. So far he has received 1228 contributions and plans to examine a sample of 50 such donations using the .05 alpha level. If he assumes an average difference of $6 one way or the other from his previous contributions, what is his probability of committing a beta or Type II error?

8. Is America's attention span getting shorter? The American Council on Journalism reported that five years ago the typical topic (e.g., political event, crime, social or economic issue, disaster, and so on) averaged 5.27 consecutive days in newspaper stories, with a standard deviation of 1.55 days. Recently, however, based upon a sample of 48 of the country's 432 daily newspapers, the general life spans of such topics averaged 4.93 consecutive days, having a standard deviation of 1.54 days. At the .05 alpha level, does the typical story endure as news for a significantly shorter period than was true five years ago? What about at the .01 alpha level?

9. According to recent polls, two-thirds (67%) of all US adults believe clergy found guilty of sexual misconduct should be dismissed from their respective orders. In a sample of 140 respondents selected from a population of about 1500 long-time church members, 82 agreed that guilty clergy should be removed. At the .05 alpha level, are long-time church members less likely to support the dismissal of offending clergy than are US adults in general?

10. About 58% of US adults say they would like to stay in the same occupation or profession all their working lives. A researcher suspects this percentage may be less among today's college seniors. In fact, he believes the figure to be about 50% among seniors. If his belief is correct, and if he intends to use the .05 alpha level and to test his hypothesis on a sample of 300 students drawn from a population of 2570 accessible college seniors, what is the probability of his committing a beta error?

11. A fairly constant situation in Illinois is that 54% of those who move do so within the same county. However, given the uncertainty of the local economy, officials are now wondering whether that figure is significantly less for the Chicago area's Cook County. With the help of area realtors, they sampled 150 of the last

1000 county families who had moved. Seventy-nine families had left the county altogether. At the .05 and .01 alpha levels, is the Cook County percentage of intracounty movers significantly less than for the state as a whole?

12. Global warming in southern Alaska? Over the last 150 years of record keeping, the average high temperature in Juneau in September has been 59°F, with a standard deviation of 2°F. However, sampling just the last 20 years, the average daily high has been 60°F. At the .05 alpha level, does this represent a statistically significant increase in the city's September highs? What about at the .01 alpha level?

13. The US Army recently ran 3500 soldiers through field trials using a new piece of equipment. Following the trials, a random sample of $n = 300$ was quizzed on the operation and maintenance of the piece. Participants took an objective test having four possible answers per question: a, b, c, or d. When the typical score turned out to be 29 correct on the 100-item quiz, training personnel were irate, claiming soldiers might just as well have rolled dice to pick their answers. At the .05 alpha level, were the trainers correct, or was the typical score significantly better than random chance?

14. Assume the average college student visits a campus library (in person or on-line) some 18.96 times per quarter with a standard deviation of 3.21 visits. From a population of approximately 700 graduate students, we randomly select 32 and find they make such visits an average of 20.42 times per quarter with a standard deviation of 5.92 visits. At the .05 alpha level, does the average among the graduate population differ at all from the university-wide average? What about at the .01 alpha level?

15. Seventy percent of US adults believe airport security personnel should have the right to hand-search all luggage and belongings. Among a sample of 120 frequent flyers selected from one airline's list of approximately 1800 such customers, 67 agreed with unlimited rights of search. At the .05 alpha level, does the percentage of frequent flyers supporting unlimited searches differ from that of the general public?

16. With students' uncertainties about the revised Scholastic Aptitude Test (SAT), a tutoring company hopes to raise clients' average scores by 80 points (on the new 2400-point scale). Early test takers averaged 1520 points on the exam, with a standard deviation of 140 points. From 415 students now eligible for the test, the company randomly selects 35, offers them its tutoring free, and plans

to record their average SAT scores. If the company assumes an 80-point average increase for these students and uses the .05 alpha level in testing for the success of its program, what risk does it run of committing a Type II error?

17. Teenagers under age 18 are sometimes tried as adults when they commit serious crimes. Nationally, 10 years ago, the average age of teens tried in adult courts was 16.23 years with a standard deviation of .60 years. Last year, 447 teens were tried as adults for serious crimes. In a sample of 28 cases, the average age of the defendants was 15.94 years, with a standard deviation of .68 years. At the .01 alpha level, does this evidence suggest that the average age of youths tried in adult courts has significantly declined in the last decade?

18. In the last census, about 33% of western states' populations were Latino. One western county suspects its current population differs significantly from this figure. In a random sample of 1100 residents drawn from its 1.1 million residents, 407 indicated Latino or Hispanic roots. At the .01 alpha level, does the percentage of the county's population identifying as Latino differ from the last census figure?

19. Are televised football games and their commercial breaks getting longer, or does it just seem that way? During National Football League games televised five years ago, the average commercial break lasted 3.12 minutes with a standard deviation of .78 minutes. Last year, in a sample of 36 games selected from the 248 broadcast, the average commercial length was 3.31 minutes with a standard deviation of .42 minutes. At the .05 alpha level, has the average length of games' commercial breaks significantly increased over the last few years?

20. To bolster its application for a lucrative contract with the state, a particular company claims to have a commendable record of gender equity. Asked to document its claim, the company randomly selects 225 of its 820 female employees and finds that 81 report instances of discrimination or harassment. Nationally, surveys reveal that about 42% of similar companies' female employees report instances of gender discrimination or outright sexual harassment. At the .05 alpha level, do the data support the company's claim of having a significantly lower rate of gender-based complaints? What about at the .01 alpha level?

21. Research has consistently shown that 41% of US adults regularly attend religious services. A rural sociologist suspects this percentage is higher in smaller cities and towns. In fact, her preliminary data suggest it may be as high as 48%. To adequately test her hypothesis, she plans to sample around 600 adults in a community of 10,000 and to use the .01 alpha level. Assuming her suspicions are correct, what risk does she run of possibly committing a beta error?

22. An anthropologist recently unearthed the fossil remains of 8 adult individuals in the tidelands near Marseille, France, and carbon dated their ages to about 60,000 years. He estimates their average height to have been 68.69 inches. He knows from hundreds of skeletal remains, moreover, that the early European *Homo sapiens sapiens* population had a mean height of 70.30 inches with a standard deviation of 2.19 inches. At the .05 alpha level, could his sample be part of the known *Homo sapiens* population in Europe? Does it deviate significantly from the normal range for that population's adult heights?

23. A research corporation's recent poll revealed that 43% of US adults believe the government should limit the profits companies are allowed to make. Among a random sample of 300 college students drawn from a population of 22,000, 147 agreed that companies' profits should be limited.

    a. At the .05 alpha level, is the proportion of students supporting limited profits significantly greater than among the general public?

    b. If we assume the percentage of college students favoring limited profits is indeed 49% or .49, what risk did we run of committing a Type II error with our test in (a)?

24. Among divorced Atlanta residents, the average first marriage lasted 7.36 years with a standard deviation of 2.10 years. For a random sample of 32 divorced marriage counselors selected from the 890 certified therapists practicing in the region, a sociologist found their first marriages to have lasted an average of 6.63 years with a standard deviation of 1.87 years.

    a. At the .05 alpha level, had the counselors' first marriages lasted for a significantly shorter time than those of Atlantans as a whole?

    b. Assuming the actual average for counselors' first marriages is 6.63 years, and given the way the sociologist conducted her H_0 test in (a), what risk did she run of possibly committing a beta error?

25. A bond issue to build a new middle-school gym is on the ballot in a certain school district. About 58% of the voters in the district are known to favor the bond issue. In the Palm Grove precinct, which has 2650 voters, however, school officials suspect the percentage favoring the bonds is significantly higher. In a random telephone survey of 400 Palm Grove voters, 260 say they intend to vote for the bonds. At the .05 alpha level, are school officials correct in stating that the level of support is significantly higher in the Palm Grove area?

26. A campus bookstore sells various clothing items bearing the university logo, available in one of the three school colors: red, yellow, or white. Every

quarter, the bookstore sells about 2750 university sweatshirts. In a sample of 250 sales, 95 were red sweatshirts. At the .05 alpha level, does the rate at which red sweatshirts are purchased differ significantly from that expected by random chance?

27. In the last census, the average age of California's Asian residents was 33.37 years with a standard deviation of 12.60 years. With recent immigration, a demographer suspects that the average age of Asian Californians is declining. His survey randomly samples 900 of the state's 4.3 million Asian residents and finds an average age of 32.21 years. At the .01 alpha level, has the average age of Asian Californians declined since the last census?

28. On a standardized math test given at the end of the sixth grade, the average score is 50 points and the standard deviation is 15 points. From a population of 430 children, a sample of 35 receives its sixth-grade math instruction almost entirely on computers rather than through traditional methods. The sample averages 54.5 points at the end of the school year. At the .05 alpha level, does the sample's average score differ at all from the average based on traditional instruction?

29. It has been claimed that about 44% of US adults believe they have had out-of-body "astral travel" experiences. At Miami's annual Psychic Jamboree, 450 people showed up for a panel discussion entitled "Gravity: Fact or Fiction?" We randomly sampled 60 people in the audience, and 33 claimed to have had astral travel experiences. At the .05 alpha level, do our data indicate a significantly higher percentage of astral travelers at the panel session than is claimed among the general public?

30. How are probability and random chance involved in hypothesis testing?

31. What is the general theory or logic of testing hypotheses with single-sample data?

32. What does it mean to say, "Hypothesis testing is a search for the rare event versus the normal occurrence"?

33. Why do we always test the null hypothesis (H_O) rather than the alternative hypothesis (H_A)?

34. Why may we never be certain or 100% sure of any H_O decision? Why may we never eliminate the possibility of error completely?

35. What are alpha (Type I) and beta (Type II) errors? Why are they important in hypothesis testing?

36. Answer the following two questions with explanations rather than simply yes/no answers:

    a. In making a decision regarding H_O, may one make both an alpha and a beta error at the same time?

    b. May one make *neither* an alpha nor a beta error?

37. What are the differences between one- and two-tailed hypothesis tests? How do they differ in terms of general theory, logic, and procedures?

38. Other things being equal, why is it harder to reject the H_O with a two-tailed test than with a one-tailed test?

39. Why do we say, "The alternative hypothesis determines the nature of the test to which we subject the null hypothesis"?

40. What are the operating characteristics and power of a hypothesis test?

41. Other things being equal, why does the power of an H_O test vary directly with sample size?

42. How are z and t tests similar? In what ways do they differ?

43. Answer the following four questions with explanations rather than single-sentence answers:

    a. If we fail to reject our H_O at the .05 alpha level, what decision would we make at the .01 alpha level?

    b. If we reject our H_O at the .05 alpha level, what decision would we make at the .01 alpha level?

    c. If we reject our H_O at the .01 alpha level, what decision would we make at the .05 alpha level?

    d. If we fail to reject our H_O at the .01 alpha level, what decision would we make at the .05 alpha level?

CHAPTER 6

Estimates and Tests with Two Samples: Identifying Differences

In this chapter, you will learn how to:

- Construct confidence interval estimates for the difference between both independent and related population averages
- Construct confidence interval estimates for the difference between both independent and related population proportions
- Test hypotheses for the difference between two sample averages
- Test hypotheses for the difference between two sample proportions

THE KEY WORD FOR THIS CHAPTER is again *difference*. So far, we have considered how one sample and population might differ. Now, we look at two samples (representing two different populations) and the sampling distribution of the difference between them. Suppose we take many similar pairs of samples and calculate the difference between each pair. Graphing the differences produces the sampling distribution of the difference. In principle, it is the same as our earlier sampling distributions. Now, however, instead of a single statistic, we have a distribution of the difference between two statistics.

As an example, assume we have samples from two populations. In Chapter 5, we had one known population and one sample, the sample possibly differing from that population. Instead, we now have two samples and usually two quite distinct populations from which they were selected. Based on the samples, we may establish a confidence interval estimate for the difference between the two populations. Or, we may test whether the two samples differ significantly and compare our particular two-sample difference to the distribution of all such differences that might occur by random chance. In either case, *whenever we have two samples we must consider whether they are related or independent.* That affects our formulas and procedures.

What is meant by matched or related samples? When we have **matched or related samples,** *we make a conscious effort to see that every person in Sample A has a counterpart in Sample B.* We might attempt to match people on several variables considered important to a study—at least things we wish to control—and then randomly assign one person of each pair to the experimental group and one to the control group. Or we may have a before-and-after design. We test people before and again after a certain treatment or experience. Every person is matched with himself or herself, so to speak. In other words, we have paired measurements on the same individual, and each person acts as his or her own control. In contrast, data may come from **independent samples,** *in which we have two quite separate and distinct random samples, drawn separately from different populations, and quite possibly of different sizes.*

This is really all that is new here: two samples and the question of whether they are related or independent. The rest is quite familiar: means and proportions, hypothetical sampling distribution curves centered around either sample statistics or known population figures, the standard error, confidence levels and alpha regions, z versus t for problems involving averages, and so on. We look first at cases of two independent samples (the more common situation) and later at procedures for related samples.

Independent Samples

Many Samples and Usual Differences

The basic concepts of estimation and hypothesis testing are already familiar, so we will simply introduce this new model of the sampling distribution of the difference and the standard error of the difference here. Thereafter, we will use these concepts in confidence interval estimation and in hypothesis testing. We imagine a sampling distribution curve of the differences between many pairs of samples. The differences might be between two sample means $(\overline{X}_1 - \overline{X}_2)$ or two sample proportions $(p'_1 - p'_2)$. The curve should be centered around the true difference between the two population means $(\mu_1 - \mu_2)$ or the two population proportions $(p_1 - p_2)$. As a measure of the spread in such a curve, we have the **standard error of the difference between two population means** $(\sigma_{\overline{X}_1 - \overline{X}_2})$ or the **standard error of the difference between two population proportions** $(\sigma_{p_1 - p_2})$. Figures 6.1 and 6.2, respectively, illustrate these two curves.

Figure 6.1 The Sampling Distribution of the Difference between Two Population

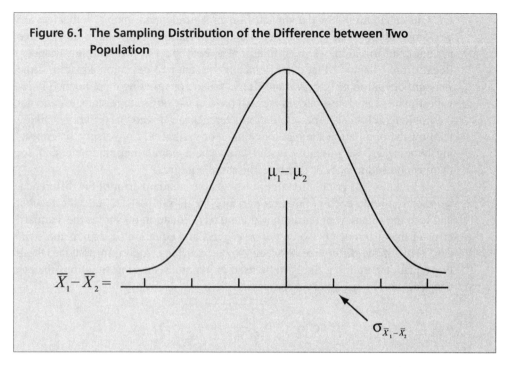

Figure 6.2 The Sampling Distribution of the Difference between Two Population Proportions

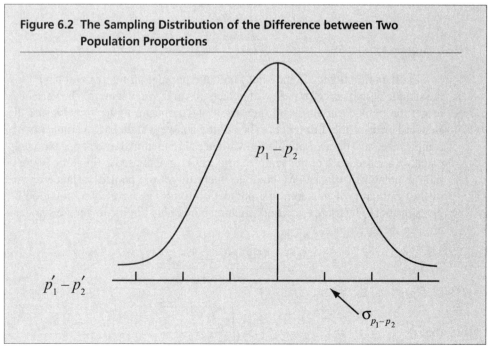

Our calculations for the standard errors here depend upon whether we are working with population or sample standard deviations. Note that if all we have is sample information, we must first pool or combine the two sample variances. Recall from Chapter 2 that the variance is the standard deviation squared; if the standard deviation is 3, the variance is 9. When constructing a confidence interval estimate, it is almost certain we will have to use our sample statistics and get the **pooled variance.** If the population averages are unknown, we are not likely to know the population standard deviations or variances. Conversely, when testing hypotheses, we probably would know these population parameters. If so, there is no need to pool or combine the sample figures.

If we know the population figures, to get the standard error of the difference, we merely divide each population variance by its sample size, add the results, and take the square root (equations 60 and 61). Equation 60 shows the standard error of the difference between two *averages*, and equation 61 depicts the standard error of the difference between two *proportions*. Again, in practice, these two equations are more likely to be used in hypothesis testing situations than to construct confidence interval estimates.

$$(60) \qquad \sigma_{\bar{X}_1 - \bar{X}_2} = \sqrt{\frac{\sigma_1^{\,2}}{n_1} + \frac{\sigma_2^{\,2}}{n_2}}$$

$$(61) \qquad \sigma_{p_1 - p_2} = \sqrt{\frac{p_1 q_1}{n_1} + \frac{p_2 q_2}{n_2}}$$

In other situations, we may not know the population figures and have only the sample variances (s^2 or $p'q'$) with which to work. We first merge the variances to get the pooled variance and then use this composite figure to solve for the standard error of the difference. When doing so, we weight each variance by its sample size, and the two population variances are assumed to be approximately equal. In a sense, each of our sample variances is an unbiased estimator of a common population variance. We combine them to get the pooled variance or one overall estimator of this common population figure. For *averages,* equation 62 gives the pooled variance, $s_p^{\,2}$, and equation 63 uses that figure to get the standard error of the difference, $s_{\bar{X}_1 - \bar{X}_2}$.

$$(62) \qquad s_p^{\,2} = \frac{s_1^{\,2}(n_1 - 1) + s_2^{\,2}(n_2 - 1)}{n_1 + n_2 - 2}$$

$$(63) \qquad s_{\bar{X}_1 - \bar{X}_2} = \sqrt{\frac{s_p^{\,2}}{n_1} + \frac{s_p^{\,2}}{n_2}}$$

In the case of *proportions,* we use equation 64 to first combine the p' values to get $\overline{P}$, read as "*P*-bar," part of the pooled variance. Again, each p' is weighted by its sample size, n. (This is not necessary when we have related or paired samples.) Next, the product of $\overline{P}$ and $\overline{Q}$ $(\overline{P} \cdot \overline{Q})$ is divided by n for the pooled variance for proportions, where $\overline{Q}$ is the complement of $\overline{P}$, or $\overline{Q} = 1 - \overline{P}$. We then use this figure to solve for the standard error of the difference in equation 65.

(64)
$$\overline{P} = \frac{p_1'(n_1) + p_2'(n_2)}{n_1 + n_2}$$

For the standard error of the difference:

(65)
$$s_{p_1 - p_2} = \sqrt{\frac{\overline{P} \cdot \overline{Q}}{n_1} + \frac{\overline{P} \cdot \overline{Q}}{n_2}}$$

Estimating Differences Between Two Populations

Population averages. Estimates of the confidence interval for the difference between two means follow the same general format of all confidence interval formulas:

(66) Confidence Interval = Point estimate $\pm$ (z or t) $\times$ (Standard error of the difference)

To estimate the difference between the means of two populations, we seek a confidence interval estimate for the difference between μ_1 and μ_2. The generic equation 66 has several parts. First, the point estimate becomes $(\overline{X}_1 - \overline{X}_2)$, the difference between our two sample means. With confidence intervals, it is somewhat arbitrary which sample and mean is designated as 1 and which as 2. While we might have some logical method for deciding this (e.g., this year's mean minus last year's mean, or simply a wish to get a difference that is positive rather than negative), in the end, our conclusions will be the same. Exactly how we get the $(\overline{X}_1 - \overline{X}_2)$ difference is largely a matter of choice. We must make sure our interpretation of the results takes into account how the difference was determined, but that is all.

Second, we apply the same criterion as before to determine whether z or t is appropriate. With averages, our sample size must be at least 30 in order to use z, and we now apply this guideline to each sample separately. If either or both samples should fall below $n = 30$, we must use one of the t curves. In this case, our t coefficient has $n - 1$ degrees of freedom associated with each sample. Our total degrees of freedom become $(n_1 - 1) + (n_2 - 1)$, but we usually write this as $n_1 + n_2 - 2$. Therefore, for t, equation 67 holds.

(67)
$$\text{d.f.} = n_1 + n_2 - 2$$

So there is no confusion here, we have only one t score, not two. We combine the degrees of freedom from both samples to look up our one t-critical value. This may lead to another small problem, however. Our total degrees of freedom might not be listed in the t table (Table C). In the usual table, we find d.f. entries from 1 to 30 and then only periodic listings. Table C is no exception. It contains t values for 1 through 30 d.f., for 40, 60, 120, and even infinity (∞, the entries for which are identical to z scores). But what do we do if $n_1 + n_2 - 2$ gives us, say, 45 d.f.? We have two options: (1) We may interpolate, which would give us at least an approximate figure, or (2) we may go to the next lowest number of d.f. on the table. The rationale here is that we have at least that many degrees of freedom; we could not use entries for the next highest number because we do not have that many degrees of freedom.

Finally, our generic formula (equation 66) includes the standard error of the difference for the two population means. The only question is (1) whether we know both population standard deviations (σ) and variances (σ^2), or (2) whether we must first combine our two sample variances (s^2) to get the pooled variance (s_p^2). In all probability, especially for confidence interval estimates, we must do the latter. As noted previously, if we are using the sample means to estimate the difference between two unknown population means, the chances of knowing the actual population variances are remote. In most estimation problems, we use our two sample variances to get the pooled variance before finding our standard error.

Putting these elements together, we get several possible formulas for estimating the difference between two independent population averages. Remember, we must choose whether we can use z or t, and then we have to see whether we know the σ^2 figures or have to use the pooled variance estimator, s_p^2. Equations 68 and 69 show the use of z for the cases where σ^2 is known and not known, respectively:

(68)   C.I. Estimate for $\mu_1 - \mu_2 = (\overline{X}_1 - \overline{X}_2) \pm z\sqrt{\dfrac{\sigma_1^2}{n_1} + \dfrac{\sigma_2^2}{n_2}}$

(69)   C.I. Estimate for $\mu_1 - \mu_2 = (\overline{X}_1 - \overline{X}_2) \pm z\sqrt{\dfrac{s_p^2}{n_1} + \dfrac{s_p^2}{n_2}}$

Equations 70 and 71 show the use of t, where t has $n_1 + n_2 - 2$ degrees of freedom, for the cases where σ^2 is known and not known, respectively. For the latter case, s_p^2 is found from equation 62.

(70)   C.I. Estimate for $\mu_1 - \mu_2 = (\overline{X}_1 - \overline{X}_2) \pm t\sqrt{\dfrac{\sigma_1^2}{n_1} + \dfrac{\sigma_2^2}{n_2}}$

(71)   C.I. Estimate for $\mu_1 - \mu_2 = (\overline{X}_1 - \overline{X}_2) \pm t\sqrt{\dfrac{s_p^{\,2}}{n_1} + \dfrac{s_p^{\,2}}{n_2}}$

1. *There has been some debate about rising college costs in different parts of the country. We wish to estimate the difference between the annual tuition costs at public colleges and universities in the Midwest versus those in the Southeast. From a random sample of 15 midwestern institutions, we find an average tuition of $6,933 per year, with a standard deviation of $210. In a similar sample of 12 southeastern colleges and universities, we find a mean of $6,416 and a standard deviation of $193. At the .95 confidence level, what is the estimate for the true difference between the midwestern and southeastern average tuitions?*

Designating the midwestern universities as sample #1 and the southeastern universities as sample #2 gives us a positive difference. Our point estimate is $\overline{X}_1 - \overline{X}_2 = \$6933 - \$6416 = \$517$. (We could have written $\overline{X}_M - \overline{X}_S$, for the midwestern minus the southeastern average; it would make identification easier.) Both sample sizes are less than 30 (15 and 12, respectively), so we will use t rather than z. If $n_1 + n_2 - 2 = 25$ d.f., $t = \pm 2.060$ at the .95 confidence level.

Note that we have the sample standard deviations, not the population values. We first calculate the pooled variance by using equation 62 and then use this figure to get the standard error of the difference:

$$s_p^{\,2} = \frac{s_1^{\,2}(n_1 - 1) + s_2^{\,2}(n_2 - 1)}{n_1 + n_2 - 2} = \frac{(210)^2(14) + (193)^2(11)}{15 + 12 - 2} = 41{,}085.56$$

We then use the standard error of the difference (equation 62) to get:

$$s_{\overline{X}_1 - \overline{X}_2} = \sqrt{\frac{s_p^{\,2}}{n_1} + \frac{s_p^{\,2}}{n_2}} = \sqrt{\frac{41{,}085.56}{15} + \frac{41{,}085.56}{12}} = \sqrt{6162.84} = 78.50$$

Then, for the actual confidence interval estimate, from equation 70, we get:

$$\text{.95 C.I. Est. for } \mu_1 - \mu_2 = (\overline{X}_1 - \overline{X}_2) \pm t\,(s_{\overline{X}_1 - \overline{X}_2})$$
$$= (6933 - 6416) \pm 2.060\,(78.50) = \$517 \pm \$161.71$$

For the C.I.'s lower limit: $517 - $161.71 = $355.29.
For the C.I.'s upper limit: $517 + $161.71 = $678.71.

Figure 6.3 shows the distribution curve for problem 1. We may be 95% confident that the true average difference is between $355.29 and $678.71 per year

Figure 6.3

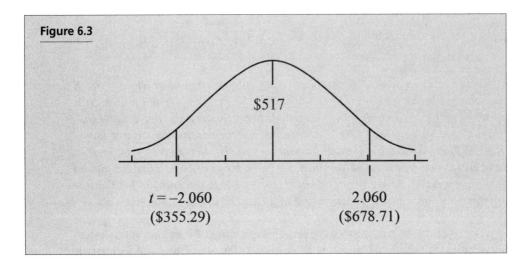

$517

$t = -2.060$ 2.060
($355.29)   ($678.71)

"favoring" the midwestern universities. That is, we are 95% certain that the average tuition at public universities in the Midwest exceeds that of similar institutions in the Southeast by anywhere from $355.29 to $678.71 per year.

2. On the average and among families with children, do urban recipients stay on public assistance longer than rural recipients? If so, by how much? In a large state whose records are available, we find an average stay on assistance of 21.32 months among a random sample of 50 former urban clients. Among a sample of 45 rural recipients we get an average stay of 17.74 months. From previous studies in the state, we know the urban standard deviation to be 4.56 months on assistance and the rural standard deviation to be about 4.11 months. What is the .90 confidence interval estimate for the average difference in urban versus rural periods on public assistance?

Since both our sample sizes are at least 30, we may use z, which is ± 1.645 for the .90 confidence level. Assuming we wish a positive point estimate, we subtract the rural average from the urban average: $(\overline{X}_U - \overline{X}_R) = 21.32 - 17.74 = 3.58$ months.

In an unusual circumstance, we know the populations' standard deviations. We do not have to pool any sample variances. The population standard deviations are 4.56 for the urban and 4.11 for the rural. We calculate the standard error of the difference by using equation 60.

$$\sigma_{\overline{X}_U - \overline{X}_R} = \sqrt{\frac{\sigma_U{}^2}{n_U} + \frac{\sigma_R{}^2}{n_R}} = \sqrt{\frac{(4.56)^2}{50} + \frac{(4.11)^2}{45}} = \sqrt{.791} = .890$$

For our final confidence interval estimate, we use equation 68:

$$.90 \text{ C.I. Est. for } \mu_U - \mu_R = (\overline{X}_U - \overline{X}_R) \pm z \, (\sigma_{\bar{x}_U - \bar{x}_R})$$
$$= 3.58 \pm 1.645 \, (.890) = 3.58 \pm 1.46$$

For the C.I.'s lower limit: 3.58 − 1.46 = 2.12 months on public assistance.
For the C.I.'s upper limit: 3.58 + 1.46 = 5.04 months on public assistance.

Figure 6.4 shows the distribution curve for problem 2. We are 90% confident that the average difference in time on public assistance is between 2.12 and 5.04 months longer for urban recipients.

Population proportions. When we estimate the difference between two population proportions, the procedures are basically the same as that for two population averages. Our confidence interval contains a point estimate plus or minus an error term. The point estimate becomes $p'_1 - p'_2$, and, as with averages, which is p'_1 and which is p'_2 is not especially important. We merely take our way of deriving $p'_1 - p'_2$ into account when we interpret our final confidence interval estimate.

The first error term component is the z coefficient. We consult Table 4.1 to see whether our two sample sizes are sufficient for using the z curve. Remember that both sample sizes must qualify. We check the table with the minimum p' or q' value for each sample separately and make sure each n meets the minimum criterion listed (or interpolated).

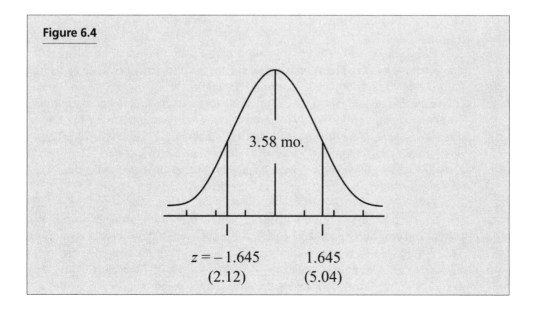

Figure 6.4

3.58 mo.

$z = -1.645$ 1.645
(2.12)   (5.04)

Assuming our sampling distribution of the difference conforms to the z curve, the other part of the error term is the standard error of the difference for proportions. Deriving this standard error is a little different than deriving the standard error of the difference for the means. In that situation, we *may* have known the population variances (although it was unlikely) and not have had to pool the two sample variances. However, it is difficult to imagine a similar situation for proportions. Logically, we must *always* pool the two sample variances when estimating the difference between two populations. We would not know the population figures (p_1 and p_2). With proportions, the variances are derived from the product of the p and q values themselves. It follows that if we know the actual population proportions, there is no need for a confidence interval estimate at all. We may easily get the true difference.

Hence, we have a situation somewhat different from that involving means, and this makes equation 61 unlikely for confidence intervals. Nevertheless, while we do not know the true p and q values for both populations and would not use this formula for estimation, equation 61 is critical for two-sample hypothesis tests in which the population values are known. In practical situations involving confidence intervals for two proportions, however, expect to first pool the two p' figures to get $\overline{P}$ (see equation 64) and then substitute this $\overline{P}$ value and its complement, $\overline{Q}$, into equation 65. This gives us at least an estimated standard error of the difference, $s_{p_1 - p_2}$.

Finally, for confidence interval estimates for the difference between two population proportions, use equation 72:

(72)   C.I. Estimate for $p_1 - p_2 = (p'_1 - p'_2) \pm z \, (s_{p_1 - p_2})$

3. A large school district is considering a switch from its traditional calendar to year-round schooling (YRS), making summer a regular school term. The district has also experienced growing numbers of immigrant students enrolling in the last few years. The school board, before studying any plan, wishes to know what percentages of immigrant versus native-born or US- citizen parents favor YRS. Among a random sample of 300 immigrant parents in the district, 102 say they favor developing such a plan. From 300 randomly selected citizen parents, the school board finds 87 in favor. What is the .99 confidence interval estimate for the difference in YRS support rates for recent immigrant parents versus citizen parents?

To set up the problem, p'_1 equals our sample proportion of recent immigrant parents in favor of YRS, and p'_C equals the proportion of citizen parents in favor. The p'_1 value for immigrant parents is 102/300, or .34. For citizen parents, p'_C is 87/300, or .29. Our point estimate, $p'_1 - p'_C$, is .34 − .29, or .05. Referring to Table 4.1, each sample size (300) meets the criterion for using the z curve, and at the .99 confidence level, $z = \pm 2.575$.

No population values are given, so we rely upon our sample numbers. Solving for the pooled variance and $\overline{P}$ (see equation 63), we have:

$$\overline{P} = \frac{p'_I(n_I) + p'_C(n_C)}{n_I + n_C} = \frac{.34(300) + .29(300)}{300 + 300} = \frac{189}{600} = .315$$

Thus, $\overline{Q} = 1 - \overline{P} = 1 - .315 = .685$.

For the standard error of the difference using equation 65, we get:

$$s_{p_I - p_C} = \sqrt{\frac{\overline{P} \cdot \overline{Q}}{n_I} + \frac{\overline{P} \cdot \overline{Q}}{n_C}} = \sqrt{\frac{.315(.685)}{300} + \frac{.315(.685)}{300}} = \sqrt{.0014} = .038$$

We solve for the confidence interval estimate itself using equation 72:

$$.99 \text{ C.I. Est. for } p_I - p_C = (p'_I - p'_C) \pm z(s_{p_I - p_C})$$
$$= (.34 - .29) \pm 2.575(.038) = .05 \pm .10$$

For the C.I.'s lower limit: $.05 - .10 = -.05$, or -5%.
For the C.I.'s upper limit: $.05 + .10 = .15$ or 15%.

Figure 6.5 shows the distribution curve for problem 3. This is an interesting result. The true difference could favor either immigrant or citizen parents. It was mentioned earlier that we must take into account how our point estimate was derived when interpreting a confidence interval. This is a good example of that rule. It is especially important when a confidence interval includes both positive and negative numbers. We determined our point estimate by using immigrant minus native citizen percentages. Therefore, any positive difference means immigrant parents are more in favor of YRS. A negative difference means the citizen parents' percentage is higher. We interpret our interval here to mean that, with .99 confidence, immigrant parents could be more in favor of YRS by as much as 15 percentage points or citizen parents might be more in favor by up to 5 percentage points. The true difference could go either way. At the 99% confidence level, therefore, we cannot be sure whether more immigrant or citizen parents would be more supportive of year-round schooling.

4. *The Insurance Institute of California recently completed a survey of retired physicians in the state. Of 380 physicians who had spent all or almost all their careers working for health maintenance organizations (HMOs), 87 had been involved in malpractice cases. Among a sample of 415 physicians who had been*

Figure 6.5

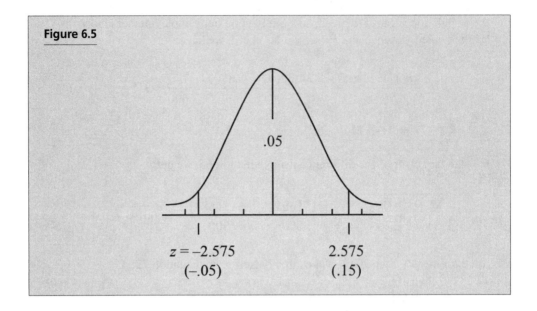

$$z = -2.575 \qquad 2.575$$
$$(-.05) \qquad (.15)$$

entirely in private practice, 79 had experienced malpractice incidents. What is the Institute's .95 confidence interval estimate for the difference in the percentages of HMO versus private practice physicians involved in malpractice cases?

To start with our p' values, we get p' = 87/380 = .23 for HMO physicians (Sample 1). Among those in private practice (Sample 2), $p'_2 = 79/415 = .19$. These two values and the sample sizes meet the requirements for using the z distribution (Table 4.1), and $z = \pm1.96$ at the .95 confidence level. Our point estimate is $p'_1 - p'_2 = .23 - .19 = .04$.

We must pool the two sample variances since we are not given population p and q values, and then get our standard error of the difference (see equation 65):

$$\overline{P} = \frac{p'_1(n_1) + p'_2(n_2)}{n_1 + n_2} = \frac{.23(380) + .19(415)}{380 + 415} = \frac{166.250}{795} = .209$$

Thus, $\overline{Q} = 1 - .209 = .791$, and

$$s_{p_1 - p_2} = \sqrt{\frac{\overline{P} \cdot \overline{Q}}{n_1} + \frac{\overline{P} \cdot \overline{Q}}{n_2}} = \sqrt{\frac{.209(.791)}{380} + \frac{.209(.791)}{415}}$$

$$= \sqrt{.0004 + .0004} = \sqrt{.0008} = .028$$

The confidence interval becomes:

$$.95 \text{ C.I. Est. for } p_1 - p_2 = (p'_1 - p'_2) \pm z \ (s_{p'_1 - p'_2}) = .04 \pm 1.96 \ (.028) = .04 \pm .05$$

For the interval's lower limit: $.04 - .05 = -.01$, or -1%.
For the interval's upper limit: $.04 + .05 = .09$, or 9%
Figure 6.6 presents the distribution curve for problem 4. Again, it appears the true difference might go either way. Given the way the point estimate was calculated, with 95% confidence, the Institute might conclude physicians in private practice are 1% more likely to experience malpractice suits. However, those in HMOs might be up to 9% more likely to do so. The true difference between the two populations is somewhat in doubt here, but it does appear that HMO physicians *may* be slightly more likely to face malpractice incidents during their careers.

Summary. Based upon two independent sample statistics, we establish a confidence interval estimate for the difference between the two unknown population parameters, either averages or proportions. Our point estimate is the difference between the two sample statistics themselves, we use z or t as the circumstances dictate, and we typically have to pool or combine variances before determining the standard error of the difference.

These elements are also part of hypothesis testing with two independent samples. They are simply used in slightly different but familiar ways.

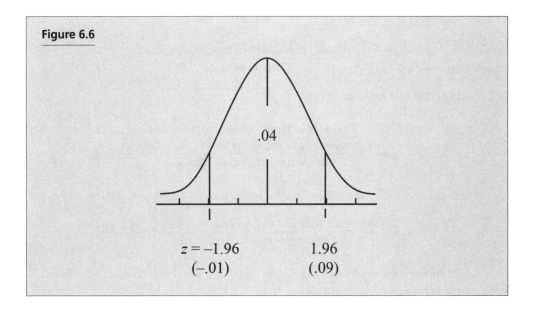

Figure 6.6

.04

$z = -1.96$
$(-.01)$

1.96
$(.09)$

Testing Differences Between Two Samples

Procedures for testing hypotheses with two independent samples are generally the same as with one sample. We have an H_O curve, conduct a z test or a t test, have one- and two-tailed tests, and must contend with the possibility of errors. All the familiar elements are here, except we now think in terms of testing for the difference between two sample statistics, using the standard error of the difference, and so on.

In terms of the actual methods, as before, we calculate the z- or t-observed and compare this figure to a critical value taken from the z or t tables. To understand how this is done, we consider a theoretical point first and then practical formulas.

The theoretical principle is that the population parameters, the μ_O or p_O terms used in the equations in Chapter 5, are now assumed to equal zero (0.00). In the single sample case, we had a known value for the H_O curve's mean or proportion. This was the population figure. Taking our alpha level into account, we then tested to see if our sample statistic differed significantly from this parameter. The H_O said there was no difference between that sample figure and the known population parameter. In the case of two samples, our null hypothesis still maintains "There is no difference," but it refers to the difference between two population averages or proportions, the populations being those from which our independent samples come. Our H_O now suggests that $\mu_1 - \mu_2$ equals zero, no difference. The same is true for proportions: $p_1 - p_2 = .000$. Since the H_O's population parameter is assumed to be zero, the population value in the numerators of our z and t formulas now equals zero. To illustrate this, first consider the single-sample generic formula in Chapter 5 for determining z- or t-observed. Using z_O as an example, we had (see equation 53):

$$z_O = \frac{\text{Sample statistic} - \text{Population parameter}}{\text{Standard error}}$$

Now, for two samples, it is:

$$z_O = \frac{\begin{array}{c}\text{Difference between two sample statistics} - \\ \textit{Difference between two population parameters}\end{array}}{\text{Standard error of the difference}}$$

According to H_O, the italicized term in the numerator equals zero, so we have:

$$(73) \quad z_O = \frac{\text{Difference between two sample statistics } (\overline{X} \text{ or } p') - 0.00}{\text{Standard error of the difference}}$$

Dropping the extraneous "0.00" term gives us revised generic formulas. For sample averages, we have equation 74:

(74)
$$z_0 \text{ or } t_0 = \frac{\text{Difference between two sample averages}}{\text{Standard error of the difference}}$$

For sample proportions, we have equation 75:

(75)
$$z_0 = \frac{\text{Difference between two sample proportions}}{\text{Standard error of the difference}}$$

Under the null hypothesis, we still assume the difference between the two populations is zero. We simply do not show that zero in our working formulas.

Sample averages. When testing for the difference between two averages from independent populations, equation 74 has both z and t versions. No matter which we use, our calculations are the same. It is a matter of whether our sampling distribution of the difference conforms to the z curve or to a t curve:

(76)
$$z_0 = \frac{\overline{X}_1 - \overline{X}_2}{\sigma_{\overline{X}_1 - \overline{X}_2}}$$

(77)
$$t_0 = \frac{\overline{X}_1 - \overline{X}_2}{\sigma_{\overline{X}_1 - \overline{X}_2}}$$

Equations 76 and 77 show the standard error terms as $\sigma_{\overline{X}_1 - \overline{X}_2}$. This presumes we know the actual population standard deviations (σ's) and have used them in equation 60 to get the standard errors. However, in cases where we do not have the population information, we may write the standard error as $s_{\overline{X}_1 - \overline{X}_2}$ and derive it and the pooled variance by using equations 62 and 63, respectively. Finally, in conjunction with equation 77, our t-critical value (t_c) has ($n_1 + n_2 - 2$) degrees of freedom.

5. *An industrial psychologist working for a Silicon Valley high-tech firm wishes to study the possibility of using visually impaired employees in a new task. The job will require unusually fine hearing and touch discrimination but no critical use of sight at all. She wishes to know whether, at the .05 alpha level, the production rates of legally blind and normally sighted employees might differ. Among a reasonably random sample of 12 potential employees having certified visual impairments, she finds a mean of 62 units inspected per day with a standard deviation of 5 units. With a sample of 16 normally sighted workers, she gets a mean of 59 units inspected per day and a standard deviation of 4 units. Based on her evidence, do the two samples differ significantly?*

First, the two samples are of different sizes and independent of each other. The psychologist also wishes to do a two-tailed test. She wishes to determine whether the sample averages "differ significantly," not whether one is greater than the other. Because each sample consists of less than 30 people, we use a *t* test rather than *z*. In setting up the problem, we let the visually impaired subjects be Sample 1 and their normally sighted counterparts be Sample 2.

$$n_1 = 12, \overline{X}_1 = 62 \text{ units}, s_1 = 5 \text{ units}$$

$$n_2 = 16, \overline{X}_2 = 59 \text{ units}, s_2 = 4 \text{ units}$$

Our hypotheses are:

$$\text{H}_\text{A}: \overline{X}_1 \neq \overline{X}_2 \,; \text{H}_\text{O}: \overline{X}_1 = \overline{X}_2 \quad \text{or} \quad \text{H}_\text{A}: \overline{X}_1 - \overline{X}_2 \neq 0.00 \,; \text{H}_\text{O}: \overline{X}_1 - \overline{X}_2 = 0.00$$

Either way of writing H_A and H_O is acceptable. Another way of expressing these hypotheses is often used as well. Since our sample statistics, $\overline{X}_1$ and $\overline{X}_2$, are unbiased estimators of their population parameters, μ_1 and μ_2, the hypotheses are often expressed in terms of these parameters. In fact, some people argue it is more proper to do it this way. Our hypotheses might therefore have been written:

$$\text{H}_\text{A}: \mu_1 \neq \mu_2; \text{H}_\text{O}: \mu_1 = \mu_2 \quad \text{or} \quad \text{H}_\text{A}: \mu_1 - \mu_2 \neq 0.00; \text{H}_\text{O}: \mu_1 - \mu_2 = 0.00$$

t-critical has $(n_1 + n_2 - 2)$ degrees of freedom = $12 + 16 - 2 = 26$ d.f. With 26 d.f., the .05 alpha level, and a two-tailed test, according to Table C, $t_c = \pm 2.056$. Figure 6.7 shows the distribution curve for problem 5.

Before getting the standard error of the difference, we must pool the two sample variances (see equation 62):

$$s_p^{\,2} = \frac{s_1^{\,2}(n_1 - 1) + s_2^{\,2}(n_2 - 1)}{n_1 + n_2 - 2} = \frac{5^2(11) + 4^2(15)}{12 + 16 - 2} = \frac{515}{26} = 19.808$$

For the standard error itself (see equation 63):

$$s_{\overline{X}_1 - \overline{X}_2} = \sqrt{\frac{s_p^{\,2}}{n_1} + \frac{s_p^{\,2}}{n_2}} = \sqrt{\frac{19.808}{12} + \frac{19.808}{16}}$$

$$= \sqrt{1.651 + 1.238} = \sqrt{2.889} = 1.700$$

Finally, we calculate t_0 (see equation 77 with $\sigma_{\overline{X}_1 - \overline{X}_2}$ replaced by $s_{\overline{X}_1 - \overline{X}_2}$):

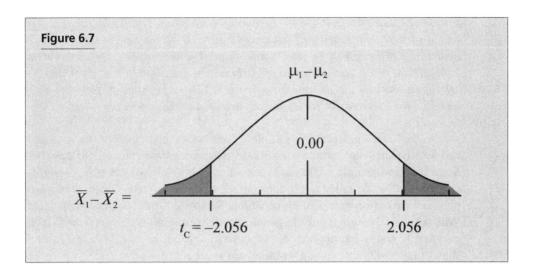

Figure 6.7

$$t_O = \frac{\overline{X}_1 - \overline{X}_2}{s_{\overline{X}_1 - \overline{X}_2}} = \frac{62 - 59}{1.700} = \frac{3}{1.700} = 1.765$$

We have insufficient evidence to reject the null hypothesis at the .05 alpha level because $t_C = \pm 2.056$, and $t_O = 1.765$. The difference between the two sample means, as reflected in t_O, falls within the bounds of normal variation or difference. At the .05 alpha level, the two sample means do not differ significantly. Visually impaired and normally sighted employees do not perform at significantly different rates.

This is another case in which the nature of the test has an important consequence. We have a two-tailed test and a t_C value of ± 2.056. With a one-tailed test, let's say an upper-tailed test, Table C shows that the t_C figure would be 1.706. We would then reject H_O and conclude that visually impaired workers inspect significantly more units per day than do normally sighted employees. If the psychologist rephrases her question and sets up her test differently, the outcome changes. This also illustrates the fact that, for a given alpha level, it is more difficult to reject H_O with a two-tailed test than with a one-tailed test.

6. *The rising cost of school books is becoming increasingly worrisome to local school districts. One school official, concerned that his state and its districts were paying more for books than comparable states and districts, examined samples of his own and other states' costs for very similar US history texts. In his state, even though the cost of its chosen texts has risen, the variation in prices has changed little: the standard deviation is about $3.20. Beyond that, in his*

sample of 42 districts in his state, he finds an average cost of $50.60 per text. For an out-of-state comparison, he samples 45 districts across the country and finds an average price of $48.44 per US history book. The standard deviation for the national text prices, according to publishers' previous statistics, is $4.08. At the .01 alpha level, is the administrator correct? Does his state pay significantly more, on the average, for its US history texts than do comparable states?

We have two independent samples. They are comparable but not matched, and we have different sample sizes: 42 districts in one case versus 45 in the other. Since the administrator wishes to know whether his state's districts pay "significantly more," a one-tailed test is appropriate. If we let his state's districts be Sample 1 and the out-of-state districts constitute Sample 2, we get an upper-tailed test: Will $\overline{X}_1 - \overline{X}_2$ be sufficiently large and positive enough to be significant? Note also that a z test is appropriate because each sample has at least 30 cases. Table A shows that $z_C = 2.33$ for an upper-tailed test at the .01 alpha level.

$$n_1 = 42, \overline{X}_1 = \$50.60, \sigma_1 = \$3.20$$

$$n_2 = 45, \overline{X}_2 = \$48.44, \sigma_2 = \$4.08$$

Our hypotheses are:

$$H_A: \overline{X}_1 > \overline{X}_2; H_0: \overline{X}_1 \leq \overline{X}_2 \quad \text{or} \quad H_A: \overline{X}_1 - \overline{X}_2 > 0.00; H_0: \overline{X}_1 - \overline{X}_2 \leq 0.00$$

$$H_A: \mu_1 > \mu_2; H_0: \mu_1 \leq \mu_2 \quad \text{or} \quad H_A: \mu_1 - \mu_2 > 0.00; H_0: \mu_1 - \mu_2 \leq 0.00$$

Figure 6.8 presents the distribution curve for problem 6. Note that, as with problem 5, four different sets of null and alternative hypotheses are possible. Each set or pair expresses the same thing, but we are overdoing hypotheses statements here just to illustrate a point: there are different ways of doing the problem. In practice, unless an instructor prefers a certain method, use whichever format makes the most sense to you.

The information given includes no sample standard deviations: s_1 and s_2. Even if we have those statistics, we use the population figures, σ, when they are available. For the standard error, we use equation 59:

$$\sigma_{\overline{X}_1 - \overline{X}_2} = \sqrt{\frac{\sigma_1^2}{n_1} + \frac{\sigma_2^2}{n_2}} = \sqrt{\frac{(3.20)^2}{42} + \frac{(4.08)^2}{45}} = \sqrt{\frac{10.240}{42} + \frac{16.646}{45}}$$

$$= \sqrt{.244 + .370} = \sqrt{.614} = .784$$

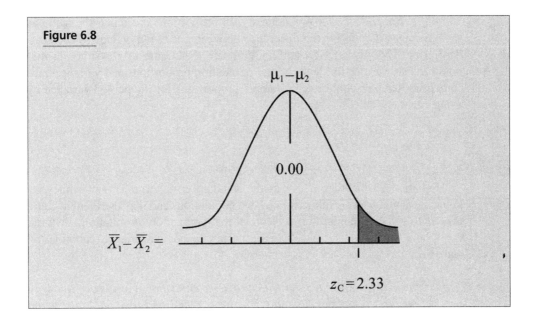

Figure 6.8

For z-observed, we use equation 76:

$$z_O = \frac{\overline{X}_1 - \overline{X}_2}{\sigma_{\overline{X}_1 - \overline{X}_2}} = \frac{50.60 - 48.44}{.784} = \frac{2.16}{.784} = 2.755$$

Thus, we may reject the null hypothesis at the .01 alpha level. The administrator was correct. School districts in his state are paying significantly more for their US history texts, on the average.

As a general note, we must set up one-tailed hypotheses tests correctly here. All the elements must form a logically consistent package. Assume we expect the mean we designated as $\overline{X}_1$ to be the larger of the two averages. If we calculate our difference as $\overline{X}_1 - \overline{X}_2$, we are obliged to do an upper-tailed test. If $\overline{X}_1$ is indeed larger, is the $\overline{X}_1 - \overline{X}_2$ difference sufficiently positive to reach the upper tail of the H_O curve? Conversely, if we calculate our difference as $\overline{X}_2 - \overline{X}_1$ (and we could), we expect a negative difference. Is it then large enough to fall into the lower extreme of the H_O curve? The important consideration is that in this sort of situation our difference of the means must be consistent with our alternative hypothesis and the kind of H_O test we perform.

We may test for the difference between sample proportions also. As in Chapter 5 and earlier in Chapter 4, however, there is only a z test (no t test) with proportions.

Sample proportions. Our procedures with proportions are much the same as with averages. We divide the difference between our two p' values by the standard error. This gives us z_0, and we compare that figure to z_c. Although it is possible that we may know the actual population proportions and variances, we sometimes have to pool our two sample proportions for the pooled variance by using equation 78.

(78)
$$z_0 = \frac{p'_1 - p'_2}{\sigma_{p_1 - p_2}}$$

Equation 78 includes $\sigma_{p_1 - p_2}$ as the standard error term. This presumes we know the actual p and q values for our two populations and may use them in equation 61 to get that standard error. Without that population information, we write the standard error as $s_{p_1 - p_2}$ and solve for it using the pooled variance (equations 64 and 65).

7. *In comparatively liberal San Francisco, does educational level correlate with opinions regarding the death penalty? From national and regional polls, about 75% of those with at least some college education are known to favor the death penalty, and the figure is about 65% for those with no college. Is this pattern true in the Bay Area? Specifically, are San Francisco's college-educated adults more favorable or less favorable to the death penalty than those with fewer years of formal education? In a local study, two random samples of adults responded to the question: "Should the death penalty be eliminated or continue to be enforced in certain cases?" Among the 273 respondents who never attended college, 175 favored retaining a death penalty. For the 231 respondents with at least some college, 164 favored capital punishment in certain cases. At the .05 alpha level, do the respective populations' opinions differ at all?*

To review the problem, we have two independent samples, one with 273 respondents and the other with 231. Among those with less education (Sample 1), the proportion supporting a death penalty (p'_1) is 175/273, or .64. Among those with more education, the proportion (p'_2) is 164/231, or .71. Referring to Table 4.1 with our minimum values (specifically, with $q'_1 = .36$, and $q'_2 = .29$), we see that both samples are large enough to permit the use of z, indicating that the sampling distribution of the difference is normal. Also, since the question is whether or not the populations "differ at all," not whether one is more favorable than the other, we must do a two-tailed test.

$n_1 = 273, p'_1 = .64, p_1$ is believed to be about .65 (or 65%), so $q_1 = 1 - .65 = .35$

$n_2 = 231, p'_2 = .71, p_2$ is thought to be about .75 (or 75%), so $q_2 = 1 - .75 = .25$

Our hypotheses are:

$$H_A: p'_1 \neq p'_2; H_0: p'_1 = p'_2 \quad \text{or} \quad H_A: p'_1 - p'_2 \neq .000; H_0: p'_1 - p'_2 = .000$$

$$H_A: p_1 \neq p_2; H_0: p_1 = p_2 \quad \text{or} \quad H_A: p_1 - p_2 \neq .000; H_0: p_1 - p_2 = .000$$

(Again, in doing the practice problems, you need present only one form of the hypotheses.)

At the .05 alpha level for a two-tailed test, $z_C = \pm 1.96$. Figure 6.9 shows this distribution curve.

Since we have information as to the two population proportions, .65 and .75, we may use them to find the standard error (equation 61). We do not need $\overline{P}$ and $\overline{Q}$ for the pooled variance first. For the standard error of the difference:

$$\sigma_{p_1 - p_2} = \sqrt{\frac{p_1 q_1}{n_1} + \frac{p_2 q_2}{n_2}} = \sqrt{\frac{.65(35)}{273} + \frac{.75(25)}{231}}$$

$$= \sqrt{.0008 + .0008} \quad \sqrt{.0016} = .04$$

Equation 78 provides z_0:

$$z_0 = \frac{p'_1 - p_2}{\sigma_{p_1 - p_2}} = \frac{.64 - .71}{.04} = \frac{-.07}{.04} = -1.75$$

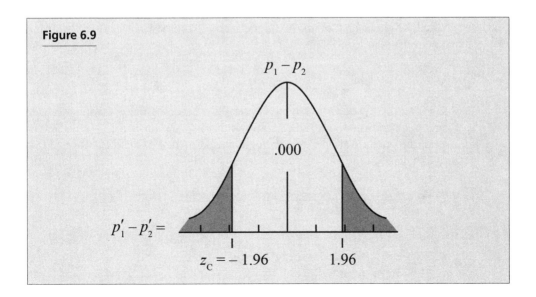

Figure 6.9

$p_1 - p_2$

.000

$p'_1 - p'_2 =$

$z_C = -1.96$ 1.96

We fail to reject the null hypothesis at the .05 alpha level because $z_0 = -1.75$, but $z_C = \pm 1.96$. Our conclusion is that, in San Francisco, educational level, at least as measured here, does not affect opinions as to whether a death penalty should be retained. A higher proportion of those with some college education does indeed favor a death penalty, but that percentage is not *significantly* higher than that found among those with no college.

8. *The National Sleep Foundation (NSF) recently estimated that 126 million Americans suffer "insomnia problems at least a few times per week." A researcher suspects insomnia may affect significantly fewer men than women. She randomly samples 360 adult males, and finds that 133 have the sort of sleeping problem described by the NSF. Among a random sample of 400 women, the figure is 176. Is the researcher correct? At the .05 alpha level, do significantly fewer males report recurrent instances of insomnia?*

The researcher has two independent samples. If males become Sample 1 and females Sample 2, we will subtract the female value (p'_F) from that of males (p'_M): $p'_M - p'_F$. If more females have trouble sleeping, the difference should be negative. Is it sufficiently negative to fall into the lower alpha region? This indicates a lower-tailed test.

Does the sampling distribution of the difference conform to the normal curve? The sample proportions are $p'_M = 133/360 = .37$, and $p'_F = 176/400 = .44$. Checking .37 and .44 in Table 4.1, we see that our sample sizes are large enough to suggest that the distribution does approximate the z curve. z_C at the .05 alpha level is -1.645. How about the standard error? We must use our sample p' values to get the pooled variance first because no separate population values are reported for either men or women. Finally, our hypotheses reflect the lower-tail nature of the test:

$$H_A: p'_M < p'_F; \; H_0: p'_M \geq p_F \quad \text{or} \quad H_A: p'_M - p'_F < .000; \; H_0: p'_M - p'_F \geq .000$$

(As before, the hypotheses may be expressed in terms of the population values, p_M and p_F, if preferred.) Figure 6.10 shows the distribution curve.

For the pooled variance and then the standard error of the difference (equations 64 and 65):

$$\overline{P} = \frac{p'_M(n_1) + p'_F(n_2)}{n_M + n_F} = \frac{.37(360) + .44(400)}{360 + 400} = \frac{309.200}{760} = .41$$

$$\overline{Q} = 1 - \overline{P} = 1 - .41 = .59$$

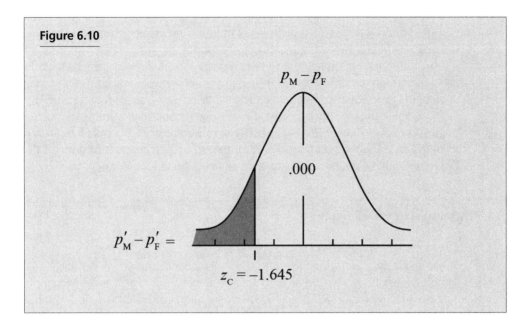

Figure 6.10

$$s_{p_M - p_F} = \sqrt{\frac{\overline{P} \cdot \overline{Q}}{n_M} + \frac{\overline{P} \cdot \overline{Q}}{n_F}} = \sqrt{\frac{.41(.59)}{360} + \frac{.41(.59)}{400}}$$

$$= \sqrt{.0007 + .0006} = \sqrt{.0013} = .036$$

Equation 78 (using $s_{p_M - p_F}$) results in:

$$z_O = \frac{p'_M - p'_F}{s_{p_M - p_F}} = \frac{.37 - .44}{.036} = \frac{-.07}{.036} = -1.944$$

We therefore reject H_O at the .05 alpha level. It appears the researcher is correct. Significantly fewer men report regular insomnia than do women. Once more, the choice of an alpha level is crucial, however. If the researcher opts for the .01 level, we fail to reject her null hypothesis because z_C then equals −2.33.

Summary. We may establish confidence intervals and test hypotheses for the difference between two independent samples. We do this by envisioning a sampling distribution of the difference between the means or proportions of the two samples and calculating a standard error of the difference. We then establish a confidence interval in terms of this hypothetical sampling distribution, using the observed difference between our two sample means or proportions as the central point estimate. In testing hypotheses, we compare our observed $\overline{X}_1 - \overline{X}_2$ or $p'_1 - p'_2$ difference, in effect, to a zero difference assumed under the null hypothesis. If the observed difference is sufficiently greater or less than zero (depending upon our H_A and how we have set up the test), we have evidence for rejecting H_O.

Conceptually, the situation is much the same for confidence intervals and hypothesis tests involving differences between related or paired samples, discussed next. We have a sampling distribution of the difference, a standard error of the difference, and so on, but we do have to adjust our equations somewhat.

Related or Paired Samples

Many Sample Pairs and Usual Differences

What about the case where samples are *not* independent, where we have two samples consisting of paired observations? For every score or X value in Sample 1, there is a counterpart in Sample 2. Every person recorded in Sample 1 also has another score or measurement recorded in Sample 2, or at least every person's matched counterpart has. An obvious example of this involves repeated measures or before-after designs. Individuals are measured or tested at Time 1, exposed to some experimental treatment or experience, and thereafter tested again at Time 2. Is there a significant difference between the Time 1 and Time 2 scores? Considering each pair of scores, what is the average or mean difference between the Time 1 and Time 2 scores? In the case of proportions, what is the difference between the Time 1 and Time 2 p' values?

Our approach and calculations here are somewhat different for means and proportions. For means, we now have **d scores,** *which are the difference between each pair of scores.* The procedures are the same as before, but we now work with d scores rather than X values. We have the mean difference, $\overline{d}$ (read as "d-bar"), and we get a standard deviation of the differences and finally a standard error of the mean difference. In the case of proportions, our approach is basically the same as it was with independent samples, but, even though we still work with $p_1 - p_2$ differences, the way in which we calculate the standard error is new.

Our curve is still the hypothetical sampling distribution of the difference. For means, on the one hand, it is the distribution of many average differences, or $\overline{d}$ scores, taken from many samples of paired differences. For proportions, on the other hand, it is the distribution of the typical differences ($\overline{P}$) found among samples of paired p' values. We imagine many sets of matched samples, each producing an average or typical difference derived from all the pairs in the sample. These average differences, when graphed, form the sampling distribution of the difference. The curve is centered around the theoretical difference between paired populations. These curves are illustrated in Figures 6.11 and 6.12, for population means and proportions, respectively. In Figure 6.11, $\overline{D}$ is the mean difference between two related populations, and $s_{\overline{d}}$ is the standard error of the mean difference. In Figure 6.12, $p_1 - p_2$ is the difference between the proportions of two related populations, and $s_{\overline{p}}$ is the standard error of the difference.

Although much of this seems similar to situations involving independent samples, related samples do require some adjustments. First, with averages, we typically use a *t* curve with related samples, rather than the normal curve. Why? With all the effort and investment necessary to find matched samples (or pair respondents), we usually have small samples, compared to how easy it is to secure random samples for just one test or measurement. For matched pairs, every person in Sample 1 needs a reasonable counterpart in Sample 2. Or, if we use people as their own controls, we must find individuals who will go through two tests or measurements (i.e., before and after, Time 1 and Time 2). For these reasons, our matched samples are typically smaller, and our resulting sampling distribution of the mean difference often conforms to a *t* distribution rather than to the normal distribution.

Figure 6.11 The Sampling Distribution of the Difference Between Two Related Population Means

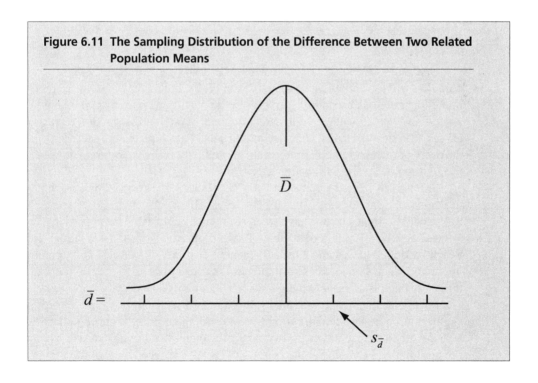

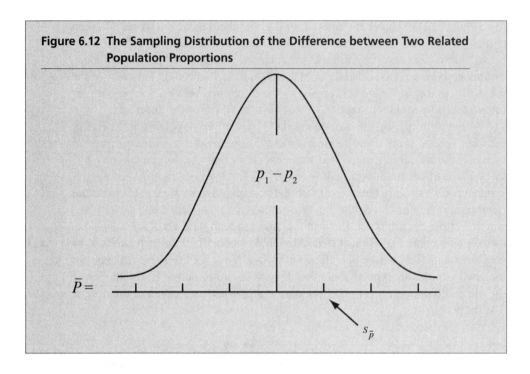

Figure 6.12 The Sampling Distribution of the Difference between Two Related Population Proportions

$p_1 - p_2$

$\bar{P} =$

$s_{\bar{p}}$

One could, of course, envision using z here, and we *must* do so with proportions. We will look at practice problems using z, but realistically, t is more appropriate for situations involving averages. In fact, many statistics texts omit z with related samples altogether. In this sense, actual sampling distribution curves, t curves, may be less peaked and have more spread than depicted in Figure 6.11. In the event of matched samples large enough to justify using the z curve, however, the appropriate z value is substituted for t.

Second, for related samples, as may be obvious, the two samples are identical in size: n_1 and n_2 are naturally the same. This is not always the case with independent samples, as we have seen. (In related samples, if some participants are lost between Times 1 and Times 2, we drop any Time 1 observations for which we do not have Time 2 measurements.) This fact of identical n's becomes important when we pool our two sample proportions to get $\bar{P}$. Since both n's are the same, there is no need to weight each p' value by its respective sample size.

Finally, as mentioned previously, the standard error terms are calculated in slightly different ways. With proportions we no longer weight our p' values by n, and with means we now work with d scores rather than X values. Also, we do not pool the two sample variances for s_p^2. There is no need to pool the two independent estimates for the unknown population variance. The standard error of the difference is calculated another way. We now use the d scores. These minor adjustments will become clearer as we consider specific examples and equations. In the meantime, our overall procedures may be represented by the following familiar generic equations.

For confidence intervals:

(79)   C.I. Est. for the Mean Difference
             = Sample averages' difference $\pm$ (z or t) $\cdot$ (Standard error)

(80)   C.I. Est. for the Difference between Related Proportions
                = Sample proportions' difference $\pm$ z (Standard error)

For hypotheses tests for means:

(81) z_0 or $t_0 = \dfrac{\text{Sample averages' difference}}{\text{Standard error of the difference}}$

For hypotheses tests for proportions:

(82) $z_0 = \dfrac{\text{Difference between sample } \textit{proportions}}{\text{Standard error of the difference}}$

Estimating Differences Between Related Populations

Pairs of population averages. We wish to estimate the difference between two related population means, or $\overline{D}$, the population mean difference. Our point estimate, the center point around which we construct the interval, is $\overline{d}$, the average difference between the paired observations in our samples. To get our standard error of the difference ($s_{\overline{d}}$) we first calculate the standard deviation for the difference scores or d scores. We then divide this standard deviation by $\sqrt{n}$. Does this sound familiar? It should—this is exactly how we got a standard error previously when dealing with single sample situations. Again, the only difference is that now our observations are d scores instead of X scores. Therefore, the confidence interval estimate for $\overline{D}$ using z is:

(83)   C.I. Estimate for $\overline{D} = \overline{d} \pm z\,(s_{\overline{d}})$

and using t, it is:

(84)   C.I. Estimate for $\overline{D} = \overline{d} \pm t\,(s_{\overline{d}})$

where t has $n - 1$ degrees of freedom, n is the number of paired observations, the **mean difference,** $\overline{d}$, is given by equation 85:

$$\overline{d} = \frac{\Sigma d}{n}$$

(85)

and the **standard error of the difference,** $s_{\overline{d}}$, is given by equation 86, with the **standard deviation of the difference,** s_d, calculated from equation 85.

(86)

$$s_d = \sqrt{\frac{\Sigma d^2 - \frac{(\Sigma d)^2}{n}}{n-1}}$$

(87)

$$s_{\overline{d}} = \frac{s_d}{\sqrt{n}}$$

9. *A psychologist advertised in the campus newspaper for subjects who thought they had unusual test anxiety. From interviews with those who responded, she selected a representative sample of 10 students. Just prior to the start of the spring semester, the students went through a series of counseling and training sessions designed to help them reduce their test anxiety. They were asked to practice what they had learned during the coming spring term. Assuming all else was reasonably equal for the two semesters, their grade point averages (GPAs) for the fall and spring terms were compared at the end of the academic year. Given the data below, what is the 90% confidence interval for the before- and after-counseling difference in GPAs?*

Subject:	1	2	3	4	5	6	7	8	9	10
Fall GPA ("Before"):	2.76	3.01	2.19	2.54	3.71	2.95	2.27	2.66	3.18	3.27
Spring GPA ("After"):	2.91	3.15	2.21	2.93	3.68	3.14	2.19	2.93	3.45	3.83

This is an obvious case of related samples. The *before* GPAs come from the fall semester, and the *after* GPAs come from the spring semester. With a sample of 10, we use t rather than z. Note that the sample size is 10, not 20, because even though we have 20 separate observations, the sample size is determined by the number of pairs, not by the total number of measurements.

That also affects the degrees of freedom. Instead of two separate and independent samples, there are two scores on the same variable per person for one sample. With one sample, our degrees of freedom are again $n - 1$, where n is the number of paired scores.

In Table 6.1, the first column is the subject number. The next two columns consist of the raw data, the fall and spring GPAs. The last two columns show the initial difference calculations needed: the d scores (Spring GPA – Fall GPA) and d^2 values.

At the .90 confidence level with $n - 1$ (or 9) d.f., $t = \pm 1.833$:
We first find the mean difference $(\bar{d})$ by using equation 85:

$$\bar{d} = \frac{\Sigma d}{n} = \frac{1.88}{10} = .188$$

Next, we solve for the standard deviation of the differences, s_d (equation 85), and then the standard error of the mean difference, $s_{\bar{d}}$ (equation 87):

$$s_d = \sqrt{\frac{\Sigma d^2 - \frac{(\Sigma d)^2}{n}}{n-1}} = \sqrt{\frac{.697 - \frac{(1.88)^2}{10}}{9}} = \sqrt{\frac{.344}{9}} = \sqrt{.038} = .196$$

$$s_{\bar{d}} = \frac{s_d}{\sqrt{n}} = \frac{.196}{\sqrt{10}} = \frac{.196}{3.162} = .062$$

Table 6.1

Subject	Fall GPA ("Before")	Spring GPA ("After")	(Spring–Fall) d	d^2
1.	2.76	2.91	.15	.0225
2.	3.01	3.15	.14	.0196
3.	2.19	2.21	.02	.0004
4.	2.54	2.93	.39	.1521
5.	3.71	3.68	−.03	.0009
6.	2.95	3.14	.19	.0361
7.	2.27	2.19	−.08	.0064
8.	2.66	2.93	.27	.0729
9.	3.18	3.45	.27	.0729
10.	3.27	3.83	.56	.3136
Totals			$\Sigma d = 1.88$	$\Sigma d^2 = .6974$

We use equation 84 for the actual confidence interval estimate:

.90 C.I. Estimate for $\overline{D} = \bar{d} \pm t(s_{\bar{d}}) = .188 \pm 1.833\,(.062) = .188 \pm .114$

For the interval's lower limit: $.188 - .114 = .074$ grade points.
For the interval's upper limit: $.188 + .114 = .302$ grade points.

The distribution curve for problem 9 is shown in Figure 6.13. Assuming other things are fairly equal for the two semesters, the psychologist may be 90% confident that her anxiety-reducing techniques result in a typical GPA increase of anywhere from .074 to .302 grade points.

10. *Do immigrants' households tend to contain more children than households of native-born adults? Thirty-five immigrant households were each recently matched with a native household regarding the age, gender, educational level, economic background, and religion of the head of the household. The number of children in each pair of households was recorded, and the following summary data were noted:*

- *Immigrant households had a total of 92 children versus 63 for native families; therefore, $\Sigma d = 92 - 63 = 29$.*
- *The average difference, $\bar{d}$, between matched immigrant and native households was $(\Sigma d)/n$ or $29/35 = .829$ children. In other words, immigrant households averaged .829 more children.*
- *The sum of the squared difference scores, Σd^2, was 127.*

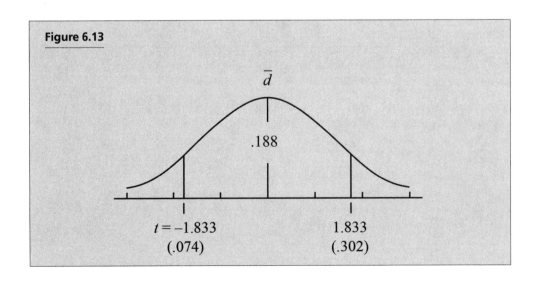

Figure 6.13

$\bar{d}$

.188

$t = -1.833$ 1.833
(.074)   (.302)

What is the .99 confidence interval estimate for immigrant-native differences in the average number of children per household?

In this example, we have a sample of 35 pairs of matched respondents or households (which indicates using z rather than t). Table A shows that the z coefficient for the .99 confidence level is $z = \pm2.575$. We are already given the summary information rather than the raw data themselves, which saves time. The differences were derived by subtracting the number of children in native households from those in immigrant households ($X_I - X_N$), so we must be aware of that when we interpret our confidence interval estimate.

Besides the information given and for $n = 35$ pairs, we need the standard deviation of the differences, s_d (from equation 86) and then the standard error of the mean difference, $s_{\bar{d}}$ (from equation 87):

$$s_d = \sqrt{\frac{\sum d^2 - \frac{(\sum d)^2}{n}}{n-1}} = \sqrt{\frac{127 - \frac{(29)^2}{35}}{34}} = \sqrt{\frac{127 - 24.029}{34}} = \sqrt{\frac{102.971}{34}} = \sqrt{3.029} = 1.740$$

$$s_{\bar{d}} = \frac{s_d}{\sqrt{n}} = \frac{1.740}{\sqrt{35}} = \frac{1.740}{5.916} = .294$$

By using equation 83, we get the actual confidence interval:

.99 C.I. Estimate for $\overline{D} = \bar{d} \pm z\,(s_{\bar{d}}) = .829 \pm 2.575\,(.294) = .829 \pm .757$

For the interval's lower limit: $.829 - .757 = .072$ children
For the interval's upper limit: $.829 + .757 = 1.586$ children

Figure 6.14 shows the distribution curve for problem 10. The study suggests that with 99% confidence, immigrant households tend to average between .072 and 1.586 more children than similar native households.

Pairs of population proportions. We may also estimate the difference for proportions from related populations. Apart from the calculation of the standard error, the situation with related samples is the same as for independent samples. We still use $p'_1 - p'_2$ as the point estimate and $p_1 - p_2$ to refer to the true difference between the population proportions. Similarly, we must also check Table 4.1 using the smallest of our p' or q' values from each sample to see whether the sampling distribution curve approximates the normal distribution. Assuming it does, the adjustment in solving for the standard error is that we no longer weight each p' value by its respective n because the two n's are identical, by definition. Our first step is to take the average of the two p' values to get $\overline{P}$ (and $\overline{Q}$).

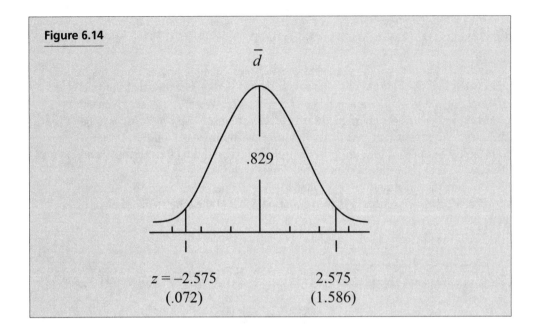

Figure 6.14

$$\overline{P} = \frac{p_1' + p_2'}{2}, \overline{Q} = 1 - \overline{P}$$

We then find the standard error of the difference by using equation 88:

(89)

$$s_{\overline{p}} = \sqrt{\frac{\overline{P} \cdot \overline{Q}}{n}}$$

Equation 89 inserts the standard error into the final step in finding the confidence interval estimate for the population proportions:

(90)   C.I. Estimate for $p_1 - p_2 = (p_1' - p_2') \pm z\,(s_{\overline{p}})$

11. *A physician was interested in whether social facilitation might influence the perceived effectiveness of a new arthritis remedy. Two groups of arthritis patients were matched as to the severity of their symptoms, general health, activity levels, age, and sex. The new medication, which was expected to take effect in a matter of minutes, was administered to one group collectively. In an osteoarthritis clinic setting, these patients stayed together for at least 30 minutes after administration of the medication and were allowed to freely socialize and interact if*

they wished. On the assumption that some patients in this sample would derive noticeable benefits from the medication, would this possibly influence some of their fellow subjects to report an improvement due to a contagion or halo effect? Subjects in the other sample were given the medication privately and never met the other patients from their group. The question concerned how much the two samples would differ in the proportions reporting improvement. Based upon examinations and interviews, the physician determined that 36 of 60 patients in the interactive sample (Sample 1) believed their conditions had improved. Of 60 patients receiving the medication alone (Sample 2), the comparable figure was 33 patients. What is the .95 confidence interval estimate for the difference between the proportions claiming improvement?

The samples were purposely matched, and $n = 60$. The p' value for Sample 1 is 36/60, or .60. For Sample 2, it is 33/60, or .55. The two q' figures are $q'_1 = 1 - .60 = .40$, and $q'_2 = 1 - .55 = .45$. Consulting Table 4.1 using these q values shows that our sampling distribution curve conforms to the normal distribution, and we may proceed. For our eventual estimate, the z coefficient will be ±1.96 for the .95 confidence level, and our point estimate for the difference becomes $.60 - .55$, or .05.

First, we need to determine $\overline{P}$ and $\overline{Q}$, and then our standard error of the difference (using equations 88 and 89, respectively):

$$\overline{P} = \frac{p'_1 + p'_2}{2} = \frac{.60 + .55}{2} = \frac{1.15}{2} = .575 \text{, and } \overline{Q} = 1 - .575 = .425$$

$$s_{\overline{p}} = \sqrt{\frac{\overline{P} \cdot \overline{Q}}{n}} = \sqrt{\frac{.575(.425)}{60}} = \sqrt{\frac{.244}{60}} = \sqrt{.0041} = .064$$

We may now establish a confidence interval for the difference by using equation 90:

.95 C.I. Est. for $p_1 - p_2 = (p'_1 - p'_2) \pm z\,(s_{\overline{p}}) = (.60 - .55) \pm 1.96\,(.064) = .05 \pm .13$

For the interval's lower limit: $.05 - .13 = -.08$
For the interval's upper limit: $.05 + .13 = .18$

Figure 6.15 shows the sampling distribution curve for problem 11. Based upon these results, the physician cannot be sure a halo effect exists. It is possible 18% more patients would report improvement if the medication is administered

Figure 6.15

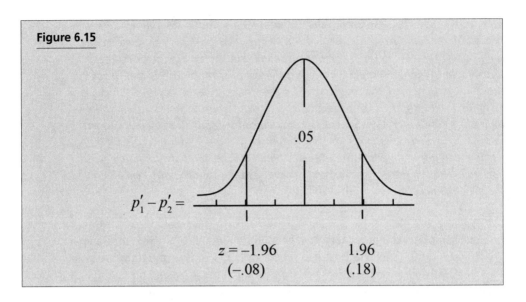

$$p'_1 - p'_2 =$$

$$z = -1.96$$
$$(-.08)$$

$$1.96$$
$$(.18)$$

.05

in a group milieu. However, the confidence interval encompasses zero or no difference, so the true difference may go the other way. In fact, up to 8% more patients might claim relief when the medication is given privately. The evidence therefore does not clearly confirm any social benefit of group treatment as opposed to private treatment.

12. *A recent survey asked respondents whether they approved of using recycled water to irrigate crops and produce intended for human consumption. From more than a thousand respondents, researchers were able to match 113 people with a connection to local community gardens, that is, those leasing plots or on waiting lists (Sample 1), with 113 counterparts uninvolved in anything but occasional and/or minor home gardening (Sample 2). Pairs of sample members were matched for approximate age, educational level, and ethnicity. Among those with garden connections, 62 approved using recycled water to grow human food. In Sample 2, 54 did so. What is the .90 confidence interval estimate for the difference in the rates at which these two populations support irrigation with recycled water?*

The question describes matched samples. Among the gardeners (Sample 1), the proportion supporting recycled water use is $p'_1 = 62/113 = .549$, and $q'_1 = 1 - .549 = .451$. For those comparatively uninvolved in gardening (Sample 2), p'_2 is $54/113 = .478$, and $q'_2 = 1 - .478 = .522$. After consulting Table 4.1 with our minimum p or q values (specifically, q'_1 and p'_2), we determine that we may use the z distribution, and $z = \pm 1.645$. For the interval, our point estimate is $p'_1 - p'_2 = .549 - .478$, which is $.071$, or $.07$.

As before, we start with $\overline{P}$ (equation 88) and then $s_{\overline{p}}$, our standard error of the difference (equation 89):

$$\overline{P} = \frac{p'_1 + p'_2}{2} = \frac{.549 + .478}{2} = \frac{1.027}{2} = .514 \ , \quad \overline{Q} = 1 - .514 = .486$$

$$s_{\overline{p}} = \sqrt{\frac{\overline{P} \cdot \overline{Q}}{n}} = \sqrt{\frac{.514\,(.486)}{113}} = \sqrt{.0022} = .047 = .05$$

For our estimation of the difference, we use equation 90:

$$.90 \text{ C.I. Est. for } p_1 - p_2 = (p'_1 - p'_2) \pm z\,(s_{\overline{p}}) = (.55 - .48) \pm 1.645\,(.05) = .07 \pm .08$$

For the interval's lower limit: $.07 - .08 = -.01$
For the interval's upper limit: $.07 + .08 = .15$

Figure 6.16 shows the distribution curve for problem 12. We may be 90% confident that the true difference in the rates of approving recycled water's use is between –1% and 15%. There is a 90% probability that nongardeners may be more approving by 1% or that gardeners may be up to 15% more approving. Our

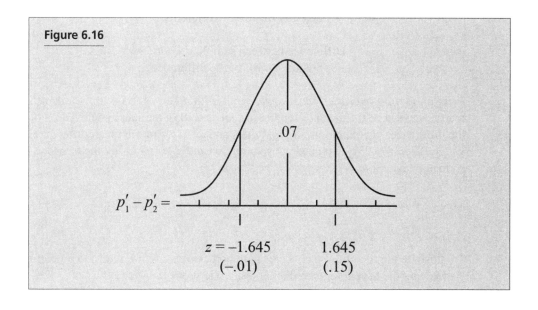

Figure 6.16

$p'_1 - p'_2 =$

$.07$

$z = -1.645$　　1.645
$(-.01)$　　　$(.15)$

results do not confirm a definite difference of opinion going either way, but, given the possibility of as much as 15% greater support among gardeners, it seems likely that they may be at least slightly more approving of recycled water than are nongardeners.

Testing Differences Between Related Samples

Besides establishing confidence intervals, we may also test hypotheses involving differences between related samples. The new point estimates and standard errors also apply to hypothesis testing. Procedures for these tests follow the basic format outlined in previous sections. According to our null hypothesis, the difference between our samples (i.e., our two samples' statistics) is zero. The numerator of our z_0 and t_0 formulas (equation 74) includes the actual difference between our sample statistics minus 0.00. Again, we drop the "minus 0.00." The standard errors of the difference are the denominators of our formulas and are calculated just as they are for confidence intervals. Finally, with means, we typically use a t test rather than z because matched samples are generally small, but each test is illustrated below. When looking up t-critical (t_C), we have $n - 1$ degrees of freedom, where n is, again, the number of paired rather than total observations. For proportions, we must consult Table 4.1 to make sure n and our population proportions permit the use of the normal distribution.

In generic form, our procedures were described in equations 81 and 82, repeated here:

$$\text{For means: } z_0 \text{ or } t_0 = \frac{\text{Sample averages' difference}}{\text{Standard error of the difference}}$$

$$\text{For proportions: } z_0 = \frac{\text{Difference between sample } \textit{proportions}}{\text{Standard error of the difference}}$$

Pairs of sample averages. Again, we work with d scores. As with confidence intervals, we need the mean difference, $\bar{d}$ (equation 85); the standard deviation of the differences, s_d (equation 86); and the standard error of the mean difference, $s_{\bar{d}}$ (equation 87). We also need our hypotheses and z_0 or t_0. Using the terms described in equations 84–86, z_0 or t_0 becomes:

$$(91) \qquad\qquad z_0 \text{ or } t_0 = \frac{\bar{d}}{s_{\bar{d}}}$$

To illustrate the use of these concepts in hypothesis testing, we return to problem 9 involving the psychologist reducing test anxiety.

13. *In addition to establishing a confidence interval, the psychologist wishes to test whether the before and after GPA differences are significant at the .05 alpha level. We have the before and after means for students' GPAs. Given her data below and using the .05 alpha level, are students' post-counseling (after) GPAs significantly higher than their pre-counseling (before) GPAs?*

Subject: *1   2   3   4   5   6   7   8   9   10*

Fall GPA ("Before"): 2.76 3.01 2.19 2.54 3.71 2.95 2.27 2.66 3.18 3.27

Spring GPA ("After"): 2.91 3.15 2.21 2.93 3.68 3.14 2.19 2.93 3.45 3.83

To start, the problem calls for a t test because n consists of just 10 cases. It also makes sense to conduct an upper-tailed test, given the psychologist's interest in "significantly higher" GPAs "after." This means subtracting the Fall GPAs from those recorded in the spring to get $\overline{d}$ (or "after" – "before"): Is the positive (+) mean difference large enough to be statistically significant? Finally, from Table C, for a one-tailed test with $n - 1 = 9$ degrees of freedom, t_c becomes 1.833 at the .05 alpha level. Figure 6.17 presents the distribution curve for problem 13.

The hypotheses are H_A: $\overline{d} > 0.00$; H_O: $\overline{d} \leq 0.00$, and H_O assumes again that the true difference is zero. If we express our hypotheses in terms of population parameters, we can write: H_A: $\overline{D} > 0.00$; H_O: $\overline{D} \leq 0.00$.

Taking advantage of earlier calculations from problem 9, we get: the mean difference is $\overline{d} = .188$; the standard deviation of the differences is $s_d = .268$; and the standard error of the mean difference is

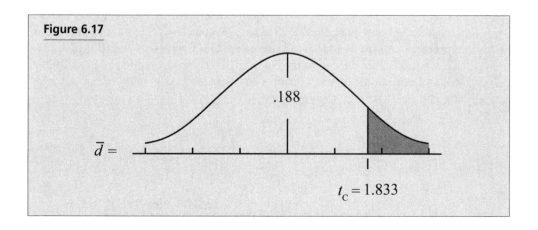

Figure 6.17

$\overline{d} =$

.188

$t_c = 1.833$

$$s_{\bar{d}} = \frac{s_d}{\sqrt{n}} = \frac{.268}{\sqrt{10}} = .085$$

For t_O, we get (see equation 90):

$$t_O = \frac{\bar{d}}{s_{\bar{d}}} = \frac{.188}{.085} = 2.212$$

We may reject H_O. The psychologist's strategies resulted in significant gains in GPAs at the .05 alpha level. She used a very small sample but got encouraging results.

14. *Compared to other global entities, US corporations are alleged to be particularly stingy regarding paid vacation time for employees. An economist wished to assess students' opinions about paid holidays and conducted a poll in a senior International Business course. As the semester began, she asked the 37 registered students how many weeks paid annual vacation each believed a middle-management five-year corporate employee should receive. At the end of the course, with 32 semester-long students still enrolled, she posed the same question. At the .01 alpha level, do the students' perceptions as to appropriate annual vacation times differ between the beginning and end of the semester? In other words, at the .01 alpha level and given the data shown, does the course appear to have changed the students' opinions at all?*

The two columns in ***bold italics*** in Table 6.2 show the actual data given. They were not presented separately as part of the question merely to save space here.

To analyze the situation described, since the question asks whether opinions simply "differ," a two-tailed test is appropriate. With 32 cases, a z test works, and Table A shows that $z_c = \pm 2.575$ at the .01 alpha level. First, we determine the hypotheses and the H_O curve (Figure 6.18). Our hypotheses are: H_A: $\bar{d} \neq 0.00$; H_O: $\bar{d} = 0.00$. Then we determine the mean difference, $\bar{d}$ (from equation 85); the standard deviation, s_d (from equation 86); the standard error, $s_{\bar{d}}$ (from equation 86); and z-observed, z_O (from equation 81):

$$\bar{d} \; (\text{After} - \text{Before}) = \Sigma d/n = 20/32 = .625$$

$$s_d = \sqrt{\frac{\Sigma d^2 - \frac{(\Sigma d)^2}{n}}{n-1}} = \sqrt{\frac{78 - \frac{(20)^2}{32}}{31}} = \sqrt{\frac{78 - 12.500}{31}} = \sqrt{\frac{65.500}{31}} = \sqrt{2.113} = 1.454$$

Table 6.2

Student	Weeks Before	Weeks After	(After − Before) d	d^2
1	4	4	0	0
2	3	5	2	4
3	2	3	1	1
4	2	4	2	4
5	1	2	1	1
6	0	1	1	1
7	4	8	4	16
8	2	2	0	0
9	3	2	−1	1
10	0	2	2	4
11	3	3	0	0
12	0	3	3	9
13	5	5	0	0
14	3	2	−1	1
15	1	3	2	4
16	0	3	3	9
17	6	6	0	0
18	0	0	0	0
19	4	5	1	1
20	3	0	−3	9
21	2	2	0	0
22	3	2	−1	1
23	4	5	1	1
24	6	6	0	0
25	3	4	1	1
26	2	2	0	0
27	5	3	−2	4
28	3	3	0	0
29	0	1	1	1
30	6	6	0	0
31	3	5	2	4
32	4	5	1	1
Totals			$\Sigma d = 20$	$\Sigma d^2 = 78$

$$s_{\bar{d}} = \frac{s_d}{\sqrt{n}} = \frac{1.454}{\sqrt{32}} = \frac{1.454}{5.657} = .257$$

$$z_O = \frac{\bar{d}}{s_{\bar{d}}} = \frac{.625}{.257} = 2.432$$

Figure 6.18

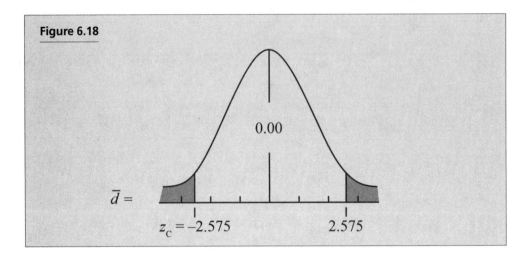

With a two-tailed test at the .01 alpha level, the economist may not reject her null hypothesis. With her test criteria ($z_c = \pm 2.575$), she concludes that students' opinions as to appropriate vacation times have not significantly changed.

However, had the economist set up her test differently, her conclusion would differ. For an upper-tailed test (such as the question "Do students subscribe to more paid holiday time after being exposed to international norms?"), $z_c = 2.33$, and she would reject her H_0. Retaining a two-tailed test but using the .05 alpha level also would have affected the outcome. At the .05 alpha level, $z_c = \pm 1.96$, and she would reject her H_0, too. So, even though the economist lacks evidence to reject her null hypothesis as the test is originally designed, she *could* reject that H_0 under other test conditions.

For related samples and the mean-difference hypothesis test, we follow the same basic rules as with independent samples. Only the degrees of freedom and the standard error terms differ.

Pairs of sample proportions. Procedures for pairs of sample proportions are much the same as previously described for related samples. With proportions, only the standard error differs from the case of independent samples. Since n is constant, we need not weight our sample proportions by that factor. We need the sample difference, $p'_1 - p'_2$, and we divide that figure by the standard error of the difference. To get that standard error, we use $\overline{P}$ and $\overline{Q}$, just as with confidence intervals. The standard error allows us to test our H_0 by solving for z-observed, z_0. As before, that null hypothesis assumes the difference between the two population proportions is zero. After consulting Table 4.1 to make sure our sampling distribution conforms to the z curve, we may solve for z_0:

(91)
$$z_O = \frac{p'_1 - p'_2}{s_{\bar{p}}}$$

Let's look at more practice problems. Problem 15 refers to the data collected in problem 11.

15. *Assume the physician administering the trial arthritis medication to two groups of patients had wished to conduct a hypothesis test. Anticipating that patients reacting to the medication in each other's company might indeed experience a halo effect and report success at a higher rate, a one-tailed test might be performed. At the .05 alpha level, are patients who are allowed to interact with others more likely to claim improvement?*

As in problem 11, Sample 1 includes those patients receiving the medication in a clinic and collective setting, and Sample 2 consists of patients receiving the medication privately. To take advantage of previous calculations in problem 11, we have:

$p'_1 = .60$, $p'_2 = .55$, $p'_1 - p'_2 = .60 - .55 = .05$, $\overline{P} = .575$, $\overline{Q} = .425$, and $s_{\bar{p}} = .064$

The hypotheses are H_A: $p'_1 > p'_2$; H_O: $p'_1 \leq p'_2$. At the .05 alpha level, $z_C = 1.645$, and the distribution curve is as shown in Figure 6.19.

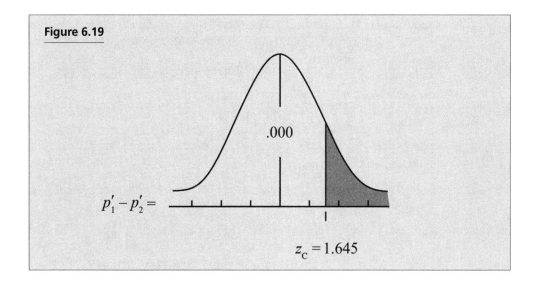

Figure 6.19

.000

$p'_1 - p'_2 =$

$z_C = 1.645$

Then, from equation 92,

$$z_0 = \frac{p'_1 - p'_2}{s_{\bar{p}}} = \frac{.05}{.064} = .781$$

Thus, the physician has insufficient evidence to reject his null hypothesis at the .05 alpha level. Patients receiving the medication in a collective setting are not significantly more likely than those receiving it alone to report improvement from the treatment. Such halo effects as may have been present do not produce significantly better results among patients in Sample 1, and this confirms the physician's earlier finding with his confidence interval.

16. *A sociologist administered a survey to incoming college freshmen and their parents during an orientation weekend. The survey included the statement, "It is important to contribute in some way to charity." Some 207 students and a parent of these students each responded to the poll. Among the students, 110 agreed with the statement, whereas 126 parents did so. At the .05 alpha level, do the students' and parents' levels of agreement differ? What about at the .01 alpha level?*

We may assume the sociologist has matched samples, with each student and parent constituting a pair. For parents, p'_1 = 126/207 = .61. For students, p'_2 = 110/207 = .53. The respective q' values become .39 and .47, and with these figures, n = 207 is adequate for assuming a normal sampling distribution (Table 4.1). A two-tailed test is appropriate because the sociologist wishes to know whether the student and parent samples "differ," not whether one percentage is larger than the other. From Table A, at the .05 alpha level for a two-tailed test, $z_C = \pm 1.96$, and at the .01 level, $z_C = \pm 2.575$. Figure 6.20 shows the distribution curves for the two alpha levels for problem 16.

Our hypotheses are $H_A: p'_1 \neq p'_2$; $H_0: p'_1 = p'_2$. For the standard error term, we need $\bar{P}$, $\bar{Q}$, and $s_{\bar{p}}$:

$$\bar{P} = \frac{p'_1 + p'_2}{2} = \frac{.61 + .53}{2} = \frac{1.14}{2} = .57 \text{ ; so } \bar{Q} = 1 - .57 = .43$$

$$s_{\bar{p}} = \sqrt{\frac{\bar{P} \cdot \bar{Q}}{n}} = \sqrt{\frac{.57(.43)}{207}} = \sqrt{.0012} = .034$$

$$z_0 = \frac{p'_1 - p'_2}{s_{\bar{p}}} = \frac{.61 - .53}{.034} = \frac{.08}{.034} = 2.353$$

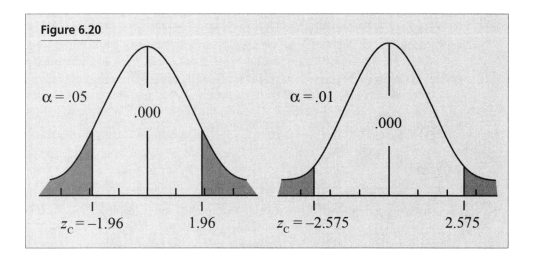

Figure 6.20

The sociologist may reject her H_O at the .05 alpha level because $z_O = 2.353$ and exceeds the critical value of ±1.96. This means that at the .05 alpha level, the rates at which students and their parents subscribe to the importance of giving to charity differ significantly. That is not the case at the .01 alpha level, however. When $\alpha = .01$, the evidence is insufficient to reject H_O ($z_C = ±2.575$). At this more stringent level, z_O falls within the bounds of normal variation. The sociologist may therefore be 95% certain that students' and parents' beliefs differ, but not 99% sure. (How might her conclusions have differed had she opted for an upper-tailed test instead?)

Summary

We have looked at problems involving two samples. The samples may be independent or they may be matched or paired in some fashion. In either case, we envision a sampling distribution of the difference. In constructing a confidence interval, this sampling distribution is centered around the difference between our two sample statistics. Given our confidence level, we establish a range within which the true difference between the two population parameters is suspected to fall. When we test hypotheses, the sampling distribution of the difference constitutes our H_O curve. The central value of the curve is assumed to equal zero. Under the null hypothesis, the difference between the two populations is assumed to be zero. If our actual or observed difference falls into a designated tail of the curve, we reject H_O and conclude our two samples (and populations) differ significantly.

Chapter 7 considers different sorts of hypotheses tests, procedures known as nonparametric tests. In this new situation, we do not use the population parameters in our calculations nor make any assumptions about populations being normally distributed. We also work with different levels of data: ordinal scores, ranked ordinal categories, and nominal category frequencies rather than numerical X values or d scores. Our new statistical tools include the Wilcoxon rank-sum test and its companion, the Mann-Whitney U test, and a statistic called chi square.

Exercises

1. In the context of a survey on ethnic relations, a poll asked students to respond to the statement, "Diversity enhances the campus' learning environment for all students." Of 342 female respondents, 264 agreed, whereas 178 of 258 male participants did so. At the .01 alpha level, do the respective percentages of female and male students supporting the diversity statement differ significantly? What about at the .05 alpha level?

2. A careful study timed the minutes waiting in checkout lines at two supermarket chains. Among 23 shoppers at Food Court, the average wait was 7.32 minutes, with a standard deviation of 2.11 minutes. At about the same time of day, 20 shoppers at JiffyShop averaged 5.24 minutes, with a standard deviation of 1.77 minutes. What is the .95 confidence interval estimate for the difference in average waiting times?

3. An insurance company matched samples of central city and suburban automobile service stations for approximate size, business capacity, union membership of employees, and traffic volume at each location. The same malfunctioning car was sent to each of the 12 city and 12 suburban stations, and estimates for repairs were received. At the .01 alpha level, is the average central city estimate significantly less than that among suburban stations?

Station pairs:	1	2	3	4	5	6	7	8	9	10	11	12
Suburban ($):	421	503	481	431	480	478	427	455	464	438	442	430
Central city ($):	398	457	420	428	441	437	451	425	415	405	411	433

4. Finalists in the Crisco National Fry-Off are teams from Oregon and Texas. Oregon's 17-person team averages 48.06 points, with a standard deviation of 12.37 points. Texas's 9 entrants, on the other hand, average 40.73 points, with a standard deviation of 10.61 points. The judges are about to award Oregon the

coveted Golden Grease Award when the Texas captain jumps to his feet. "No way, no way!" he shouts. "The rules say it has to be 95% certain the winning team is better than the runner-up. Oregon's score is not significantly better." Is the Texas captain correct, or has Oregon really won?

5. Two minor league baseball franchises in the same league have been matched for market size, number of season ticket holders (about 450 in each case), recent records, and stadium sizes. Team A, after mounting an aggressive sales campaign, manages a 70% renewal rate for season tickets. For Team B, the renewal rate is 64%. Has Team A's effort paid off? Is its renewal rate significantly higher than that of Team B at the .05 alpha level?

6. Nationwide, a political science Ph.D. candidate was able to match 300 Democratic with 300 Republican candidates for tenure in office and political level (county, state, Congressional, etc.). All were eligible for re-election. Seventy-two Republicans were not re-elected, nor were 63 Democrats. What is the .99 confidence interval estimate for the difference in re-election failure rates?

7. In different districts, two community college faculties were surveyed as to whether they favored unionization, that is, the introduction of a faculty union on their respective campuses. At Briarcliffe, 303 of 453 faculty favored union representation. At Lafayette, 248 of 427 faculty members did so. What is the .90 confidence interval estimate for the difference in the percentages favoring faculty unions?

8. A university researcher was able to track the remittance records of undocumented workers in the United States. Matching migrant workers for approximate age, education, English language ability, and for gender, family status, and country of origin, she was able to track the remittances of 18 undocumented workers in the Southwest and their 18 counterparts in the Northeast. What is the .90 confidence interval estimate for the difference in the average monthly amounts remitted from these two regions of the country?

Pair	SW	NE	Pair	SW	NE	Pair	SW	NE
1	96	90	7	115	107	13	109	103
2	123	114	8	118	86	14	95	111
3	97	99	9	88	101	15	106	93
4	127	121	10	103	80	16	117	100
5	113	112	11	120	104	17	110	96
6	140	109	12	124	98	18	98	89

9. Senior citizens' living arrangements are varied. A senior-living consultant conducted a survey matching widowed female and male retirees for age, income, current family and living situations, and general health. Among 75 female seniors, 30 indicated a preference for some sort of group living environment, whereas 39 of their male counterparts did so. What is the .90 confidence interval estimate for the percentage difference in group residential preferences?

10. A market researcher examined the effectiveness of people going door-to-door selling coupon books to aid local charities. She matched 35 single-person and 35 two-person sales teams for approximate age, gender, and experience, and recorded the number of books sold by each. Given the data below, are the two-person sales teams significantly more successful than single sales people at the .05 alpha level? What about at the .01 alpha level?

Match	Solo	Pair	Match	Solo	Pair	Match	Solo	Pair
1	7	9	13	10	12	25	19	22
2	14	15	14	9	14	26	16	15
3	18	23	15	15	11	27	17	18
4	10	14	16	7	10	28	13	13
5	15	11	17	5	9	29	14	16
6	16	8	18	4	11	30	7	13
7	23	20	19	13	9	31	15	11
8	14	17	20	17	14	32	10	11
9	12	12	21	6	12	33	16	12
10	7	13	22	3	8	34	11	18
11	4	10	23	11	5	35	14	12
12	11	17	24	14	17			

11. Does being popular in high school translate into later success? Employed young adults in their mid-twenties, none of whom had attended college, were surveyed about popularity in high school and also about present incomes. Among the 70 people who described themselves as being "popular or very popular," the average gross income was $5726 per month, with a standard deviation of $485 per month. For the 115 who reported they were "not generally popular," the average was $5950 per month, with a standard deviation of $512 per month. At the .01 alpha level, do the current monthly salaries differ significantly?

12. Nutritionists and psychologists combined to evaluate free weight-loss programs offered for middle school children who were at least 15 pounds heavier than the typical weights for their respective genders and heights. In the summer between the seventh and eighth grades, the 43 children on Program A lost an average of 9.83 pounds, with a standard deviation of 2.16 pounds. At the same

time, the 48 students on Program B averaged an 11.91-pound loss, with a standard deviation of 3.60 pounds. What is the .90 confidence interval estimate for the difference in average weight losses between the two programs?

13. A campus poll measured the support for an ethnic studies course requirement for graduation. Of the 522 minority students responding to the survey, 344 supported such a requirement. Of the 680 white respondents, 279 were in favor of the requirement. What is the .99 confidence interval estimate for the percentage difference in support for an ethnic studies requirement?

14. Using the data in Exercise 13, and using the .01 alpha level, is the percentage of white students' support of an ethnic studies requirement significantly less than that of minority students?

15. Part of a state's Department of Health Services annual survey wished to identify smoking rates in gay, lesbian, bisexual, and transgender (G-L-B-T) populations versus heterosexual populations. Of 1900 respondents, 174 identified themselves as G-L-B-T, and 53 were smokers. Among the remaining 1726 respondents, 266 were smokers. At the .95 confidence level, what is the estimate for the difference in smoking rates between the state's G-L-B-T and heterosexual populations?

16. Samples of people 60 years of age and above were asked, "Is your quality of life better now than when you were growing up?" All had been lifelong residents of the United States, and pairs of individuals in the respective samples were matched for approximate age, marital status, educational level, and current economic circumstance. The individuals in one sample ($n = 120$) were all still employed, at least part-time, whereas all 120 participants in the other sample considered themselves to be fully retired. In the working sample, 78 responded to the question in the affirmative, and 66 of the retired respondents did so. At the .01 alpha level, are retired seniors less likely to report their quality of life better now than were their working counterparts?

17. Two samples of high school students were asked whether "intelligent design" arguments, besides the theory of evolution, should be included in required biology classes. Of a group of 200 students who reported that either they or at least one immediate family member regularly attended formal religious services, 78 thought intelligent design should be taught. Among another group of 246 students who reported that neither they nor their families attended religious services, 64 believed intelligent design should be covered in classes. Does the support for inclusion of intelligent design in the classroom differ between these two sample groups at the .01 alpha level? Please conduct a one-tailed test of your null hypothesis.

18. At a particular university, the College of Engineering's alumni organization conducted a study in which 30 computer engineering alumni working in high-tech positions were each matched with a non-engineering ("other") alumnus for age, gender, overall grade point average, marital status, and number of children. All participants in the survey had graduated 15 years ago, and each was asked how many job-related residential moves he or she had made since graduation. Their responses are shown below. What is the .95 confidence interval estimate for the difference in the average number of job-related moves since graduation?

Pair	Engineering	Other	Pair	Engineering	Other	Pair	Engineering	Other
1	3	1	11	3	1	21	3	0
2	2	2	12	4	3	22	4	3
3	4	2	13	4	1	23	1	2
4	0	3	14	3	3	24	1	1
5	2	1	15	6	3	25	3	1
6	5	3	16	1	1	26	2	0
7	1	0	17	2	2	27	4	2
8	2	2	18	0	1	28	3	3
9	3	1	19	2	3	29	2	0
10	5	2	20	3	1	30	1	2

19. The United Nations' Economic Commission for the Americas recently updated information regarding Mexican and US farmers. As part of that update, 20 farmers from each country were matched as to age, size of holding, and general economic and family situations. The Commission recorded the number of years (rounded to the nearest whole year below) that each had maintained his or her present farm property. What is the Commission's .99 confidence interval estimate for the difference in average years of continuous farm maintenance between Mexican and US farmers?

Pair	Mexican	US	Pair	Mexican	US	Pair	Mexican	US
1	18	16	8	17	15	15	12	5
2	21	14	9	19	13	16	31	24
3	15	17	10	25	28	17	23	18
4	23	19	11	20	17	18	21	27
5	42	36	12	14	11	19	23	18
6	31	33	13	9	8	20	28	22
7	22	19	14	15	16			

20. In a confidential survey conducted for a large government agency, more than one thousand employees were asked a battery of questions, which included agreeing or disagreeing with the item, "I would be willing to take a lie detector test if required for a job promotion." As part of the analysis, the responses of 150 employees with five years or more of service were matched with those of 150 employees having two years or less of service. The samples of workers were matched with regard to marital status, gender, approximate salary, and job description's general level of responsibility. Among those with longer periods of service, 87 agreed that a lie detector test would be acceptable. Of those with two years or less on the job, 102 did so. At the .01 alpha level, do the two samples differ in their willingness to take a lie detector test?

21. A county social services outreach worker conducted a study of local day care use among single parents and among parents with an adult partner in the home. Each parent had just one child and was in his or her mid to late twenties. The 50 single parents averaged 27.16 day care hours per week. The 80 participants with partners in their homes averaged 21.63 day care hours per week. The outreach worker took her respective standard deviations from known statewide figures for her two populations: 3.52 hours per week for single parents and 2.91 hours per week for those with partners. What is the .99 confidence interval estimate for the difference in average weekly day care use between these two populations of parents?

22. A researcher compared social work departments from public and private universities. Not counting electives, the 20 departments at public universities averaged 6.03 required core courses for the bachelor's degree, with a standard deviation of 1.09 courses. The core requirements at 17 private institutions averaged 5.22 courses, with a standard deviation of .94 courses. At the .05 alpha level, is the average number of core courses required for undergraduate social work degrees at private universities significantly less than that required at public institutions? What about at the .01 alpha level?

23. In a particular tri-county metropolitan area, more than one thousand domestic violence offenders (mostly male) had undergone a court-ordered one-year treatment and counseling program during the last five years. Researchers were able to isolate 252 cases of repeat offenders having charges against them both before and after the treatment program, and had access to Department of Probation records for all 252 individuals. Before the treatment program, alcohol was involved in 61% of the sample's domestic violence incidents. Following treatment, alcohol was involved in 70% of all repeat incidents. What is the .95 confidence

interval estimate for the percentage difference in alcohol-related incidents before and after the intervention? At the .01 alpha level, is alcohol significantly more likely to be involved in domestic violence incidents following exposure to the treatment and counseling program?

24. A national realty company compared sales agents in two large regional offices serving similar markets. Fifteen agents from Office A were generally matched for age and experience with 15 agents from Office B. Given the data below, do the number of sales made over a month's time in the two offices differ significantly at the .05 alpha level?

Pair	A	B	Pair	A	B	Pair	A	B
1	4	3	6	3	0	11	2	4
2	5	5	7	5	3	12	0	2
3	2	1	8	4	1	13	3	5
4	3	4	9	3	2	14	4	2
5	1	2	10	3	3	15	3	1

25. In your own words, what is the sampling distribution of the difference between two sample averages? Two sample proportions?

26. What is the pooled variance? Why do we sometimes pool the variances from two samples?

27. What is the difference between pairs of independent versus related samples? Please give original examples of research situations involving each.

CHAPTER 7

Exploring Ranks
and Categories

In this chapter, you will learn how to:

- Test hypotheses using categorical data
- Test hypotheses using nominal and ordinal data
- Use measures of association to assess the strength of relationships between pairs of variables

WHAT ABOUT INSTANCES IN WHICH the data are *not* interval-ratio or continuous or for which there are more categories than can be legitimately reduced to a dichotomy? What about ordinal and nominal data? For these cases, we should not calculate averages, although it may at times be possible to calculate p and q values. We may still, however, conduct hypotheses tests using data ranks or category frequencies. Some of these tests use the ranks among a set of scores rather than the actual X values themselves. Others use the numbers of cases or frequencies falling into each category of the variable(s). When given rankable numerical scores from independent samples, we compare two groups using the Wilcoxon rank-sum test or its close companion, the Mann-Whitney U test. Do the data ranks of the respective sample groups differ significantly? With categorized data, either nominal or ordinal, we use a popular statistic known as chi (pronounced *ki*) square and measures of association. Do category frequencies differ significantly from those expected by random chance?

Although these tests differ in the kinds of data used, they do share a fundamental feature: We do not use known population parameters such as μ_O or p_O, in the tests themselves. Even though we may assume populations 1 and 2 are similar, we make no other assumptions about their general features, normality, or skew. For these tests, we rely strictly on statistical theory, probability, and random chance, not on known population parameters; hence they are called **nonparametric.**

Tests for Ranks

The **Wilcoxon rank-sum** test and the **Mann-Whitney U test** *allow us to test for differences between two samples when we have ordinal (rankable) data.* These procedures do not require normally distributed populations. As the "rank-sum" in the Wilcoxon test's name suggests, these procedures use the ranks of the original scores rather than the X scores themselves. For this test, we rearrange our X values in ascending (or descending) order and assign a rank to each one (e.g., rank #1 for the highest score). Our calculations then use the ranks rather than the initial X values themselves. This is necessary when we have ordinal data and possibly some imprecision in our original measurements, such as when we have sets of numbers possibly subject to interpretation, somewhat lacking in precision, and in which we do not have the confidence reserved for true interval-ratio measurements. (Recall the discussion of the level of measurement in Chapter 1.)

The Wilcoxon and Mann-Whitney tests are very similar. There is a minor difference in procedure, but both use independent samples, combine the scores of those samples to get one set of overall ranks, and then test to see whether the sums of the samples' respective ranks differ. The tests do permit samples of different sizes, but, allowing for n_1 and n_2 differences, the null hypothesis assumes that the sums of the ranks from the two samples are about equal. Upper- or lower-tailed tests are possible if we hypothesize that one sample should have significantly higher or lower ranks than the other. Both tests are typically conducted using z, although small-sample versions are available, as we shall see presently. Given the tests' similarities, we will first look at the large-sample case using the Wilcoxon test, and then at the small-sample case illustrating the Mann-Whitney procedure.

Large Samples: Wilcoxon's Rank-Sum Test

For larger samples, when n_1 and n_2 are each 10 or more, we sum the ranks, get a mean and standard error for the sampling distribution of the summed ranks, and finally conduct a z test. Scores from the two samples are pooled, and one overall set of ranks is assigned, typically from the highest score to the lowest. We separate the two groups again, and sum the ranks of the smaller sample, n_1, obtaining a value designated as W. If both samples are the same size, either may be used for n_1. Also, if we have tied or duplicate X scores, we give each the average rank of all ranks they would otherwise occupy. For example, with ordinal measurements of 48, 43, 43, and 40, we assign rank #1 to 48. The values of 43 occupy ranks 2 and 3, but we have no justification for making one of them 2 and the other 3. We therefore give them each the rank of 2.5. With ranks 2 and 3 accounted for, the score of 40 assumes rank #4.

One caution is that statistical operations may be distorted with large numbers of tied scores and shared ranks. This is a common problem with statistics using ranked data. Therefore, avoid using such tests when 20% or more of the original scores are part of tied or duplicated values. Assuming each n is at least 10 and that less than 20% of our scores are tied with other scores, however, we first get W by using equation 93:

$$(93) \qquad\qquad W_1 = \Sigma \ (\text{Ranks of } n_1),$$

where n_1 is the smaller of the two independent samples.

Then, for the mean and standard error of the sampling distribution of the ranked sums, we get equations 94 and 95, respectively:

$$(94) \qquad\qquad \mu_{W_1} = \frac{n_1(n_1 + n_2 + 1)}{2}$$

$$(95) \qquad\qquad \sigma_W = \sqrt{\frac{n_1 n_2 (n_1 + n_2 + 1)}{12}}$$

And finally, z-observed is found from equation 96:

$$(96) \qquad\qquad z_O = \frac{W_1 - \mu_{W_1}}{\sigma_w}$$

In the now-familiar procedure, z_O is compared to z_C and the null hypothesis either rejected or retained.

The hypotheses may be one- or two-tailed. We are comparing the summed ranks of the smallest sample to a population figure that, under H_O, assumes the summed ranks of both samples should be about the same—that each sample's ranks should add to about the average figure for *all* the ranks. We test the smallest sample's ranks against this average-for-all-possible-ranks, our hypothetical figure under H_O. There are variations in just how we express our hypotheses, of course, but in symbolic form and for two-tailed tests, we write: H_A: $W_1 \neq \mu_w$; H_O: $W_1 = \mu_w$.

Similarly, one-tailed tests express an expected directional difference. If we expect our smallest sample to include X values assigned the lowest numbered ranks (e.g., ranks 1, 2, 3, etc.)—even though they may be the highest X scores—we conduct a lower-tailed test: *we expect the sum of that sample's ranks to be less than the average rank sum.* A lower-tailed test includes directional hypotheses: $H_A: W_1 < \mu_W; H_O: W_1 \geq \mu_W$.

Conversely, if we expect the X values of the smaller n_1 to get the highest-numbered ranks, we formulate hypotheses suitable for an upper-tail test: $H_A: W_1 > \mu_W; H_O: W_1 \leq \mu_W$.

1. *A survey regarding sociocultural and ethnic relations on campus asked several questions about feelings of acceptance in classes; treatment by professors and by peers; and receptions accorded minority viewpoints, dress and apparent lifestyles, and sexual orientations. From a larger random sample, we abstracted the responses of students identifying themselves as gay, lesbian, bisexual, or transgender (G-L-B-T). Among these respondents (n = 23), 13 further identified themselves as white and 10 as other than white, or minority. With opinions measured on 5-point agree/disagree scales and with feelings of acceptance positively scored and ranging from an overall high of 35 to a low of 7 (over seven items), white and minority G-L-B-T students' responses were compared. At the .05 alpha level, do minority G-L-B-T students report significantly more problems regarding acceptance on campus than their white G-L-B-T counterparts?*

For minority G-L-B-T students (n_1 = 10):

$$X_1 = 35 \quad 15 \quad 23 \quad 25 \quad 21 \quad 27 \quad 13 \quad 12 \quad 20 \quad 11$$

For white G-L-B-T students (n_2 = 13):

$$X_2 = 26 \quad 20 \quad 10 \quad 28 \quad 18 \quad 34 \quad 31 \quad 22 \quad 19 \quad 21 \quad 24 \quad 14 \quad 29$$

As with other tests, we first diagnose and set up the problem. Each n is at least 10, so the large-sample Wilcoxon procedures apply. Our hypotheses reflect the expectation that minority students will have lower perceived-acceptance scores than white students. But the higher the score, the lower the numbered rank: the highest acceptance score ranks 1 and the lowest ranks number 23 (or the $n_1 + n_2$th rank). We would therefore expect the minority students' *ranks* to be numbered higher. Since minority G-L-B-T students constitute the smaller sample, or n_1, we expect the sum of the n_1 ranks to be *greater than* the average of the summed ranks. This suggests an upper-tailed test with $z_C = 1.645$ at the .05 α level: The hypotheses are therefore $H_A: W_1 > \mu_W; H_O: W_1 \leq \mu_W$.

The data from both samples are pooled and ranked as follows (Table 7.1):

Table 7.1

Pooled X Scores	Race/ Ethnicity	Rank	Pooled X Scores	Race/ Ethnicity	Rank
35	M	1	21	W	12.5
34	W	2	20	M	14.5
31	W	3	20	W	14.5
29	W	4	19	W	16
28	W	5	18	W	17
27	M	6	15	M	18
26	W	7	14	W	19
25	M	8	13	M	20
24	W	9	12	M	21
23	M	10	11	M	22
22	W	11	10	W	23
21	M	12.5			

We are thus able to calculate the parameters to test our hypotheses:

$$W_1 = \text{sum of minority students' ranks} =$$
$$1 + 6 + 8 + 10 + 12.5 + 14.5 + 18 + 20 + 21 + 22 = 133$$

$$\mu_{W_1} = \frac{n_1(n_1 + n_2 + 1)}{2} = \frac{10(10 + 13 + 1)}{2} = \frac{10(24)}{2} = \frac{240}{2} = 120$$

$$\sigma_W = \sqrt{\frac{n_1 n_2 (n_1 + n_2 + 1)}{12}} = \sqrt{\frac{10(13)(10 + 13 + 1)}{12}} = \sqrt{\frac{3120}{12}} = \sqrt{260} = 16.125$$

$$z_0 = \frac{W_1 - \mu_{W_1}}{\sigma_W} = \frac{133 - 120}{16.125} = \frac{13.000}{16.125} = .81$$

Figure 7.1 depicts the distribution curve for problem 1.

Since $z_c = 1.645$ at the .05 alpha level, with $z_0 = .81$ we have insufficient evidence to reject H_0. Minority students' rankings are not higher than those of white students, meaning their feelings of acceptance on campus are not appreciably lower. We conclude that among a sample of gay, lesbian, bisexual, and transgender students, race/ethnicity does not affect perceptions of acceptance in the classroom and on campus.

Smaller Samples: The Mann-Whitney U Test

While small, the samples in problem 1 meet the minimum size ($n_{1,2} \geq 10$) for using the normal curve, but that is not always the case. Both the Wilcoxon and

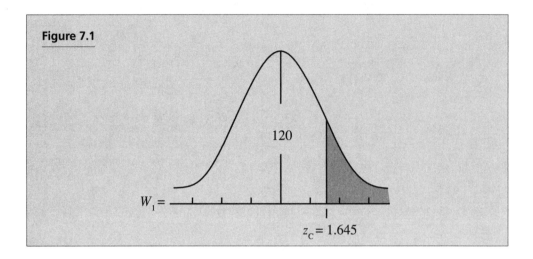

Figure 7.1

Mann-Whitney U tests offer small-sample alternatives; here we focus on the latter. We calculate U values based upon our samples' sizes and the sums of ranks in our two samples. For U-observed:

(97) $$U_O = \text{the smaller of} \begin{pmatrix} n_1 n_2 + \dfrac{n_1(n_1+1)}{2} - W_1 \\[2em] n_1 n_2 + \dfrac{n_2(n_2+1)}{2} - W_2 \end{pmatrix}$$

Again, W refers to the sum of the samples' respective ranks after we have pooled all the X scores and determined one common set of overall ranks. We then compare the smaller figure for U_O to a critical value by using Table D. If our observed value U_O is equal to or less than the value in Table D, we reject our null hypothesis.

2. *Thirteen police sergeants are being groomed for possible promotions. As part of their overall evaluation process, each candidate's job performance is rated by supervisors on a 0–100 scale, with higher scores representing more favorable reviews. At the .01 alpha level, do the ratings given female and male sergeants differ significantly?*

Females' ratings ($n_F = 5$): 73 87 94 66 80
Males' ratings ($n_M = 8$):   71 96 81 63 77 88 60 91

To set up the problem, since it is doubtful the supervisors' evaluations represent interval-ratio measurements, we turn to the Mann-Whitney procedure. Our samples are also too small (<10) to justify using the normal curve, so we will compare U_O to the critical value (U_C) rather than using a z test. In Table D at the .01 alpha level, we find a critical value of 4 where the $n_F = 5$ and $n_M = 8$ entries intersect. Our hypotheses and test are two-tailed because we are asked whether females' and males' ratings differ, not whether one tends to rank higher than the other. Our hypotheses are: H_A: $U_O \neq \mu_U$; H_O: $U_O = \mu_U$, where μ_U is the mean of the ranks based upon the sample sizes and the number of ranks possible. This expected mean is reflected in the critical values of U (or U_C) we looked up by using n_F and n_M in Table D.

For the actual procedures, all the scores are first pooled, ranked, and then separated into female and male ranks to calculate W_F and W_M:

Score (X):	96	94	91	88	87	81	80	77	73	71	66	63	60
Female/Male:	M	F	M	M	F	M	F	M	F	M	F	M	M
Rank:	1	2	3	4	5	6	7	8	9	10	11	12	13

$$\Sigma \text{ (Female ranks)} = W_F = 2 + 5 + 7 + 9 + 11 = 34$$

$$\Sigma \text{ (Male ranks)} = W_M = 1 + 3 + 4 + 6 + 8 + 10 + 12 + 13 = 57$$

Equation 97 for U_O calls for the smaller of two values:

$$n_F n_M + \frac{n_F(n_F + 1)}{2} - W_F = 5(8) + \frac{5(5+1)}{2} - 34 = 40 + 15 - 34 = 21$$

$$n_F n_M + \frac{n_M(n_M + 1)}{2} - W_M = 5(8) + \frac{8(8+1)}{2} - 7 = 40 + 36 - 57 = 19$$

Therefore, $U_O = 19$, and we see from Table D that $U_C = 4$ at the .01 α level. Since $U_O > U_C$, we may not reject H_O. The evaluations given to female and male promotion candidates do not differ significantly when $\alpha = .01$.

One final note regarding the Mann-Whitney procedure: one-tailed tests are possible, and their hypotheses are similar to those for the one-tailed Wilcoxon rank sum test. For a lower-tailed U test, we have: H_A: $U_O < \mu_U$; H_O: $U_O \geq \mu_U$. For the upper-tailed alternative, we get: H_A: $U_O > \mu_U$; H_O: $U_O \leq \mu_U$.

When we have two independent samples and ordinal, numerical measurements, we may test for differences using the Wilcoxon rank sum test or the Mann-Whitney U test. With related samples, we would turn to the **sign test.** We encounter data suitable for the sign test less frequently, however, and it is not covered here. Instead, we now turn to a widely used statistic: chi square. Chi square is used to answer questions such as: Do sample frequencies differ from random chance? Are two variables independent or statistically related to each other?

Frequencies, Random Chance, and Chi Square

The chi square test differs from the previous hypothesis tests in that it uses category frequencies and both ordinal and nominal data, but there are also similarities. For example, we have an alpha level and conduct the test in terms of probabilities. We have null and alternative hypotheses. And just as before, we test and reject or fail to reject H_O. We calculate chi square-observed (χ^2_O) and compare this to a chi square-critical value (χ^2_C) taken from Table E. The critical value is derived by using degrees of freedom. The difference we test, however, does not involve ranks or population parameters.

With chi square testing, we work with the differences between observed frequencies (O) and expected frequencies (E). **Observed frequencies** *are the actual category frequencies in the data; they are the frequencies we observe in our sample.* **Expected frequencies** *are those we would expect to get in our categories based upon random chance or according to a criterion such as normality (a normal distribution).* For example, how many cases *should* we have in each category if only random chance determines the frequencies? Or if our distribution is normal? We calculate a series of observed minus expected frequency differences ($O - E$), one for each category, and these differences become the basis for our chi square-observed statistic (χ^2_O).

A brief and simple example may illustrate the comparison of expected and observed frequencies. For any 100 flips of a normal coin, the laws of probability suggest 50 heads and 50 tails. According to random chance, these are the *expected frequencies.* Assume we actually do flip a coin 100 times and get 60 heads and 40 tails. These are the *observed frequencies,* the numbers we observed in our sample of 100 flips. The chi square statistic uses the difference between the observed and the expected frequencies: Do they differ enough in this example to be statistically significant? Does our coin deviate *significantly* from random chance at the .05 alpha level? What about at the .01 level? What if the frequencies are 52 heads and 48 tails, or 55 of one and 45 of the other? Are those differences from 50-50 significant? Do the frequencies differ from chance rather than from a population parameter?

Like the Wilcoxon, Mann-Whitney, and sign tests, the chi square test is non-parametric because it does not rely upon or include a population parameter. We compare data to what we would expect by random chance or to some hypothetical distribution of data suggested by a particular hypothesis or assumption. We are not comparing our sample data to a known parameter. We are instead putting our sample data side-by-side with what might have occurred by random chance or according to some other criterion (e.g., what we might get if a population is normally distributed). This feature of chi square is reflected in its hypotheses.

Chi square's hypotheses are somewhat different from those of earlier examples. We do not consider one- or two-tailed tests or upper- and lower-tailed tests with chi square. Our hypotheses are always the same, and they always address the observed versus expected differences. The null hypothesis is consistent with previous examples and states: There is no difference between our observed and expected frequencies. According to H_0, observed category frequencies do *not* differ significantly from those expected by random chance or according to an external criterion we have established. It says, in essence, there is nothing going on, there is no factor operating that makes our observed frequencies differ significantly from the expected numbers. The null hypothesis is always written as "$O = E$" or that observed and expected frequencies are equal—at least within the bounds of normal variation. The alternative hypothesis states just the opposite. According to H_A, there *is* something going on, there *is* a factor or variable skewing the observed frequencies away from random chance. It says the observed and expected frequencies do differ significantly, and it is always written as "$O \neq E$."

There actually are two chi square tests. Both incorporate the general logic above and use $O - E$ differences. The **one-way chi square test,** sometimes used as a **goodness-of-fit test,** *tests for significant differences among the category frequencies of a single variable.* The **two-way chi square test,** more commonly called the **chi square test of independence,** *tests for an association between two variables.* Are two variables statistically related or independent of each other?

Chi Square with One Variable

This test starts with the frequency distribution for a single categorical variable. Our purpose is to see whether the category frequencies are significantly different. Do some categories appear to be *over-chosen* and some *under-chosen,* so to speak, at least in terms of what we would expect? Are some category frequencies significantly greater than expected and some significantly smaller?

We first calculate chi square-observed, χ^2_O. There are several steps in this procedure; we go through them one by one here and eventually put them all together in a sample problem:

• Step 1. We start with the observed category frequencies, which come from the sample data. They are the category counts or tallies observed in our sample.

• Step 2. We calculate the expected frequencies.

(98)
$$E = \frac{n}{k}$$

where n is the total sample size, and k is the number of categories. On occasion, the expected frequencies may be determined by a particular hypothesis regarding ideal or normal numbers, as in the goodness-of-fit test. In the one-way chi square, however, all the expected frequencies are the same. How many cases do we have, and how many categories are there?

• Step 3. For each category, we subtract the expected frequency (E) from each observed frequency (O). The sum of the $O - E$ differences should always be zero, allowing for rounding error. (We encountered a similar situation before for deviations around the mean. The positive and negative deviations around the mean canceled each other out, and the sum of any $X - \overline{X}$ differences was always zero.) For $O - E,$ we have both positive and negative quantities. However, the $O - E$ differences must cancel each other out over all the categories. After all, we have the same finite number of cases (n) spread over both the expected and observed distributions. If the $O(f)$ is larger for some categories, then the $E(f)$ must necessarily be larger for others.

• Step 4. We square the $O - E$ differences: $(O - E)^2$. This is the equivalent of $(X - \overline{X})^2$ when we calculated standard deviations. Since a column of numbers always adding to zero did not give us a practical measure of spread around the mean, we squared the deviation scores. Because our $O - E$ values always total to zero, they also do not give us a usable measure of how much the observed and expected distributions differ. So we resolve this by squaring the $O - E$ differences.

• Step 5. Divide the $(O - E)^2$ values by the expected frequencies: $(O - E)^2 /E$. We do this because we do not want χ^2_o to unduly reflect our sample size. χ^2_o is based upon $O - E$ differences. It stands to reason that, with larger sample sizes, we have potentially larger $O - E$ values. For instance, with a sample of 50 cases spread across several categories, the individual $O - E$ differences would be small. In contrast, with a sample of several hundred, the individual $O - E$ differences could be quite large. We want to control against getting a significant chi square merely due to a large sample. Statistical significance should reflect legitimate differences between the observed and expected distributions, not differences inflated simply by large samples and calculations involving big numbers. To control for this, we divide $(O - E)^2$ by the expected frequencies. The $E(f)$'s reflect the sample size. The larger the sample size, the larger the $E(f)$. In a sense, we divide by $E(f)$

to bring the $O - E$ differences back into perspective. We normalize them, in a way, so they will not unfairly reflect our sample size.

• Step 6. Calcluate χ^2_O, given here by equation 99:

(99) $$\chi^2_O = \Sigma \left(\frac{(O - E)^2}{E} \right) = 5.24; \ \chi^2_C = \ 9.49$$

Before making our decision regarding H_O, we need one final bit of information, and that is our chi square-critical value: χ^2_C. To compare χ^2_O to a χ^2_C value, we need the degrees of freedom. When we worked with t curves, our degrees of freedom reflected the sample size $(n - 1)$. With the one-way χ^2 statistic, our degrees of freedom are based upon the number of categories, or k. Instead of having $n - 1$ individual scores free to vary, we now have $k - 1$ category frequencies:

(100)   Degrees of freedom = d.f. = $k - 1$

Table E lists the critical values for chi square. We find χ^2_C where our degrees of freedom and alpha level intersect. For the chi square test, if χ^2_O is equal to or greater than χ^2_C, we reject our null hypothesis. Otherwise, we fail to do so. The chi square distribution and H_O curve are illustrated in Figure 7.2.

Some features of the curve are reminiscent of the z and t curves, whereas other features are clearly different. The vertical axis and height of the curve still depict relative frequency and probability. The higher the curve, the greater the probability that particular $O - E$ differences will occur. There is still an alpha area and a critical value. We still reject H_O when χ^2_O falls into the alpha region, but some aspects of the curve are new. The horizontal axis now represents the $O - E$ differences as measured by χ^2_O. That axis starts at zero, where the observed and expected frequencies would be identical. That is unlikely, and the curve is low at that point. The curve peaks where the $O - E$ differences would be normal or expected and due merely to random chance. This is the area of normal variation, where the observed and expected frequencies vary but do so within a normal range. The tail of the curve, in contrast, represents the extreme situation in which $O - E$ differences are comparatively large, unusual, and statistically significant. Here, the observed frequencies differ markedly from random chance or from the frequencies expected. The χ^2 curve generally assumes the shape shown in Figure 7.2, although it does tend to become more normal and symmetrical as n increases.

There is more to our analysis when we reject our H_O. We are obligated to describe which categories appear to be "over-chosen" $(O > E)$ and which appear

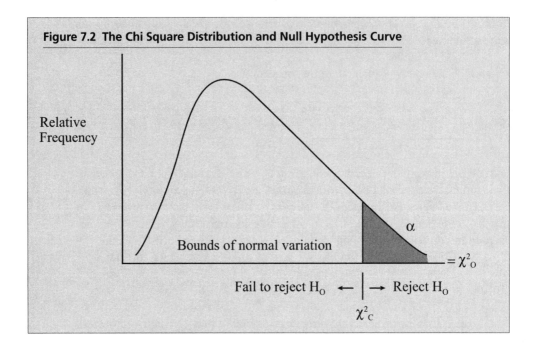

Figure 7.2 The Chi Square Distribution and Null Hypothesis Curve

to be "under-chosen" ($O < E$). We examine each category's $O - E$ differences to determine in which categories we appear to have significantly more cases than expected, and in which we have noticeably fewer, as well as where the $O - E$ discrepancies are the greatest, either positive or negative. This type of analysis is not required with the z and t tests because for those tests our results are quite clear and unambiguous. With χ^2, however, statistical significance simply tells us there are $O - E$ differences *somewhere;* we do not know exactly where just by virtue of rejecting H_O. We have to go back to our original data and look for $O - E$ patterns. We might naturally have a particular interest in the over-represented categories: What seem to be the significantly favored or chosen responses? Which have unusually high frequencies or counts? We then describe these patterns in a paragraph to accompany our claim rejecting the null hypothesis.

3. *The contract between a teachers' union and the local school district is about to expire, and negotiations on a new contract are not going well. The union's leaders decide to ask a random sample of teachers what should be the foremost issue in contract talks—the one absolutely nonnegotiable demand. The results of the responses from 105 teachers are shown below. At the .05 alpha level, do the frequencies with which teachers have chosen certain nonnegotiable issues differ significantly from random chance?*

Salaries: *f = 29 votes*   *Tenure, promotion policies:*   *f = 18*
Fringe benefits: f = 21   *Classroom working conditions: f = 22*
Other: *f = 15*

Our null hypothesis for chi square always states that, allowing for normal or random error, the observed and expected frequencies do not differ: The alternative and null hypotheses may be written out in more extended form but are usually abbreviated as: $H_A: O \neq E$; $H_O: O = E$.

With five response categories ($k = 5$), we have $k - 1 = 4$ degrees of freedom (see equation 100). From Table E, χ^2_C is 9.49 at the .05 alpha level. The distribution curve is shown in Figure 7.3.

The calculations to determine χ^2_O for this problem are presented in Table 7.2.

From equation 99, we get the following value for χ^2_O:

$$\chi^2_O = \Sigma \left(\frac{(O - E)^2}{E} \right) = 5.24; \ \chi^2_C = 9.49$$

Because $\chi^2_O < \chi^2_C$, we have insufficient evidence to reject H_O at the .05 alpha level. The observed frequencies do not differ significantly from random chance. No demands (categories) were chosen with significantly greater or lesser frequency than others. This makes things a little more difficult for the union negotiators because the teachers appear not to have any statistically significant priorities among the issues.

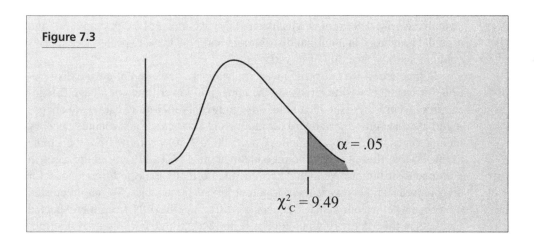

Figure 7.3

$\alpha = .05$

$\chi^2_C = 9.49$

Table 7.2

Contract Issue	(O)	Obs f (E)	Exp f (O – E)	(O – E)²	(O – E)²/E
Salaries	29	21	8	64	3.05
Tenure/promotion policies	18	21	–3	9	.43
Fringe benefits	21	21	0	0	0.00
Classroom working conditions	22	21	1	1	.05
Other	15	21	–6	36	1.71
Totals	$\Sigma(O) = 105$	$\Sigma(E) = 105$			$\Sigma(O - E)^2/E = 5.24$

Before moving on to the chi square test of independence, the one-way chi square test has another fairly common application, at least when we have interval-ratio measurements. That is the goodness of fit. The calculations and procedures are no different, but the test does serve a slightly different function.

A Goodness-of-Fit Test: Are the Data Normal?

We sometimes use chi square to compare our categorized frequencies to a particular set of hypothesized frequencies or to those of a normal distribution. In other words, we test whether a distribution differs significantly from normality. Each data category has a certain observed frequency (O). The expected frequencies (E) become the numbers likely to occur in each category in the case of a normal distribution, and we determine them by consulting the normal curve table. Essentially, we are determining what percentage or number of cases should fall into each data category in a normal distribution, and how these expected frequencies compare to those actually observed.

In practice, if we know the mean and standard deviation of an age distribution, for instance, we may translate the upper and lower limits of all age categories into z scores. (Again, recall that this requires interval-ratio data if we are going to use an average, a standard deviation, and z scores.) If a particular category includes ages with z scores ranging from 1.00 to 1.50, we know from the z table (Table A) that this category should contain about 11% (.1119) of all the cases in a normal distribution. We then determine the normally expected frequencies for all categories and compare them to the observed frequencies. We are interested in seeing whether our actual frequencies differ significantly from the expected numbers at the .05 or .10 alpha levels. If not, we assume our observed distribution is essentially normal. (Given that we accept approximations to the normal curve, an alpha level of .01 or less might be too demanding or rigid here.) This is an important function of the one-way chi square test, sometimes called the **chi square test of normality.** Its only difference from other one-way χ^2 procedures involves the expected frequencies, as problem 4 shows.

4. *In a particular state, drivers with repeated offenses are required to pass a lengthy and difficult objective test to retain their licenses. The test produces a wide range of scores, with a mean of about 70 points and standard deviation of 20 points. Given the results below for 300 randomly selected offenders, are the test scores normally distributed? Do they depart from normality at the .05 alpha level?*

Score	f	Score	f	Score	f
< 30	13	50–69	93	90–110	49
30–49	48	70–89	87	> 110	10

Assuming interval-ratio measurements with $\mu = 70$ and $\sigma = 20$, the breakdown of test scores shown conforms to convenient segments of a normal distribution. Scores of 70–89, for example, cover the interval from the mean to about one standard deviation above the mean. Those in the 30–49 range fall between one and two standard deviations below the mean. We transpose the scores onto an *X*-axis and, by consulting the normal curve table (Table A), note the proportion of cases falling into certain intervals when the distribution is normal (see Figure 7.4).

Chi square consists of a comparison between observed frequencies and those expected or predicted by a normal curve model. Therefore, we translate the proportions of cases into expected frequencies when $n = 300$. Between 70 and about 89, for instance, we expect 34.13% of all cases: .3413(300) = 102.39 cases. We compare our actual 87 cases to this expected number of 102.39 and do the same for the other categories.

To complete the picture, our hypotheses remain the same, $H_A: O \neq E$; H_0: $O = E$, and χ^2_c is 11.07 with $k - 1 = 5$ degrees of freedom and a significance level of .05. Figure 7.5 shows the chi square distribution curve for practice problem 4.

The calculations to obtain χ^2_O are presented in Table 7.3. Therefore,

$$\chi^2_O = \Sigma \left(\frac{(O - E)^2}{E} \right) = 13.12; \ \chi^2_C = 11.07$$

Because $\chi^2_O > \chi^2_C$ there is sufficient evidence to reject the null hypothesis. At the .05 alpha level, the distribution of drivers' test scores differs significantly from normality.

Figure 7.4

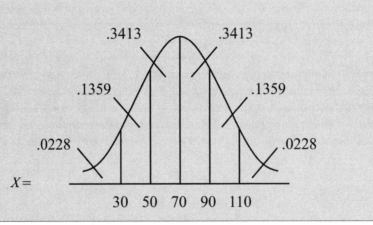

.3413 .3413

.1359 .1359

.0228 .0228

$X =$

30 50 70 90 110

Figure 7.5

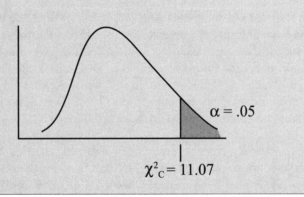

$\alpha = .05$

$\chi^2_C = 11.07$

Table 7.3

Score	(O)	Obs f (E)	Exp f (O − E)	(O − E)²	(O − E)²/E
< 30	13	6.84	6.16	37.95	5.55
30–49	48	40.70	7.23	52.27	1.28
50–69	93	102.39	−9.39	88.17	.86
70–89	87	102.39	−15.39	236.85	2.31
90–110	49	40.77	8.23	67.73	1.66
> 110	10	6.84	3.16	9.99	1.46
Totals	$\Sigma(O) = 300$	$\Sigma(E) = 300.00$			$\Sigma(O - E)^2/E = 13.12$

χ^2_O tells us that differences exist, but it does not by itself tell us where they are. We next examine Table 7.3 to note which categories appear to be over- and under-represented. Comparing the observed and predicted frequencies, there appear to be more scores than expected in the extremes of the distribution and fewer in the two central categories. Specifically, drivers scored ≤49 and ≥90 more often than expected. Despite the comparatively high frequencies between 50 and 89, fewer scores fell there than expected. Our drivers' test scores are not normally distributed.

In summary, the one-way chi square test examines the frequency distribution of a single variable. Do the category frequencies differ significantly from random chance or from other expected numbers? The chi-square statistic, however, merely tells us whether such differences exist. It does not in itself tell us where they are, that is, *which* categories are over- or under-represented. To determine this, we return to the original data and to the respective $O - E$ differences in particular. Finally, we describe any significant response patterns in terms of unusually high or low category frequencies.

The next version of chi square, with two variables, is quite similar to that with one variable. Both versions are based upon $O - E$ differences and go through the same basic steps to derive χ^2_O. There are some differences, however, because now we have two variables and wish to see whether they are statistically related to each other. Does one vary in relation to the other, or do they vary independently of one another?

Chi's Test of Independence: Are Two Variables Related?

With two variables, we work with contingency tables, or cross-tabulation tables, that we construct from the data. We wish to see to what extent responses on one variable are contingent on those of another. In other words, **contingency tables** *cross-classify the categories of two variables*. When we do so, we generate a series of cells in a table. These cells become the equivalent of the categories of the one-way χ^2. In the chi square test of independence, we calculate the $O - E$ difference for each cell.

Before introducing the chi square test of independence by way of an example, note that, despite the similarities to the one-way version of chi square, there are two important differences. First, we calculate our expected frequencies by a different method. Each cell in the χ^2 table now represents the intersection of a certain column and row. For each cell, we ask: Given the cell's column total, its row total, and the grand total, how many responses should we expect? The grand total is our sample size or n, that is, the total number of cases in our table. We use the marginal totals to get the expected frequency for a particular cell, given by equation 101:

(101) $$E(f) = \frac{(\text{Row total}) \, (\text{Column total})}{\text{Grand total}}$$

Second, the degrees of freedom are calculated differently. They are based upon the size of our table, and given by equation 102.

(102)   Degrees of freedom = (Number of rows – 1) (Number of columns – 1)

For the chi square test of independence, we are testing to see whether two variables are statistically related to each other or statistically independent. Our null hypothesis says the two variables are independent. According to H_0, our observed cell frequencies do not differ significantly from the expected or random chance frequencies. It states that there is no association between the variables to make the cell frequencies deviate significantly from random chance.

5. *A campus survey polled college students' opinions of television content. Their self-reported liberalism or conservatism was cross-tabulated with whether they believed the sexual content of TV to be excessive. The data are shown in Table 7.4. Are the two variables related or statistically independent at the .01 alpha level? If they are related, please describe the nature of their association.*

First, let's look at the elements of Table 7.4. The predictor or independent variable (IV), liberalism-conservatism, is the row variable; it runs across the rows. We suspect it will influence the dependent (column) variable (DV) opinions regarding TV's sexual content. The alternative arrangement for such tables is also possible (i.e., with the IV being the column variable and the DV running across

Table 7.4 Sexual Content Excessive by Liberalism-Conservatism

Liberalism–Conservatism	Agree	DK/NS	Disagree	Total
Liberal	[a] 287	[b] 60	[c] 193	540
Moderate/Centrist	[d] 366	[e] 60	[f] 172	598
Conservative	[g] 121	[h] 12	[i] 27	160
Total	774	132	392	1298

Note: DK/NS stands for "Don't Know/Not Sure," i.e., ambivalent or undecided.

the rows. In fact, many analysts prefer that format). The cells are lettered for easier reference. It is customary to letter them across the rows. We then calculate the row, column, and grand totals. For χ^2_O, we calculate the $O - E$ difference for each cell, a through i. The cells contain our O values. We then use equation 101 to derive each expected frequency (E)—and each is likely to be different this time. For example, the expected frequencies for cells a and e are:

$$E(f)_a = \frac{(540)(774)}{1298} = 322.00 \qquad E(f)_e \frac{(598)(132)}{1298} = 60.81$$

Now, as noted above, the degrees of freedom require a different procedure, using equation 102. They are based upon the size of the table, and the frequency table for problem 5 includes 3 rows and 3 columns. Note that we do not count the marginal totals as part of the rows and columns. The degrees of freedom here are $(3 - 1)(3 - 1) = (2)(2) = 4$, and with 4 d.f., $\chi^2_C = 13.28$ at the .01 alpha level, according to Table E.

The hypotheses remain the same as in the one-way version of χ^2: $H_A: O \neq E$; $H_O: O = E$. The distribution curve is shown in Figure 7.6.

Now, as we did for the one-way version, we calculate χ^2_O (Table 7.5). Note that the cells in Table 7.4 become the observed frequencies (O) and that the expected frequencies (E) for each cell are calculated from equation 101.

So we have:

$$\chi^2_O = \Sigma\left(\frac{(O-E)^2}{E}\right) = 27.82; \; \chi^2_C = 13.28$$

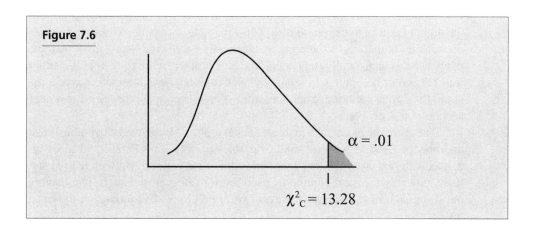

Figure 7.6

$\alpha = .01$

$\chi^2_C = 13.28$

Table 7.5

	Cell	(O)	Obs f (E)	Exp f (O – E)	(O – E)²	(O – E)²/E
Liberal	a.	287	322.00	–35.00	1225.00	3.80
	b.	60	54.92	5.08	25.81	.47
	c.	193*	163.08	29.92*	895.21	5.49
Moderate	d.	366	356.59	9.41	88.55	.25
	e.	60	60.81	–.81	.66	.01
	f.	172	180.60	–8.60	73.96	.41
Conservative	g.	121*	95.41	25.59*	654.85	6.86
	h.	12	16.27	–4.27	18.23	1.12
	i.	27	48.32	–21.32	454.54	9.41
Totals		$\Sigma(O) = 1298$	$\Sigma(E) = 1298.00$			$\Sigma(O – E)^2/E = 27.82$

We therefore reject the null hypothesis because χ^2_0 exceeds the critical value. The cell frequencies deviate from random chance at the .05 alpha level.

There is an association between the two variables, which we must now describe by examining the $O – E$ differences. It appears that liberal students tend to disproportionately disagree with the statement that there is too much sex on television, whereas conservative students tend to agree. For liberals, almost 30 more people than expected disagree. Among conservatives, almost 26 more students than expected agree regarding excessive sex on TV. (These cases are denoted by asterisks in Table 7.5.) Moderate respondents tend to agree and disagree as random chance would suggest. These tendencies account for our statistically significant χ^2_0.

Notice also that we describe the general *direction* of the association, but we do not address the *strength* of it. Chi square, as a hypothesis test, is a yes or no statistic. Is there or is there not a statistically significant association? Are the two variables statistically dependent upon each other or are they not? We either reject H_0 or we accept it. Chi square by itself does not give us an indication of the strength of an association. Such measures of association are discussed in a later section of this chapter.

These are the basic chi square statistical tools. Chi square is a hypothesis test and may be used with either single variables or, as a test of statistical independence, with two cross-tabulated variables. It is a very useful and widely used statistic, probably the most commonly used test for categorical data. Before leaving our discussion of chi square, however, we should look at variations on the basic tests above.

Nuances: Chi's Variations

There are two elaborations often encountered with chi square. The first involves Yates' correction. We apply a correction factor when (a) we have categories or cells with very small (<5) expected frequencies; and/or (b) we have only one degree of freedom and a very small number of categories or cells. One degree of freedom means we have just two categories for a one-way χ^2 (a 2×1 situation) or only two rows and two columns (a 2×2 table) for a two-way χ^2. With a small 2×2 table, there is also a simplified formula that includes Yates' correction. Second, as noted earlier, the chi square test of independence is a yes or no statistic: we either reject H_0 or we do not. χ^2_0 indicates the existence, but not the *strength* of an association between two variables. However, χ^2_0 may be used as the basis for measures of association, statistics that do indicate the degree of association. Two such measures are phi and Cramér's V.

A Correction for Small Cases

Chi square is based upon empirical frequency counts or tallies. By their very nature, these measurements are discrete. Our data or observed frequencies are recorded in whole numbers. The curve or distribution of critical chi square values is determined mathematically and is continuous, like the z and t curves, whereas the empirical distribution is discrete. If our empirical frequencies are large, the discrepancy between the two distributions is negligible. When our frequencies are small or there are not many of them (as under 2×1 or 2×2 conditions), however, there may be some difference between the discrete empirical or observed distribution and the smooth, continuous, mathematically determined distribution of critical values. **Yates' correction,** *when used in the case of small expected frequencies or small tables, makes the empirical distribution a better approximation of the mathematically generated χ^2 curve.* (You may recall that we applied a similar correction factor when using the normal curve for binomial probabilities and had $X \pm .5$.)

Low frequencies. We apply Yates' correction to the two-way χ^2 whenever we have one or more expected frequencies less than 5. Moreover, we apply the correction factor to *all* cells, not merely those with a low $E(f)$. There is a unique formula for 2×2 tables, but with larger tables we simply subtract .5 from the absolute value of our $O - E$ differences. Amending equation 99 with the Yates' correction, we get:

(103) $$\chi^2_0 = \Sigma \left(\frac{(|O - E| - .5)^2}{E} \right)$$

Our calculations are much the same as before. Previously, once we had our $O - E$ differences, we squared them. Now, we have an intermediate step. We take the absolute value for each $O - E$ figure, $(|O - E|)$, and subtract .5. The intermediate step becomes $(|O - E| - .5)$. This is the quantity we then square and divide by the $E(f)$. Our comparison of χ^2_O and χ^2_C is just the same as before, as is the interpretation of our H_O decision.

We use equation 103 with the one-way χ^2 situation less often because there is another rule. (Isn't there always a rule?) Despite the correction factor, statisticians are very wary of circumstances in which 20% or more, or one-fifth or more, of the categories or cells suffer from low expected frequencies. But by definition, with the one-way χ^2, all the categories have the same expected frequency. (Recall this chapter's first example.) If one expected frequency is less than 5, *all* are less than 5, and that renders χ^2 inappropriate. The only exception may occur when conducting the χ^2 goodness-of-fit test in which the $E(f)$'s differ. One or two of numerous categories may have low expected or predicted numbers of cases. That would be an unusual circumstance, however.*

Typically, when low frequencies occur, we encounter them in the two-way situation and have our sample spread over a larger number of cells. This is the case in problem 6.

6. *An anonymous campus survey asked students how often they have lied to a professor about the reason for late or missed work. The responses of graduate students were cross-tabulated with gender, as shown in Table 7.6. Are the two variables statistically related at the .05 alpha level? If so, please describe the nature of their association.*

Table 7.6 Lied To Professor about Reason for Late/Missed Work, by Gender

Gender	Sometimes	Rarely	Never	Totals
Female	[a]7	[b]15	[c]25	47
Male	[d]7	[e]7	[f]11	25
Totals	14	22	36	72

---

*When we do encounter very low frequencies, another option may be **recoding** or combining categories. For example, in a Likert scale of measurement, if we have a small expected frequency for the Strongly Disagree category, we may combine the Strongly Disagree and Disagree responses into one category, thus creating a larger frequency and a more general "Disagree" measurement. Recoding has both advantages and drawbacks, but this is a debate more suited to a research methods text and course.

Our hypotheses remain the same, H_A: $O \neq E$; H_0: $O = E$, and, according to Table E, $\chi^2_C = 5.99$ with 2 degrees of freedom at the .05 alpha level. One low expected frequency (4.86 for cell d) triggers the inclusion of the correction factor. The distribution curve is shown in Figure 7.7, and the calculations for χ^2_0, including the correction factor, are presented in Table 7.7.

Thus,

$$\chi^2_0 = \Sigma \left(\frac{(|O - E| - .5)^2}{E} \right) = .97, \text{ and } \chi^2_C = 5.99$$

We are unable to reject the null hypothesis at the .05 alpha level. Female and male graduate students do not differ significantly in their tendency to lie about the reasons for late or missed work.

Small (2 × 1 or 2 × 2) tables. Yates' correction is also used when we have only one degree of freedom and a small number of categories or cells. This means just two

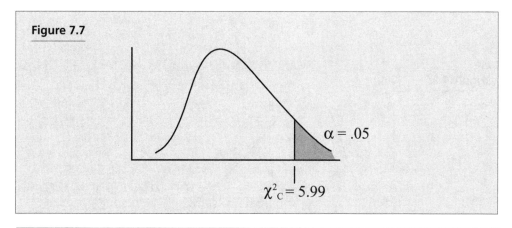

Figure 7.7

$\alpha = .05$

$\chi^2_C = 5.99$

Table 7.7

	Cell	Obs f (O)	Exp f (E)	$(|O - E| - .5)$	$(|O - E| - .5)^2$	$(|O - E| - .5)^2/E$
Female	a.	7	9.14	1.64	2.69	.2943
	b.	15	14.36	.14	.02	.0014
	c.	25	23.50	1.00	1.00	.0426
Male	d.	7	4.86*	1.64	2.69	.5534
	e.	7	7.64	.14	.02	.0026
	f.	11	12.50	1.00	1.00	.0800
Totals		72	72.00			.9743

categories for a one-way chi square or having only four cells for the two-way chi square. Equation 103 remains the procedure for the one-way chi square, as shown in practice problem 7.

7. As part of a homework assignment, a student rolls a six-sided die 75 times. The number 3 comes up 18 times. Does this frequency differ from random chance at the .01 alpha level?

Since the number 3 is but one of six numbers on the die, random chance suggests it should appear on one-sixth of the tosses. With $n = 75$, $75/6 = 12.5$ times is the hypothetical expected frequency (E). A number other than 3 should come up $75 - 12.5 = 62.5$ times, theoretically.

Our hypotheses are H_A: $O \neq E$; H_O: $O = E$, and $\chi^2_C = 6.64$ at the .01 alpha level with 1 degree of freedom (Table E). The distribution curve is shown in Figure 7.8, and the calculations for χ^2_O, including the correction factor, are presented in Table 7.8.

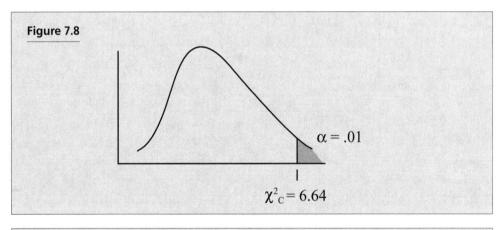

Figure 7.8

$\alpha = .01$

$\chi^2_C = 6.64$

Table 7.8

Number	Obs f (O)	Exp f (E)	(|O − E| − .5)	(|O − E| − .5)²	(|O − E| − .5)²/E
3	18	12.5	5.00	25.00	2.00
Not 3	57	62.5	5.00	25.00	.40
Totals	6	75	75.0		2.40

Therefore,

$$\chi^2_O = \Sigma \left(\frac{(|O-E|-.5)^2}{E} \right) = 2.40, \text{ and } \chi^2_C = 6.64$$

The student fails to reject her H_O at the .01 alpha level. Her outcomes are within the bounds of normal variation. Even though 3 has obviously come up more times than expected, its frequency is not enough to deviate significantly from random chance at the .01 α level. (Nor at the .05 α level, for which χ^2_C = 3.84 with 1 d.f.)

When conducting a two-way chi square test but still having one degree of freedom and working with a 2 × 2 table, a simplified formula for calculating χ^2_O allows us to use the cell and the marginal totals (the row and column totals) directly rather than having to calculate the $E(f)$'s and $O - E$ differences. In writing the formula, we refer to various frequencies by letter. For the 2 × 2 table, see Table 7.9.

Equation 104 shows the alternative procedure for χ^2_O. The correction factor remains in the numerator but instead takes the form $n/2$.

$$(104) \qquad \chi^2_O = \frac{n \left(|ad-bc| - \frac{n}{2} \right)^2}{(a+b)(c+d)(a+c)(b+d)}$$

Once we have χ^2_O, our steps are the same as before. We compare our observed value to χ^2_C taken from Table E, accept or reject our null hypothesis accordingly, and describe the nature of any significant differences revealed.

8. The survey referred to in problem 6 also asked graduate students whether they had children. The question here is whether gender correlates with having children. At the .05 alpha level, are female or male graduate students more likely

---

Table 7.9

a	b	$a+b$
c	d	$c+d$
$a+c$	$b+d$	N

to have children, or are gender and parenthood independent of each other? The data are presented in Table 7.10.

The hypotheses remain: H_A: $O \neq E$; H_O: $O = E$, and from Table E we have $\chi^2_C = 3.84$ at the .05 alpha level with 1 degree of freedom. The distribution curve is shown in Figure 7.9.

Using the alternative procedure for χ^2_O (equation 103) with the data in Table 7.10, we get:

$$\chi^2_O = \frac{n\left(|ad-bc|-\dfrac{n}{2}\right)^2}{(a+b)(c+d)(a+c)(b+d)} = \frac{72\left[|(35)(5)-(12)(20)|-\dfrac{72}{2}\right]^2}{(47)(25)(55)(17)}$$

$$= \frac{72(|175-240|-36)^2}{1,098,625} = \frac{72(65-36)^2}{1,098,625} = \frac{72(841)}{1,098,625} = \frac{60,552}{1,098,625} = .06$$

Our observed χ^2 statistic is a miniscule .06 compared to $\chi^2_C = 3.84$ at the .05 α level. We may not reject H_O. Gender and parental status are independent of

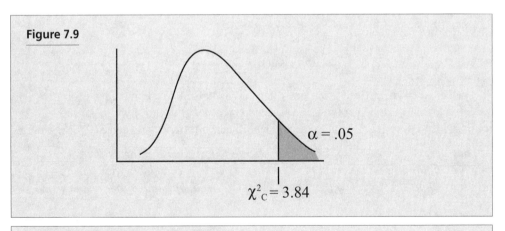

Figure 7.9

$\alpha = .05$

$\chi^2_C = 3.84$

Table 7.10 Have Children by Gender

Gender	No	Yes	Total
Female	[a]35	[b]12	47
Male	[c]20	[d]5	25
Total	55	17	72

each other among this graduate student sample. The percentages of female and male students who have children is 25.5% (12/47) and 20.0% (5/25), respectively—not a significant difference in a fairly small sample.

Whether calculated with Yates' correction or not, the two-way χ^2_O may be used to generate measures of association, statistics that give us an idea of the strength of the association between two variables.

Measures of Association: Shades of Gray

Measures of association are generally comparable but are used under different circumstances. They all vary between .00 and 1.00. The closer a measure is to 1.00, the stronger the association between the two variables. A result of 1.00 means a perfect association and predictability, or always being able to predict responses on Variable B when knowing the response on Variable A. Results of .00 mean just the opposite: there is no association between the variables whatsoever. There are guidelines for interpreting our actual outcomes between these two extremes: A result below .20 indicates a weak association. A coefficient in the .20 to .49 range is generally interpreted as a moderate degree of association. Anything higher, .50 or above, suggests a fairly strong to very strong association. There are no tables to consult for the measures of association; we merely interpret the calculated statistic.

Despite their similarities, the two chi square-based measures of association are used in different situations. The **phi coefficient,** or simply phi (φ), *is used only with 2 × 2 tables.* χ^2_O using Yates' correction and derived from a 2 × 2 table becomes the basis for the phi coefficient. **Cramér's V** *may be used with χ^2_O values based upon tables of any size.* We will look briefly at each of these measures, using chi square results from previous examples.

We typically use a measure of association only when chi square is statistically significant. There is little point to measuring the strength of a nonexistent association, as when χ^2_O is not significant. We *may* include a measure of association to confirm the absence of a relationship between two variables, but we routinely wish to assess the degree of association only when things are statistically significant. We will break that rule here in the interests of demonstrating phi, but Cramér's V, or simply *V*, will clarify earlier significant results.

2 × 2 Tables: Phi (φ). To review, the phi coefficient varies between .00 and 1.00 and is used exclusively for 2 × 2 tables. It is simply the square root of χ^2_O divided by the sample size, *n* (equation 105):

(105)
$$\varphi = \sqrt{\frac{\chi^2_O}{n}}$$

Problem 8 looked at a sample of graduate students ($n = 72$) and the possible association between gender and whether respondents had children, a 2×2 cross-tabulation. χ^2_{o} was a nonsignificant .06, very low. Nevertheless, if we take the additional step of calculating φ, we get:

$$\varphi = \sqrt{\frac{\chi^2_{\mathrm{o}}}{n}} = \sqrt{\frac{.06}{72}} = \sqrt{.0008} = .029 = .03$$

This result reaffirms the lack of an association between gender and having children among graduate students. φ is very close to .00, a practically nonexistent degree of association.

Other Tables: Cramér's V. With larger tables, we substitute Cramér's V for phi. *V* uses χ^2_{o}, *n*, and the number of rows or columns (equation 106). However, since the number of rows and columns will not be equal as a rule, *V* incorporates the minimum or smaller of two values: the number of rows in the chi square cross-tabulation minus 1 ($r - 1$) or the number of columns minus 1 ($c - 1$).

$$(106) \qquad V = \sqrt{\frac{\chi^2_{\mathrm{o}}}{n(\text{Minimum of } r - 1 \text{ or } c - 1)}}$$

Cramér's V works with practice problem 5, which cross-tabulated liberalism-conservatism with responses to a question about television's sexual content. The data were presented in Table 7.4, with three rows and three columns. χ^2_{o} was a significant 27.82, which indicated that self-defined conservative students tended to disproportionately agree, and liberals to disagree, regarding TV's sexually oriented programming being excessive. The 3×3 table allows us to use Cramér's V, and with equal numbers of rows and columns, we may use either ($r - 1$) or ($c - 1$) in Equation 106. Arbitrarily here, we use ($r - 1$) in the denominator:

$$V = \sqrt{\frac{\chi^2_{\mathrm{o}}}{n(r-1)}} = \sqrt{\frac{27.82}{1298(3-1)}} = \sqrt{\frac{27.82}{2596.00}} = \sqrt{.0107} = .104$$

V tempers the earlier chi square result. On a scale of .000 to 1.000, it registers only .104. The association between liberalism-conservatism and views of TV's sexual content is significant but weak. The significant chi square may derive in part from the large sample size (almost 1300). Among that many respondents, even fairly small differences may prove statistically significant. Cramér's V helps us to put our earlier chi square results into better perspective and refines our analysis. We may treat our results as "significant" but not as conclusive as the chi square outcome alone initially suggested.

Summary

We have looked at several hypotheses tests for ordinal and nominal data. For large or small two-sample situations with numerical and ordinal or ranked measurements, the Wilcoxon rank-sum test and the Mann-Whitney U test are options. When using categorical data with single-sample situations and either ordinal or nominal measurements, chi square is a common statistic. Chi square also permits follow-up analyses in the form of measures of association, statistics offering indicators of the strength of the association between two variables.

Chapters 5, 6, and now 7 have looked at the one- and two-sample cases for hypothesis testing. What if we have more than two samples? Comparing numerous majors, how many quarters or semesters do students need to graduate, on the average? Comparing several different makes of compact cars, which get the best gas mileage? Which campus parking area averages more available spaces when you get to school, and is the difference statistically significant? Do various weight-loss programs differ in the average number of pounds lost? These kinds of questions may involve respondents from more than just one or two samples. A popular statistic for testing hypotheses involving multiple samples is the analysis of variance, commonly referred to as ANOVA. There are numerous extrapolations of ANOVA. Chapter 8 looks at its most basic version.

Exercises

1. Does the degree to which public university students are self-supporting differ from what we would expect by random chance? A survey asked students approximately what part of their monthly expenses they personally paid. The results are shown in Table 7.11. Do students' responses differ from random chance at the .05 alpha level? What about at the .01 alpha level?

Table 7.11

Proportion of Expenses Paid Personally	Frequency (f)
Over half	160
About half	123
Less than half	120

2. Following a particularly destructive hurricane season in the southeastern United States, a local official in a small midwestern town thought he should assess residents' emergency preparedness in his own community. On a 0- to 45-point scale, his 15-item questionnaire assigned higher scores to those more capable of weathering emergencies. Part of his data (shown below) permitted comparisons of preparedness ratings provided by homeowners and renters:

Homeowners (X_O, $n = 15$): 40 31 19 22 9 4 16 27 32 14 39 20 8 17 29
Renters (X_R, $n = 12$):   23 29 18 30 7 6 24 10 33 21 5   15

At the .01 alpha level, do homeowners' and renters' preparedness self-evaluations differ?

3. A confidential and random campus survey asked high school students whether they had ever driven while alcohol- or drug-impaired. Cross-tabulated with gender, their responses are as shown in Table 7.12. At the .05 alpha level, are the two variables related or statistically independent? What about at the .01 alpha level? If they are related, please describe the nature of the association and include a measure of its strength.

Table 7.12 Ever Driven While Under the Influence of Alcohol or Illegal Drugs, by Gender

Gender	Yes	No	Total
Female	211	343	554
Male	245	273	518
Total	456	616	1072

4. A survey asked college students whether they had children and whether they knew what they wanted to do after graduation. The two variables are cross-tabulated in Table 7.13. Are the variables related or statistically independent at the .05 alpha level? If related, please describe the nature of their association and include a measure of that association's strength.

Table 7.13 Know What Want to Do After Graduation, by Parental Status

Have Children	Yes	Not Sure	No	Total
Yes	20	5	0	25
No	234	138	52	424
Total	254	143	52	449

5. A poll in a large California city revealed the responses shown in Table 7.14 for years of formal education. Using the chi square goodness-of-fit test and the .01 alpha level, are the data normally distributed? (*Hint:* As a first step, create categories for years of education that conform to known segments of a normal distribution. The sample average is 14.36 years of education, with a standard deviation of 2.88 years.)

Table 7.14 Years of Formal Education Completed

Years	Frequency	Percentage	Valid Percentage	Cumulative Percentage
0	3	.3	.3	.3
2	3	.3	.3	.6
3	3	.3	.3	.9
4	2	.2	.2	1.1
6	10	1.0	1.0	2.1
7	3	.3	.3	2.3
8	11	1.1	1.1	3.4
9	8	.8	.8	4.2
10	15	1.5	1.5	5.7
11	26	2.5	2.5	8.2
12	175	16.9	17.1	25.3
13	89	8.6	8.7	34.0
14	160	15.5	15.6	49.6
15	126	12.2	12.3	61.9
16	239	23.1	23.3	85.3
17	38	3.7	3.7	89.0
18	51	4.9	5.0	93.9
19	7	.7	.7	94.6
20	55	5.3	5.4	100.0
Total	1024	99.1	100.0	
Missing	9	.9		
Total	1033	100.0		

6. Following the first midterm examinations in her senior seminar, a professor compared the essay exam scores given by each of her two teaching assistants (TAs). The tests were graded on a 0- to 50-point scale. Do the TAs' evaluations differ at the .05 alpha level?

TA_1:   45 37 40 49 33 43 38 31
TA_2:   49 39 48 28 36 46 41 34 41

7. A poll of college students asked about their biological parents' marital statuses. The condensed data are given in Table 7.15. Do the frequencies differ from random chance at the .01 alpha level?

Table 7.15

Marital Status of Biological Parents	Frequency (f)
Married and living together	280
Other	165

8. On a 50- to 100-point scale, with higher scores indicating more correct information, samples of adults and teenagers were compared as to their knowledge of current events. At the .05 alpha level, please set up a one-tailed test to determine whether one sample's scores indicate significantly greater or lesser knowledge than the other's.

Adults: 82 63 70 54 60 84 71 72 66 73 69 56 90 81 74 91 57 67
Teenagers: 64 92 59 88 80 51 76 53 68 75 65 58 83 52 61 64 71 94
           79 77 55 62 50 78

9. As part of a poll on viewing habits, young adults were asked whether they ever watched sexually explicit shows or movies on television. Do their responses, shown in Table 7.16, differ from random chance at the .05 alpha level? What about at the .01 alpha level? Please describe the nature of any differences revealed.

Table 7.16

Watch sexually explicit shows/movies on TV?	Frequency (f)
Yes	715
No	600

10. A survey on consumers' views of possible uses for recycled water included a question about the potential costs. Respondents' expressions of concern about the cost of supplying recycled water are shown in Table 7.17. Does the distribution of their views differ significantly from random chance at the .05 alpha level? If it does differ, please describe the nature of the difference(s).

Table 7.17

Level of concern over cost of RW?	Frequency (*f*)
Slight/no concern	217
Somewhat concerned	290
Seriously concerned	234
Extremely concerned	281

11. A campus survey asked students to rank their immediate family's standard of living when they were 16, with thumbnail guidelines provided, and also to indicate whether they "had selected (or would select) a major because of its potential for a financially rewarding career." The results are cross-tabulated in Table 7.18. Are the two variables related or independent at the .01 alpha level? What about at the .05 alpha level? Please describe any association and include a measure of its strength.

Table 7.18 Choose Major for Rewarding Career, by Standard of Living at Age 16

Standard of Living When 16	Yes	Not Sure	No	Total
Poor/working class	146	21	109	276
Lower middle class	143	21	126	290
Upper middle class/wealthy	82	7	57	146
Total	371	49	292	712

12. A large supermarket chain's executives evaluated local store managers in a particular district. On a 5- to 20-point basis, managers were rated on profits made, employee relations, customer service, and personal demeanor. The separate rankings given female and male managers are shown below. Do they differ significantly at the .05 alpha level?

Female managers: 17 15 10 11 12
Male managers: 18 14 9 13 8 12 16

13. Despite consumers' charges of long waits for responses, a large corporation dealing in retail sales claims that the numbers of available customer service telephone representatives are adequate to meet all demands. In fact, the company claims the numbers of available representatives tend to be normally distributed throughout the week based upon the previous several months' demands. Research consultants recently found the following numbers of service representatives available at 20 different and randomly selected times during a three-month period. At the .05 alpha level, are the numbers normally distributed, as claimed?

Representatives
available: 12 28 21 18 15 12 23 24 4 10 19 22 6 14 11 9 17 13 16 8

14. A survey asked students about their academic majors and also to describe their general views, choosing from very liberal to moderate to very conservative. The collapsed data are cross-tabulated in Table 7.19. Are the two variables related or statistically independent at the .01 alpha level? What about at the .05 alpha level? If they are related, please describe the nature of their association and include an appropriate measure of its strength.

Table 7.19 Academic Major, by Self-Report of Liberalism/Conservatism

Self-Report Liberalism/Conservatism	Applied Arts, Business, Education	Humanities, Liberal Arts	Sciences, Engineering	Total
Liberal	96	87	86	269
Moderate	161	78	103	342
Conservative	35	15	32	82
Total	292	180	221	693

15. Does US citizenship status correlate with opinions on immigration restrictions? Following the attacks on the World Trade Center and Pentagon, the subsequent debate about immigration policies, and the introduction of various national security measures, college students were asked how they felt about restricted immigration. Their opinions were cross tabulated with their US citizenship statuses, as shown in Table 7.20. Are the two variables correlated at the .05 alpha level? What about at the .01 alpha level? Please describe any association that exists and include the appropriate measure of association.

Table 7.20 Opinions on Whether Immigration Policies Should Be More Restrictive, by US Citizenship Status

US Citizenship Status	Agree	Disagree	Total
US Citizen	402	192	594
Non-Citizen	57	47	104
Total	459	239	698

16. A cardiologist was asked to rate her patients according to how well each maintained a post-heart attack regimen of diet, exercise, and medication. Her patients' maintenance or compliance scores varied along a 25- to 75-point continuum, the higher scores indicating greater adherence to prescribed routines. She compared evaluations for patients age 60 and older with ratings for those under 60 years of age. Given the data below, do the two samples' evaluations differ significantly at the .05 alpha level?

Under
  age 60: 48 73 62 50 47 38 60 43 52 36 70 65 56 58 49 59
Age 60
  or older: 72 66 55 71 51 61 32 56 53 67 45 40 74 46 54 64 68 70 57 69 63

17. A poll asked students whether they would like to see greater diversity among a university's faculty and staff. Their answers were cross-tabulated with self-reported race and ethnicity, as shown in Table 7.21. Are the two variables associated at the .01 alpha level? If so, please describe the nature and strength of their association.

Table 7.21 Preference for a More Diverse Faculty and Staff, by Self-Reported Race/Ethnicity

Race/Ethnicity Self-Report	Yes	Not Sure	No	Total
African American	20	6	1	27
Asian	97	52	4	153
Hispanic/Latino	59	17	7	83
White	71	88	32	191
Other	36	28	9	73
Total	283	191	53	527

18. In an open-ended question, college students indicated their favorite television shows, if any. They are categorized by type In Table 7.22. Do their preferences differ from random chance at the .001 alpha level? Please describe any significant deviations from random chance.

Table 7.22

Favorite Show Type	Frequency (f)
Comedy, comedy/drama	231
Animated comedy	123
Non-crime drama/adventure	213
Crime drama/adventure	103
Specialty: cooking, travel, etc.	97
News, documentaries	101
Reality	126
Sports, sports talk	105
No favorite or type	108

19. Regular users of public transportation were asked to rate commuter bus and train service. Questions asked about punctuality, comfort, reliability, fare legitimacy, perceived safety, and so on. Commuters' ratings formed a 15- to 45-point evaluation, with higher scores representing more positive appraisals. Do the bus and train ratings shown below differ significantly at the .05 alpha level?

Bus: 39 37 29 34 36 40 31 41 42 30
Train: 45 38 33 44 41 35 32 43

20. Why are the Wilcoxon, Mann-Whitney, and chi square tests referred to as nonparametric?

21. What is the difference between interval-ratio data and the ordinal data used with nonparametric tests?

22. In what ways are the Wilcoxon, Mann-Whitney, and chi square tests similar to the z and t tests of earlier chapters?

23. What does it mean when we reject the null hypothesis with the Wilcoxon and Mann-Whitney tests?

24. What is the general rationale or logic underlying the chi square test?

25. Why does the chi square distribution tend to take on a different shape than the z or t distributions?

26. How are the one-way and two-way tests different? How are they alike?

27. What are the expected frequencies in the chi square test?

28. What does it mean when we reject the null hypothesis with the chi square test?

29. Why do we sometimes use Yates' correction?

CHAPTER 8

Analysis of Variance:
Do Multiple Samples Differ?

In this chapter, you will learn how to:

- Test for significant differences among three or more sample means
- Determine which specific sample averages differ when all you know is that some of them do

THE z, t, WILCOXON, AND MANN-WHITNEY tests work well for comparing two samples. If we have three or more samples, though, we must use a procedure called the analysis of variance, or ANOVA, also known as the F test. We calculate a statistic called the F ratio and compare this figure to F-critical (F_C) taken from a table. As with z_O or t_O, if our observed F ratio is as large or larger than the table's critical value, we reject our null hypothesis. Our null hypothesis is that all the respective sample means are equal (within the bounds of normal variation). If we reject our H_O, the sample means differ significantly.

Although there are numerous and more complicated versions of ANOVA, as well as other tests for situations involving three or more samples, this chapter presents only ANOVA in its most basic form. It is far from a comprehensive review of the F tests, but you should gain a sense of how hypotheses tests involving three or more samples are conducted. We start with ANOVA's logic or reasoning, look at the actual F test procedures, and finally consider a way to decipher the results when an F test is statistically significant.

The Logic of ANOVA

The **analysis of variance,** or **ANOVA,** *is a test for the differences among sample means.* The respective samples in ANOVA are commonly referred to as **treatment**

groups. (ANOVA parlance uses the words *samples* and *groups* interchangeably.) The assumption is that we have a research design in which several groups receive different treatments or experimental conditions or have had different experiences. Even though our H_O refers to the means of the respective groups, the logic of the test and our calculations focus on variations found in the data. Overall, there is the **total variation** that exists across all samples. Without regard to individual treatment groups, how much variation exists in the data as a whole? If we pool *all* the X scores, how much variation is there from the one overall **grand mean**? Further, this total variation is partitioned into two parts or **sources.** It is the sum of two separate kinds of variation, between groups and within each group.

Between-group variation *is directly attributable to the experimental treatment or group differences.* We compare each of the samples against the others. How much do the individual treatment groups vary from each other? If the experimental treatment or independent variable has an effect, we expect differences here. Another name for this effect is **treatment variation.**

Within-group variation *is a total of the variation found within each of the separate treatment groups.* Within each treatment group or individual sample, there is variation around that sample's average. (Recall the standard deviation measuring variation around a sample's mean.) Since all the subjects in any single sample, or group, have had the same experimental conditions or share an identical characteristic, this variation is due to random individual differences, *not* to specific conditions of the independent variable. Being due to random individual factors, within-group variation is also called the **error variation.** It is variation in individual X scores for which our experimental treatment cannot account.

The delightfully simple rationale behind the *F* test is that, if our experimental treatment or independent variable has an effect, the *between-group variation will be larger than the within-group or error variation.* Our treatment, or the different experiences of the respective groups, should produce larger and more systematic differences in the X scores than should mere individual or random factors. The between-group variations should be relatively large and the within-group variations relatively small.

The *F* test uses the ratio between these two sources of variation. First, we get a measure of each kind of variation. Next, *F*-observed (F_O), the ***F* ratio,** *is the between/within ratio, or the variation between the groups divided by the variation within the groups.* If we wish to reject our null hypothesis of no difference between the groups, we naturally hope F_O is a number larger than one, and the larger, the better. The more the variation *between* our samples exceeds that *within* them, the more evidence we have of appreciable differences between the respective treatment groups. The final step of ANOVA compares F_O to an *F*-critical value found in Table F. If the between/within ratio (F_O) is statistically significant, we reject our null hypothesis. There is significantly more variation between the sample means themselves than *around them individually.*

When the F test tells us several means are different, how do we know exactly *which* sample averages vary significantly from each other? All of them or just certain ones? A subsequent statistic untangles that for us: Tukey's HSD, or the test for *honestly significant differences*.

Do Samples Differ? The *F* Ratio

The obvious way to get the F ratio is to calculate both the between- and within-group variations directly, and this may certainly be done. Actually calculating the within-group variation is a tedious process, though (like getting the standard deviation for each separate sample), so we take a shortcut and do it an easier way. We calculate the total variation and also the variation between groups. Then, by definition, the variation that exists within the respective groups must be the difference between these two other sources of variation (equations 107 and 108).

(107) Total variation = Between-groups variation + Within-groups variation

(108) Within-groups variation = Total variation − Between-groups variation

Before we get to the formulas themselves, let us recall the importance of the sum of squares presented in Chapter 2. We measured variation by getting the standard deviation. As this concept was introduced, the key component of our formula was $\Sigma(X - \overline{X})^2$, the sum of the squared deviations from the mean. Under the square root, this was the numerator of our standard deviation formula and was abbreviated then as the *sum of squares* or SS, with a hint that you would see it later on. Well, here it is. $\Sigma(X - \overline{X})^2$ measures the variation in any set of data, and we use it—or at least its equivalent—in ANOVA. We measure our sources of variation by getting the total sum of squares [SS_T], the between-group sum of squares [SS_B], and the within-group sum of squares [SS_W].

Just as we had a quick, raw-score procedure for the standard deviation (equation 4 in Chapter 2), we now use the same method for deriving the SS figures. We do not actually go through the $(X - \overline{X})$ steps. Equation 8 for the standard deviation used the terms ΣX and $(\Sigma X)^2$ for the sum of squares, and ANOVA uses these same expressions. So for the total sum of squares, we have:

(109)
$$SS_T = \Sigma X^2 - \frac{(\Sigma X)^2}{n}$$

For the sum of squares between groups, where k = the number of treatment groups, we have:

(110) $$SS_B = \left(\frac{(\Sigma X_1)^2}{n_1} + \frac{(\Sigma X_2)^2}{n_2} + \cdots + \frac{(\Sigma X_k)}{n_k} \right) - \frac{(\Sigma X)^2}{n}$$

And for the sum of squares within the groups, we have:

$$(111) \qquad \text{SS}_W = \sum X^2 - \left(\frac{(\sum X_1)^2}{n_1} + \frac{(\sum X_2)^2}{n_2} + \cdots + \frac{(\sum X_k)^2}{n_k} \right)$$

Similar to the relationship given by equation 108, for the sum of squares, we have:

$$(112) \qquad \text{SS}_W = \text{SS}_T - \text{SS}_B$$

As a final note, in getting our F_o ratio and once we have our SS_B and SS_W figures, we derive measures called the **mean squares.** We get the mean square between groups (MS_B) and the mean square within groups (MS_W). In a way, these are the averages for our respective sums of squares.

We want to answer the question: On the average, what is the squared deviation between groups versus the squared deviations within groups? To get these average SS figures, we divide the SS_B and SS_W values by their respective degrees of freedom. For the SS between groups, we have $k - 1$ degrees of freedom, the number of groups minus 1 (equation 113).

$$(113) \qquad \text{Degrees of freedom for } \text{SS}_B = k - 1,$$

where k is the number of treatment groups. The degrees of freedom associated with the SS within groups is $n - k$, our total n minus the number of groups (equation 114).

$$(114) \qquad \text{Degrees of freedom for } \text{SS}_W = n - k,$$

where k is again the number of treatment groups, and n is the total number of cases over *all* samples or groups. Next, for the mean square between groups, we have equation 115:

$$(115) \qquad \text{MS}_B = \frac{\text{SS}_B}{k - 1}$$

And for the mean square within groups, we have equation 116:

$$(116) \qquad \text{MS}_W = \frac{\text{SS}_W}{n - k}$$

To complete this stage of ANOVA (and, yes, there is an eventual end), our observed F ratio (F_o) is the MS_B divided by the MS_W:

(117)
$$F_O = \frac{MS_B}{MS_W}$$

There are two different degrees of freedom associated with our F ratio. The numerator (MS_B) has $k - 1$ degrees of freedom, and the denominator (MS_W) has $n - k$ degrees of freedom. When we look up our critical value (F_C) in Table F, we use both. The table cross-classifies the d.f. for the numerator and that for the denominator. We find the F_C value where these two d.f. figures intersect. To interpret the F ratio, if F_O is equal to or greater than F_C, we reject the null hypothesis. If we do so, we conclude the group means differ significantly at our particular alpha level. Our hypotheses are always: H_A: Any two or more means are different; H_0: $\mu_1 = \mu_2 = \mu_k$.

When we have a significant F ratio and reject H_0, we know statistically significant differences are present between the group means, but not where such differences lie. When we find a statistically significant F ratio, we must re-examine the group means and describe the differences that have produced this result. Specifically, we are asking *which* group means are statistically different and hence mainly responsible for the significant F ratio and our rejecting H_0. In contrast, what group means are comparatively similar and not significantly different?

Which Averages Differ? Tukey's Test

When the F ratio is significant, our group or sample means differ at our particular alpha level. As with chi square in Chapter 7, we must then discover where the true differences lie. For chi square, it was a matter of comparing the observed and expected frequencies. With ANOVA, we turn to a guideline provided by Tukey's test of honestly significant differences (or HSD).

Tukey's HSD *allows us to determine a minimum or threshold inequality for significant differences between our group means.* Once we have this yardstick for honestly significant differences, we examine the actual mean differences in our samples and determine which exceed the bounds of normal variation, that is, which differences between means are statistically significant and which are not.

We determine our marker or threshold for an HSD by using equation 118 and Table G. For both the .05 and .01 alpha levels, Table G provides the q_α values used in the equation. Table G shows values for q_α where k number of means (or samples) intersects with ($n - k$) degrees of freedom for MS_W. The appropriate q_α value is inserted into equation 118:

(118)   Minimum significant difference = HSD = $q_\alpha \sqrt{\dfrac{MS_W}{n}}$

To further interpret our significant ANOVA results, and to use Tukey's HSD, we rank our sample mean differences from largest to smallest. Which differences (e.g., $\overline{X}_1 - \overline{X}_2$) exceed the HSD threshold and which do not? Which differences between sample averages are significant and which are not? Which mean differences are responsible for our significant ANOVA results?

If we have unequal sample sizes, and no single value for *n*, we take an "average," or a *weighted mean, ñ* (equation 119), also called a **harmonic mean,** for all the separate sample sizes, and we substitute this value for *n* in equation 119:

$$(119) \qquad \tilde{n} = \frac{k}{\dfrac{1}{n_1} + \dfrac{1}{n_2} + \cdots + \dfrac{1}{n_k}}$$

In the following problems, note the basic ANOVA elements: the sums of squares, the mean squares, the *F* ratio, the H_O decision, and Tukey's procedures.

1. *The Disney organization conducted a small comparative study examining the times spent at its theme parks. It tracked randomly selected visitors at its parks in California, Florida, Paris, and Tokyo, and recorded the numbers of hours spent in the parks. At the .01 alpha level, do the data collected suggest significant differences in the average times spent per daily visit at the four locations? The data, based upon a total of 40 visitors to the parks, are rounded to the nearest whole hour.*

California (C):	*7 12 6 5 10 9 13 10*	*(n = 8)*
Florida (F):	*9 5 7 11 12 5 8 4 8 7*	*(n = 10)*
Paris (P):	*4 5 8 3 7 10 9 6 5 5*	*(n =10)*
Tokyo (T):	*9 10 13 14 12 8 7 11 12 10 8 13*	*(n = 12)*

After stating the hypotheses, we calculate the information necessary for our sums of squares. These procedures go back to Chapter 2. After the sums of squares, we calculate the mean squares and finally the *F* ratio. We have $k = 4$ samples of unequal sizes and $n = 40$ cases altogether. That gives us $k - 1 = 3$ degrees of freedom associated with our numerator (MS_B) and $n - k = 36$ d.f. for the denominator (MS_W). Consulting Table F with 3 and 36 d.f., we find the *F*-critical value at the .01 alpha level is $F_C = 4.38$. The hypotheses are: H_A: Any two means are unequal; H_O: $\mu_{California} = \mu_{Florida} = \mu_{Paris} = \mu_{Tokyo}$. Figure 8.1 presents the distribution curve for problem 1.

Table 8.1 presents the calculations needed to provide the parameters for the ANOVA summary.

Figure 8.1

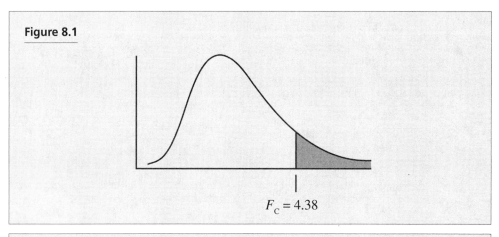

$F_C = 4.38$

Table 8.1

	ΣX	ΣX^2	$\overline{X}$	n
California	71	679	8.875	8
Florida	76	638	7.600	10
Paris	62	430	6.200	10
Tokyo	127	1401	10.583	12
$\Sigma =$	336	3148		40

For the sum of squares, from equation 109, we get:

$$\text{Between groups SS (SS}_B) = \left(\frac{(\Sigma X_C)^2}{n_C} + \frac{(\Sigma X_F)^2}{n_F} + \frac{(\Sigma X_P)^2}{n_P} + \frac{(\Sigma X_T)^2}{n_T} \right) - \frac{(\Sigma X)^2}{n}$$

$$= \left(\frac{(71)^2}{8} + \frac{(76)^2}{10} + \frac{(62)^2}{10} + \frac{(127)^2}{12} \right) - \frac{(336)^2}{40}$$

$$= (630.13 + 577.60 + 384.40 + 1344.08) - 2822.40$$

$$= 2936.21 - 2822.40 = 113.81$$

and from equation 110, we get:

$$\text{Total SS (SS}_T) = \Sigma X^2 - \frac{(\Sigma X)^2}{n} = 3148 - \frac{(336)^2}{40}$$

$$3148 - 2822.40 = 325.60$$

Therefore, according to equation 112:

Within groups SS (SS_W) = $SS_T - SS_B = 325.60 - 113.81 = 211.79$

For the mean squares, from equations 115 and 116, respectively, we get:

$$\text{Mean square between groups (MS}_B) = \frac{SS_B}{k-1} = \frac{113.81}{4-1} = \frac{113.81}{3} = 37.94$$

$$\text{Mean square within groups (MS}_W) = \frac{SS_W}{n-k} = \frac{211.79}{40-4} = \frac{211.79}{36} = 5.88$$

Finally, we get F_O from equation 117:

$$F_O = \frac{MS_B}{MS_W} = \frac{37.94}{5.88} = 6.45$$

The next step for our ANOVA is a summary table (Table 8.2).

We see that the critical value is F_C = 4.38 at the .01 level, and F_O = 6.45. We have sufficient evidence to reject H_O at the .01 α level. The sample means differ significantly. Tokyo visitors average a high of 10.58 hours at their park, whereas those in Paris average a low of just 6.20 hours, by comparison. But is that the only significant difference? Tukey's test for honestly significant differences helps to clarify such situations.

Since we have unequal sample sizes, we must first get a harmonic mean, or $\tilde{n}$ (equation 119), and then determine the minimum significant mean difference, or HSD, by using equation 118. For the q_α value from Table G, we have k = 4 samples and $n - k$ = 36 degrees of freedom associated with our MS_W. Even though $n - k$ is actually 36 d.f., Table G gives values for 30 and 40 only, so we must use the nearest entries below that number, or 30 d.f. In addition, to demonstrate its use, Tukey's HSD is shown for both the .05 and .01 alpha levels. We need a value for $\tilde{n}$, which we get from equation 119.

Table 8.2 Summary Table for ANOVA

Source of Variation		SS	d.f.	MS	F_O
(numerator)	Between Groups	113.81	3	37.94	6.45
(denominator)	Within Groups	211.79	36	5.88	

$$\tilde{n} = \frac{k}{\dfrac{1}{n_C} + \dfrac{1}{n_F} + \dfrac{1}{n_P} + \dfrac{1}{n_T}} = \frac{4}{\dfrac{1}{8} + \dfrac{1}{10} + \dfrac{1}{10} + \dfrac{1}{12}} = \frac{4}{.125 + .10 + .10 + .083} = \frac{4}{.408} = 9.804.$$

At the .05 alpha level:

$$\text{HSD}_{\alpha=.05} = q_\alpha \sqrt{\frac{\text{MS}_W}{n}} = 3.85 \sqrt{\frac{5.88}{9.804}} = 3.85\sqrt{.600} = 3.85(.775) = 2.984$$

At the .01 alpha level, all values are the same except the q_α figure, which from Table G, is 4.80:

$$\text{HSD}_{\alpha=.01} = q_\alpha \sqrt{\frac{\text{MS}_W}{n}} = 4.80(.775) = 3.720$$

Returning to our sample averages, we construct a hierarchy of their differences. Any difference of 3.720 hours or more is significant at the .01 alpha level, and any discrepancy of 2.984 hours or greater will meet the .05 alpha criterion for significance. Our differences are shown in Table 8.3.

It appears that only the Tokyo versus Paris difference is statistically significant and therefore responsible for our original *F* ratio's significance. That particular mean difference contributes significantly to the variation between the group averages. A little more exploration shows that another difference also plays a noteworthy role, however. The Tokyo versus Florida mean difference (2.983 hours) is very close to being statistically significant at the .05 alpha level. In fact, it *is* significant if we round our HSD criterion down to two decimal places. Our critical HSD and our observed difference each become 2.98. We then conclude that, even though all the mean differences contribute somewhat to the overall

Table 8.3

Samples	Averages	Differences
Tokyo versus Paris	10.583 – 6.200	4.383*
Tokyo versus Florida	10.583 – 7.600	2.983
California versus Paris	8.875 – 6.200	2.675
Tokyo versus California	10.583 – 8.875	1.708
Florida versus Paris	7.600 – 6.200	1.400
California versus Florida	8.875 – 7.600	1.275

*Significant at the .01 alpha level.

variation, the differences in average times between Tokyo and Florida and especially between Tokyo and Paris are the principal and significant ones.

2. A company wished to discover what effect a pay incentive program might have on the productivity of assembly-line employees. Twenty-one newly hired people were randomly assigned to different pay plans: Plans A and B were incentive options, and Plan C was the company's regular wage schedule. After a month on the job, each employee's output on a given day was recorded, and the data, in units output per day, are shown below. At the .05 alpha level, do the incentive plans significantly affect employees' productivity rates?

Plan A: *39   51   58   61   65   72   86*

Plan B: *22   38   43   47   49   54   72*

Plan C: *18   31   41   43   44   54   65*

We start with our hypotheses and the preliminary figures necessary for the sums of squares. Next, we derive the total and between-groups sums of squares (equations 109 and 110, respectively), and the mean squares (equations 115 and 116). We have 7 cases in each treatment group, giving us $n = 21$, and $k = 3$ groups. With 2 degrees of freedom for the F ratio numerator and 18 associated with the denominator, our F table (Table F) shows a critical value of 3.55 at the .05 alpha level. Our hypotheses are: H_A: Any two means are unequal; H_O: $\mu_A = \mu_B = \mu_C$. The distribution curve is shown in Figure 8.2.

Again, we tabulate the calculations of the parameters, as shown in Table 8.4.

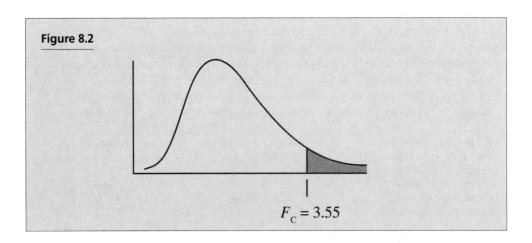

Figure 8.2

$F_C = 3.55$

Table 8.4				
	ΣX	ΣX^2	$\overline{X}$	n
Plan A:	432	28,012	61.714	7
Plan B:	325	16,487	46.429	7
Plan C:	296	13,892	42.286	7
Totals	$\Sigma X = 1053$	$\Sigma X^2 = 58,391$		21

The sums of squares and mean squares are calculated as follows:

$$\text{Total SS (SS}_\text{T}) = \ \Sigma X^2 - \frac{(\Sigma X)^2}{n} = 58,391 - \frac{(1053)^2}{21} = 58,391 - 52,800.43 = 5590.57$$

$$\text{Between-groups SS (SS}_\text{B}) = \left(\frac{(\Sigma X_\text{A})^2}{n_\text{A}} + \frac{(\Sigma X_\text{B})^2}{n_\text{B}} + \frac{(\Sigma X_\text{C})^2}{n_\text{C}} \right) - \frac{(\Sigma X)^2}{n}$$

$$= \left(\frac{(432)^2}{7} + \frac{(325)^2}{7} + \frac{(296)^2}{7} \right) - \frac{(1053)^2}{21}$$

$$= (26,660.57 + 15,089.29 + 12,516.57) - 52,800.43$$

$$= 54,266.43 - 52,800.43 \ = \ 1466.00$$

$$\text{Within-groups SS (SS}_\text{W}) = \ \text{SS}_\text{T} - \text{SS}_\text{B} = 5590.57 - 1466.00 = 4124.57$$

$$\text{Mean square between groups (MS}_\text{B}) = \frac{\text{SS}_\text{B}}{k-1} = \frac{1466.00}{3-1} = \frac{1466.00}{2} = 733.00$$

$$\text{Mean square within groups (MS}_\text{W}) = \frac{\text{SS}_\text{W}}{n-k} = \frac{4124.57}{21-3} = \frac{4124.57}{18} = 229.14$$

For the ANOVA summary table, we have Table 8.5:

Table 8.5				
Source of Variation	SS	d.f.	MS	F_O
Between groups	1466.00	2	733.00	3.20
Within groups	4124.57	18	229.14	

As we saw in Table F, F_c is 3.55, and we see here that F_O is only 3.20. Therefore, we fail to reject our H_O. The means of our three groups do not differ significantly at the .05 alpha level. Employees under Plan A average a higher production rate than their Plan B and C counterparts, but the difference is not statistically significant. With the failure to reject H_O, we go no further. We do not examine specific mean differences using Tukey's HSD procedure. With nonsignificant results, there is no justification for doing so.

This is the basic analysis of variance test. ANOVA, however, is a varied statistical tool. There is considerably more to it than covered above. Indeed, it is not uncommon for universities to focus graduate statistics courses on *just* complex ANOVA tests. They are beyond this text, however, and we turn in the last chapter to measures of correlation and regression.

Exercises

1. A sociologist compared two different organizational types (high-tech and music) with academia as to the percentages of top administrators, executives, and deans or above who were at least 50 years old. She randomly sampled 10 high-tech companies, 8 music companies, and 14 four-year colleges. For each, she recorded the percentage of top officials who were at least age 50. Her data are shown below. At the .01 alpha level, do the organizational types differ as to the typical percentages of executives age 50 or older?

High-Tech: 42 51 55 60 40 38 70 57 62 41
Music: 41 29 43 27 66 40 39 26
Academia: 63 72 87 59 69 74 83 91 84 81 77 93 68 79

2. The Advisory Group on Native American Sentencing claimed that Native Americans receive longer sentences than other offenders and offered data regarding aggravated assault convictions as evidence. Examining random selections of cases from several western states, the organization found the following

sentences (in months) given to first offenders when found guilty of aggravated assault. The data are broken down by ethnic derivation. Is the organization's claim supported at the .05 alpha level? Do Native American prisoners average significantly longer sentences for aggravated assault?

Native American: 38 31 33 39 29 34 35 37
Other minority: 37 32 38 28 37 31 33 30
Anglo/White: 34 28 27 29 31 32 26 30

3. Among adults claiming to be members of religious congregations, does attendance at religious services correlate with parental status? A researcher asked samples of congregation members to estimate how many times they had attended religious services in the last six months. His data are given below. At the .01 alpha level, do congregation members with young children attend fewer services than other adults?

No Children: 28   4 15 12   0   2 35 10 20 31
Children under age 12:   0   2   9   5 15 16   8 21 12 10
Children 12 or older: 18 22   1 17 20 30 23 16 20   0

4. For a class research project, a team of high school students measured the decibel (dB) levels* of car radios and stereos being played by samples of drivers. Randomly selecting drivers at intersections and in parking lots and estimating their ages by appearance, they collected the data shown below. At the .05 alpha level, is their teacher's suspicion correct? Do teens and young adults play car radios and stereos at significantly higher decibel levels? What about at the .01 alpha level?

Teens, young adults: 52 69 72 66 81 74 84 57 90 65 70 73
Ages 25 to 60:   56 62 57 63 70 58 60 55 63 67
Seniors, above age 60:   59 66 69 73 65 64 74 63 70 71 58 68 73

5. A family therapist randomly selected samples of patients she had classified according to their recollections of childhood as being mainly positive, mainly negative, or mixed. She also recorded their views as to the ideal numbers of children for families to have. At the .05 alpha level, do opinions of ideal numbers of children vary according to recalled childhood experiences?

---

*Most sounds are in the 50–70 dB range. A normal conversation is about 60 dB. Rustling leaves are 30 dB, while a jet taking off is about 120 dB. Hearing loss may eventually occur with repeated exposure to 85 dB or more.

Childhood mainly positive:	3	2	1	4	3	4	2	0	1	2	4	5
Childhood mixed:	0	1	4	3	2	2	1	0	1	3	2	1
Childhood mainly negative:	1	2	1	3	1	4	2	1	0	2	0	2

6. An automobile insurance company examined the one-way commuting distances of random samples of students who reportedly drove to their respective campuses. The company's researcher wondered how much commuting distances varied by academic level. At the .01 alpha level, do the average commuting distances shown below (rounded to the nearest whole mile) differ significantly by academic level of institution?

High school:	3	5	2	4	7	1	2	4	4	6	12	3	5	8	2
Community college:	7	12	18	5	9	6	10	15	11	9	14	7	13		

Four-year college,
undergraduates: 18 20 9 15 21 63 19 17 22 10 14 8 26 21
18 17 14 41 27 23

Research university,
graduate students: 25 40 15 3 29 43 32 24 4 31

7. Random samples of undergraduates were asked how often, over typical two-week periods, they had at least three alcoholic drinks at one sitting. Samples included students in campus residence halls, in fraternity or sorority houses, and in personal off-campus housing. The collected data are presented below. At the .05 alpha level, does students' drinking vary by their residential type?

Residence halls:	2	3	0	1	5	4	0	1	2	1
Fraternity/sorority houses:	4	3	3	1	2	4	3	4	0	3
Personal off-campus housing:	2	3	1	4	2	0	0	2	1	2

8. Following a steep rise in regional housing costs, an economist asked random samples of local residents what percentages of their before-tax family incomes were devoted to housing costs, either rent or mortgage payments. All respondents were part of two-income families, had two dependent children at home, and none lived in subsidized housing. She also categorized her samples by family income levels: upper, middle, and lower. At the .01 alpha level, do the average percentages of income devoted to housing vary inversely with income level?

Upper income:	21	25	19	26	20	22	23	24		
Middle income:	26	25	30	32	24	26	29	28	31	
Lower income:	29	34	37	42	33	36	38	31	35	40

9. In your own words, what is the logic or rationale of ANOVA and the *F* test?

10. How does ANOVA differ from the two-sample *t* and *z* tests?

11. In your own words, what are the mean square between groups (MS_B) and the mean square within groups (MS_W)?

12. What does it mean if we reject our null hypothesis with the *F* test?

13. When do we use Tukey's test for honest significant differences (or HSDs), and what does that procedure tell us?

CHAPTER 9

X and *Y* Together: Correlation and Prediction

In this chapter, you will learn how to:

- Calculate Pearson's r_{XY} for interval-ratio data
- Calculate Spearman's r_s for ordinal data
- Test whether r_{XY} and r_s are statistically significantly
- Use significant r_{XY} correlations to make statistical predictions

MEASURES OF CORRELATION TELL US ABOUT shades of gray; that is, they tell us not just whether a relationship exists, but about the strength of an association between two variables. They are both similar to and different from the statistics in previous chapters. They are clearly reminiscent of the chi-square-based measures of association encountered in Chapter 7. They vary between .00 and 1.00, although now there are negative values extending down to –1.00 as well. The closer to 1.00 (or to –1.00), the stronger the correlation. The closer to .00, the weaker the association. Our interpretations of specific correlations are generally the same also. We regard associations in the .00 to ±.20 range as weak, those in the ±.20 to ±.49 ranges as moderately strong, and correlations beyond +.50 or –.50 as strong to very strong. Now, however, sample sizes also influence how we interpret our correlations, and this suggests other differences from Chapter 7.

The correlation coefficients in this chapter differ from the measures of associations in Chapter 7 in two key ways. First, we reverse the previous sequence of doing the hypothesis test and then finding an association. We now establish a correlation between two variables, typically labeled *X* and *Y*, as the initial step, and then test whether it is statistically significant. In Chapter 7, for example, only if we got a significant chi square did we measure the strength of the association between two variables. Now, we get the correlation first, have a general idea as to its strength from the guidelines above, but always subject it to a significance test just in case. The "just in case" precaution comes from the fact that even unimpressive correlations in the apparently weak-to-moderate range *may* be statistically significant if based upon large enough sample sizes. For example, an *XY* correlation of .23 based upon a sample of 10 cases probably wouldn't seem to be significant. But that same .23 *is* statistically significant ($\alpha = .05$) in a sample of 80 cases. So, even though we have guidelines for interpreting our initial correlation coefficients, we always subject them to significance tests—just in case.

Second, our current measures of correlation differ from Chapter 7's associations in the kinds of data with which we work. *Now, we require numbers rather than categories and frequencies.* For example, chi square, the basis for earlier measures of association, was calculated by comparing observed and expected frequencies, but now we work with the actual *X* values (and *Y* values) that constitute precise and reliable interval-ratio measurements. Or we may work with the *ranks* of the original scores if we have ordinal data, just as we did with Chapter 7's Wilcoxon and Mann-Whitney tests. In either case, no more categories. We require: (1) *at least* ordinal, rankable measurements; and (2) numerical, not just categorical, data.

With these considerations in mind, our first correlational statistic and probably the most popular is Pearson's *r*. **Pearson's *r*** *measures the correlation between two interval-ratio variables.* We have two sets of observations. The independent or predictor variable's scores are labeled *X*, and the dependent variable's observations are labeled *Y*. Each person or element in the sample requires a score on the *X* variable and a score on the *Y* variable. Pearson's statistic calculates the degree of association between the *X* and *Y* scores, and, this being the case, it is commonly written as r_{XY}. Essentially, we are asking: Do the *Y* scores vary directly with the *X* scores (a positive *XY* correlation), vary indirectly (a negative *XY* correlation), or is there no apparent connection between the *X* and *Y* scores at all?

Having derived a correlation coefficient, or Pearson's r_{XY}, up to three more steps are possible. The second step tests the null hypothesis that the r_{XY} observed in our sample could have come from a population in which the actual correlation between *X* and *Y* is zero. Could our observed r_{XY}, assuming it appears to indicate an *XY* association, be something of a fluke? Under a null hypothesis, would it be within the bounds of normal variation for a true population correlation of zero (no association)? Put another way, does our sample r_{XY} differ significantly from an *assumed XY* correlation of zero in the population as a whole? As with previous H_0 tests, we usually wish to reject this particular null hypothesis of no true correlation. We typically wish to assert that our Pearson's r_{XY} is inconsistent with a zero correlation, thereby suggesting that there is, in fact, an *XY* association.

Next, assuming our H_0 test does suggest a relationship between X and Y that is not zero, we may use the correlation to make predictions. **Regression** *is a more formal term for a procedure that amounts to prediction.* Our third step, then, based upon X and Y being correlated, predicts a Y value (written as $\hat{Y}$ and read as "Y-hat" or "Y-predicted") for any given value of X. For instance, if we first establish a definite correlation between the College Board SAT scores (X) and subsequent grade point averages (GPAs) in college (Y), we may then use that information to predict a certain GPA for someone who achieved any particular SAT score.

We realize, of course, that our predictions will not be 100% accurate. The correlation between X and Y, at least for social and behavioral scientists, is never perfect. The fourth and final step based upon Pearson's r_{XY} establishes a confidence interval for our $\hat{Y}$ prediction. For example, given a certain SAT score, we predict a certain GPA. But this predicted GPA is an average of sorts. What GPA would the typical person with this SAT score tend to earn? To give us a little more assurance, we construct a confidence interval, a certain plus-or-minus margin of error around our $\hat{Y}$ prediction.

Related to this prediction procedure is the concept of **explained variation.** Given the correlation between X and Y, how much of the variation in Y (the dependent variable) do differences or variations in X (the independent variable) explain statistically? We may have a correlation but find that variation in the X scores accounts for only a small part of the total variation in the Y scores (a weak association). For instance, SAT scores may correlate with subsequent GPAs, but SAT test scores by themselves may account for only a small portion of the total variation in GPAs. In contrast, with a very strong XY association, differences in the X scores could account (statistically) for a major part of the variation in the Y scores. **Pearson's** r_{XY}^2, *the coefficient of determination,* measures this explained variation.

A second measure of correlation is designed for ordinal or ranked data. Instead of using actual X and Y scores, do our respondents' ranks on the X variable correlate with their ranks on the Y variable? Does someone ranking high on X, for instance, also tend to rank high on Y? Or perhaps low on Y? Our *statistical tool for ranked data is* **Spearman's rank-order correlation coefficient:** r_S.

Numbers: Pearson's Correlation (r_{XY})

As previously stated, Pearson's r_{XY} correlates two interval-ratio variables. Each respondent or element in the sample has two measurements, one on the X variable and the other on the Y variable.

A familiar way to introduce r_{XY} is to consider its relationship to z scores. In Chapter 3, we saw that if we have reliable quantitative data, a mean, and a standard deviation, we may convert any X score into a z score. The same is true of any quantitative score labeled Y, of course; simply labeling measurements X or Y does not change the basic nature of the data. For each case in a sample, we may calculate a z score for the X value (and label it z_x) and a z score on the Y variable (z_y). That locates each respondent in both the X and Y distributions of z scores. Next, we consider whether z scores on the Y variable tend to correlate at all with the z scores on the X variable. Do respondents with higher and positive z_x scores also tend to have higher and positive z_y scores? Or do those with positive z scores on one variable (above the mean) tend to have negative z scores for the other variable (below the mean)?

The **average product of the z scores** *measures the XY correlation.* For each case, we calculate a $z_x z_y$ quantity, that is, the product of the two z scores. Pearson's r_{XY} is actually the average $z_x z_y$ product: we add the $z_x z_y$ values and divide the total by n, the number of paired observations in our sample, as shown in equation 120.

$$(120) \qquad r_{XY} = \frac{\Sigma(z_X z_Y)}{n}$$

Equation 120 gives us Pearson's r_{XY}, but it requires several steps. First, we need the mean and standard deviation of the X scores ($\overline{X}$ and s_x), and the same for the Y scores ($\overline{Y}$ and s_y). We then calculate z scores for all X and Y observations and then products for all respondents' two z scores. The final step derives the average $z_x z_y$ product. This is possible, of course, but tedious, especially if our sample is fairly large. Introducing Pearson's r_{XY} in this way, as a function of z scores, anchors this statistic in earlier work (Chapters 2 and 3), but there is an easier procedure.

The **raw score formula for r_{XY}** uses the original, or "raw," X and Y values, without the need to calculate any means or standard deviations. Given the relationship of Pearson's r_{XY} to z scores and the reliance of z scores on the standard deviation, it is not surprising that terms in the raw score formula are reminiscent of an earlier standard deviation formula (equation 8). Despite looking complicated, the operations in equation 121 are more economical than those required by equation 120, as we shall see in the problem that follows

$$(121) \qquad r_{XY} = \frac{\Sigma XY - \dfrac{(\Sigma X)(\Sigma Y)}{n}}{\sqrt{\Sigma X^2 - \dfrac{(\Sigma X)^2}{n}} \ \sqrt{\Sigma Y^2 - \dfrac{(\Sigma Y)^2}{n}}}$$

Before we get to problem 1, however, we should discuss the interpretation of Pearson's r_{XY}. Two features of r_{XY} are important: (1) its size, and (2) its sign. We want to see how large r_{XY} is, and whether it is positive or negative. First, r_{XY}

varies only between -1.00 and $+1.00$. (Any r_{XY} value outside this limited $+1.00$ to -1.00 range means an error.) r_{XY} values close to -1.00 or to 1.00 indicate strong relationships, and the relationship lessens toward the middle of the range, until a value of $.00$ tells us X and Y are not related at all.

Second, the sign of r_{XY} tells us the nature of any XY association. A mathematically positive $(+)r_{XY}$ indicates a **positive or direct relationship,** *which means as the values of X increase, the values of Y also increase.* A high score on X tends to mean a high score on Y, or vice versa (low on X and low on Y). A negative $(-)r_{XY}$ value, in contrast, indicates a **negative, indirect, or inverse relationship:** *There is a tendency for the high scores on X to be associated with the low scores on Y, and vice versa* (low on X and high on Y).

Putting the size and sign together, we can interpret a few illustrative values of r_{XY}. For example, $r_{XY} = -.11$ indicates a weak negative XY relationship, a very slight tendency for Y values to decrease as the X values increase, and vice versa. $r_{XY} = .81$ describes a strong direct relationship; as X increases, there is a strong likelihood that Y will increase also. $r_{XY} = -.47$ indicates a moderate indirect association; there is a moderate tendency for Y to decrease as X increases.

When using equation 121, the numerator determines the sign of r_{XY}. It may be either positive or negative. The denominator, on the other hand, is always positive (similar to the standard deviation). The discussion following equation 8 in Chapter 2 provided an alternative to the denominator in equation 120. Specifically, the following equality exists for the sum of squares:

$$\Sigma\left(X - \overline{X}\right)^2 = \Sigma X^2 - \frac{(\Sigma X)^2}{n}$$

Therefore, the "raw score" procedure used in equation 120 can be replaced with the "deviation" procedure, and you may wish to use that method (equation 121) in solving for Pearson's r_{XY}:

(122)
$$r_{XY} = \frac{\Sigma XY - \dfrac{(\Sigma X)(\Sigma Y)}{n}}{\sqrt{\Sigma(X - \overline{X})^2}\ \sqrt{\Sigma(Y - \overline{Y})^2}}$$

With these guidelines in mind, and recalling those introduced with Chapter 7's measures of association, we turn to examples of actual calculations. In the following sections, problems 1 through 4 examine four procedures for the same data set.

1. *Administrators at Alameda University suspect that the more calculus or statistics courses an undergraduate takes, the more job offers he or she tends to receive upon graduation. The administrators randomly select 15 recent graduates and gather the data below. If there is an XY relationship, how would you describe it? Are the administrators correct in their expectation?*

Student:	1	2	3	4	5	6	7	8	9	10	11	12	13	14	15
Calculus or statistics															
courses (X):	2	0	7	4	1	1	1	0	3	0	5	3	2	4	2
Job offers (Y):	4	3	9	5	4	1	3	2	1	0	4	3	2	6	3

The number of calculus or statistics courses taken is X because it is the independent or predictor variable, presumed to affect the number of job offers, Y. We also recognize that both X and Y are interval-ratio measurements, justifying Pearson's r_{XY}. We first need a worksheet for sums required by equation 120. The information given is on the left and the required calculations on the right of Table 9.1.

From equation 120,

$$r_{XY} = \frac{\sum XY - \dfrac{(\sum X)(\sum Y)}{n}}{\sqrt{\sum X^2 - \dfrac{(\sum X)^2}{n}}\sqrt{\sum Y^2 - \dfrac{(\sum Y)^2}{n}}} = \frac{165 - \dfrac{(35)(50)}{15}}{\sqrt{139 - \dfrac{(35)^2}{15}}\sqrt{236 - \dfrac{(50)^2}{15}}}$$

$$= \frac{165 - 116.667}{\sqrt{57.333}\sqrt{69.333}} = \frac{48.333}{(7.572)(8.327)} = \frac{48.333}{63.049} = .767$$

Table 9.1

Student	Courses (X)	Offers (Y)	X^2	Y^2	XY
1	2	4	4	16	8
2	0	3	0	9	0
3	7	9	49	81	63
4	4	5	16	25	20
5	1	4	1	16	4
6	1	1	1	1	1
7	1	3	1	9	3
8	0	2	0	4	0
9	3	1	9	1	3
10	0	0	0	0	0
11	5	4	25	16	20
12	3	3	9	9	9
13	2	2	4	4	4
14	4	6	16	36	24
15	2	3	4	9	6
Totals	$\sum X = 35$	$\sum Y = 50$	$\sum X^2 = 139$	$\sum Y^2 = 236$	$\sum XY = 165$

Alameda University officials are correct. At least based on their 15 cases, there appears to be a strong direct *XY* relationship. There is a strong tendency for the number of job offers to increase as the number of calculus or statistics courses taken increases. (Remember that for indirect *XY* associations, the numerators and eventual answers automatically result in negative numbers.)

Before we do any additional analyses based on Pearson's r_{XY}, there is an important caution to keep in mind. The correlation coefficient measures only linear association or correlation. It assumes there is a consistent or linear (or straight line) association between *X* and *Y*. In the case of a strong positive association, for instance, it assumes that *Y* keeps on increasing at a constant and steady rate as long as *X* increases. Obviously then, Pearson's r_{XY} is not suitable for *all* circumstances requiring measures of correlation. We should therefore limit our use of Pearson's r_{XY} to situations in which we believe this *is* the actual nature of the *XY* relationship. For instance, suppose we wish to correlate length of time on the job (*X*) with an interval-ratio measure of job-related morale (*Y*) among teachers, nurses, or police officers. These professions are said to require unusual dedication but also are described as producing occupational burnout. The association between job tenure and morale therefore may be curvilinear. Up to a point, morale increases as does time on the job (a direct *XY* relationship). However, if feelings of burnout or demoralization develop, further years on the job may exacerbate the situation, resulting in a negative association between tenure and morale: as more time goes by, morale deteriorates. This oversimplifies a complex situation, but the example does suggest that Pearson's correlation coefficient is not always appropriate. If we used r_{XY} in such a case, it would underestimate the true *XY* association. Because r_{XY} is able to measure only a constant and linear association, it would average out the initially positive *XY* relationship and the subsequent negative relationship. We may end up with an r_{XY} value close to zero, and this would certainly be misleading. (Curvilinear associations require the correlation ratio, or eta, but that is beyond the scope of this book.)

Even granting this reservation, however, the correlation coefficient remains a very popular measure of association. In part, this is due to its forming the basis for linear regression. That topic follows a discussion of a method to determine whether a Pearson's r_{XY} is legitimate, that is, that it is not a fluke and *does* come from a population in which *X* and *Y* are correlated.

Is r_{XY} Significant?

Pearson's r_{XY} is a statistic based upon only sample data. We do not know the *true XY* correlation for the whole population (called **rho:** ρ_{XY}). However, we can test whether r_{XY} could have come from a population in which the true correlation (ρ_{XY})

is zero. If r_{XY} is consistent with or does not differ significantly from a rho value of zero, we conclude there is no real *XY* correlation. No matter what r_{XY} we get from our sample, if it falls within the bounds of normal variation around an assumed rho of .00, we conclude ρ_{XY} could equal .00 and that there is no *XY* correlation in the population. In other words, r_{XY} does not differ significantly from $\rho_{XY} = .00$, and a ρ_{XY} of .00 means no association. However, if our sample r_{XY} differs from $\rho_{XY} = .00$, we have evidence of a statistically significant correlation (at a given alpha level). We then conclude r_{XY} comes from a population in which $\rho_{XY} \neq .00$.

The null hypothesis (H_O) is $\rho_{XY} = .00$. We assume a value of .00 for the population parameter. Rho becomes the equivalent of μ_O or p_O in earlier hypothesis tests. (Pearson's r_{XY} is equivalent to our previous $\overline{X}$ or p'.) We typically deal with a two-tailed test (with H_A: $\rho_{XY} \neq .00$). This is common because r_{XY} may vary in either direction from .00. However, there is no logical reason why one-tailed tests could not be conducted. For example, we may test that r_{XY} comes from a population in which rho is significantly greater than .00. We express that directionality in our H_A: $\rho_{XY} > .00$. The H_O becomes: $\rho_{XY} \leq .00$, and we conduct an upper-tailed test.

Statistics texts usually prescribe *t* tests for this procedure, assuming that our samples will be small when we measure two variables per respondent. However, given the popularity of large-sample surveys and the ease of computer-assisted analyses, and since we are dealing with quantitative data for which means and standard deviations are possible, here we apply the earlier criterion as per the suitability of a *t* test or a *z* test: if $n < 30$, use *t;* otherwise, use *z*. (Remember that the sample size is the number of *paired* observations, *not* $n_x + n_y$.) In problem 1, $n = 15$ cases, so a *t* test is illustrated below. In actual working situations, a *t* test may not always be suitable, and when using *z*, you need not be concerned with degrees of freedom.

When using *t* in these cases, the degrees of freedom differ slightly from those appropriate for previous one-sample tests. We now have $(n - 2)$ d.f. There is one degree of freedom for the *X* measurements and one associated with the *Y* measurements. *n* is the same for both variables, of course, so we get $(n - 2)$ d.f. altogether.

As before, we calculate t_O (equation 123) or z_O (equation 124) and compare our results to a critical value from the *t* or z table, respectively.

(123)
$$t_O = r_{XY} \sqrt{\frac{n-2}{1-r_{XY}^2}}$$

where t_c has $(n - 2)$ degrees of freedom, and

(124)
$$z_O = r_{XY} \sqrt{n-1}$$

Recall that $r_{XY} = .767$ in problem 1, indicating a strong direct relationship between the number of calculus or statistics courses taken (X) and the number of job offers received upon graduation (Y).

2. *At Alameda University, does the XY relationship between the number of calculus or statistics courses taken and the number of subsequent job offers differ significantly from .00 at the .01 alpha level? In other words, does r_{XY} come from a population in which ρ_{XY} differs significantly from .00 at the .01 alpha level?*

We have fudged a little here and changed the question to illustrate the most common way of testing hypotheses for r_{XY} and ρ_{XY}. Problem 1 described a suspected positive *XY* association: more calculus or statistics courses implied more job offers. That appeared to be confirmed with r_{XY} equal to .767. Problem 2 should technically reflect this directionality, ask whether r_{XY} comes from a population in which ρ_{XY} is *greater than* .00, and prescribe an upper-tailed test. However, as indicated and as problem 2 illustrates, it is much more common to pose the r_{XY} test in two-tailed terms: Could our correlation coefficient come from a population in which ρ_{XY} *differs from* .00?

So problem 2 is written in two-tailed form to make a point; there is an inconsistency between it and problem 1. (Perhaps one may rationalize the inconsistency further by noting that the two-tailed test makes it more difficult to reject our H_0 than a one-tailed test. Recall the discussion in Chapter 5.)

To complete the solution to problem 2, our *n* of 15, a *t* test with $(n - 2) = 13$ degrees of freedom, and a two-tailed test means $t_c = \pm 3.012$ at the .01 alpha level (from Table C). From problem 1, $r_X = .767$. Therefore, our hypotheses are H_A: $\rho_{XY} \neq .00$; H_0: $\rho_{XY} = .00$. The distribution curve is shown in Figure 9.1.

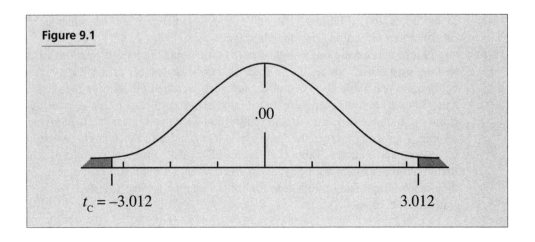

Figure 9.1

.00

$t_c = -3.012$ 3.012

We calculate t_O by using equation 123:

$$t_O = r_{XY} \sqrt{\frac{n-2}{1-r_{XY}^2}} = .767 \sqrt{\frac{13}{1-(.767)^2}} = .767 \sqrt{\frac{13}{.412}} = .767(5.617) = 4.308$$

With $t_C = \pm 3.012$ and $t_O = 4.308$, we have sufficient evidence to reject the null hypothesis at the .01 alpha level. Our sample statistic, r_{XY}, comes from a population in which the true XY correlation, ρ_{XY}, differs significantly from .00 at the .01 alpha level. We may therefore be 99% sure there is a correlation between X and Y. There is only a 1% chance we have committed an alpha error.

Having gone through all that, it is also true that a reference table allows us to look up critical values for r_{XY} itself. We may bypass the t test (or z test) altogether and consult a table to see whether $r_{XY} = .767$ is statistically significant. Instead of converting r_{XY} into a t score, we look up the critical value of r_{XY}: or r_{XY}-critical. Along the horizontal axis of our H_O curve, this is the r_{XY} value (r_{XY_C}) falling at the critical H_O acceptance/rejection point, the point where our alpha regions begin. If r_{XY}-observed is equal to or greater than r_{XY_C}, it falls farther away from the mean (.00) of our H_O curve and into an alpha region, and we reject our null hypothesis. Using r_{XY_C} is the equivalent of determining $\overline{X}_C$ or p'_C in Chapter 5 (see equations 56 and 57, respectively). This was an option instead of instead of converting $\overline{X}$ and p' into t or z scores. The only difference now is that we look up r_{XY_C} in a table rather than calculating it.

Table H presents one- and two-tailed critical values for r_{XY}. In a familiar procedure, we look up an r_{XY}-critical figure where the sample size (n) and alpha level intersect. For the present problem, $r_{XY_C} = \pm .641$ ($n = 15$, $\alpha = .01$). Since our observed or calculated r_{XY} is .767, we may reject our H_O, just as we did with our earlier t test. The sampling distribution curve using r_{XY_C} (Figure 9.2) is identical to the one using t_C (Figure 9.1). t and r_{XY} represent different and interchangeable measurement units along the same horizontal axis.

Notice that we use our sample size (n) rather than degrees of freedom when looking up a critical value of r_{XY}. There is a good reason for this. We may have occasion to use Table H in lieu of a z test, that is, when our sample includes at least 30 cases and the sampling distribution curve conforms to the normal distribution. In those cases, we are not concerned with the degrees of freedom in a sample. We merely refer to the sample size itself, n. Rather than having separate tables for t and z curves, Table H collapses all critical values into one table. The first column dispenses with degrees of freedom and identifies all samples by n, their actual size. (The same is true for Table I and the critical values of our next statistic, Spearman's r_s).

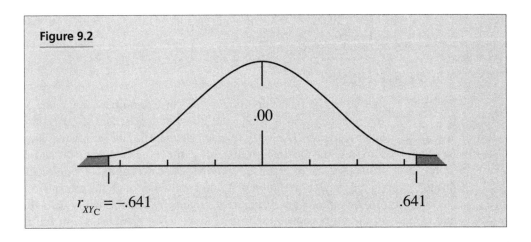

Figure 9.2

.00

$r_{XY_C} = -.641$ .641

You may wonder why we bother with equation 123 or possibly equation 124 at all. Why do we need t-observed or z-observed at all if we have Table H? There are two reasons. First, you *should* know the conceptual arguments surrounding the H_0 test. Merely looking up a number should not substitute for understanding the test's rationale and mechanics. What happens if no such table is available? You should be able to conduct the test in the absence of Table H. Knowing the test procedure is a good idea, good statistical practice, and is handy—just in case. Second, statistical programs and software usually report t (or z) values and their significance levels for this test, rather than the r_{XY_C} criterion. You should be familiar with the test to appreciate those results.

Before turning to regression, a caution may be appropriate here. Please note the two distinct and separate ways of testing the significance of r_{XY}. We test the null hypothesis that $\rho_{XY} = .000$ using either the t test or by consulting Table H for r_{XY}-critical values. In doing the t test, we compare t_0 to t_C. Alternatively, we compare r_{XY} to r_{XY_C}. But we never mix and match. We do not evaluate t-observed by looking at r_{XY_C}. Nor do we compare r_{XY} to t_C. Your test statistic (t_0 or r_{XY}) *must* be consistent with the critical value (t_C or r_{XY_C}) used to determine its significance. To do otherwise is to compare the proverbial apples and oranges, and that, of course, is unacceptable and tells us nothing.

So we have established an XY correlation and found it to be statistically significant. We now use our knowledge of that association to predict a Y value ($\hat{Y}$) for a given X value. This extrapolation of Pearson's r_{XY} involves two steps. First, given a particular X score, we predict a $\hat{Y}$ value. Then we establish a confidence interval estimate around our $\hat{Y}$ figure, just as we constructed confidence intervals around $\overline{X}$ or p' in Chapter 4. However, if r_{XY} is not statistically significant, our analysis ends here. Unless we reject the null hypothesis that $\rho_{XY} = .00$, we may not use r_{XY} as a basis to predict $\hat{Y}$ values.

Using X to Predict Y $(\hat{Y})$

Pearson's r_{XY} measures the linear or straight-line correlation between X and Y. On a graph, the X values appear along the horizontal axis, and the Y values appear on the vertical axis (Figure 9.3). Each of the 15 data points (or dots) in Figure 9.3 represents the intersection of a single respondent's X and Y scores from problem 1. Such graphs are known as **scatterplots.** A single best line to represent the *linear* correlation of X and Y drawn through the graph (from the lower left to the upper right in Figure 9.3) goes by several names. It is the **line of best fit,** the **regression line,** or the $\hat{Y}$ **line.** If the Y values rise or fall in perfect and constant relation to the X values, we get a perfect r_{XY} equal to 1.00 or to –1.00, and all the X, Y data points fall on the line.

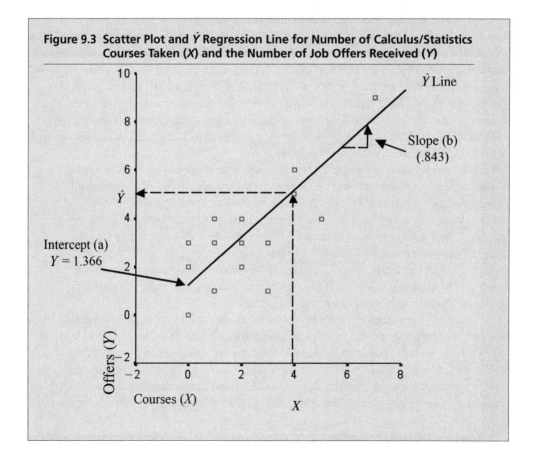

Figure 9.3 Scatter Plot and $\hat{Y}$ Regression Line for Number of Calculus/Statistics Courses Taken (X) and the Number of Job Offers Received (Y)

We could, of course, take a ruler and draw the line of best fit by eye or freehand. But since we need to be precise with this line, we determine exactly where to draw it mathematically, that is, by using certain formulas. The line has two mathematical features of note. One is the **intercept,** labeled *a, which is the Y value at which the line crosses (intercepts) the vertical Y axis when X = 0.00.* This locates a point of origin for the line in vertical space, the Y dimension. The other feature of the line is the **slope,** labeled b, *which is the angle or slant of the line.* Specifically, it tells us how much the line goes up or down (in the vertical Y dimension) for every unit increase along the horizontal X axis. With a slope of 1.22, for instance, the line goes up 1.22 units of measurement in Y for every unit increase in X along the lower axis. A slope of –.90 means the line goes down as it progresses from left to right (indicated by a negative number): a .90 decrement in Y units for every one-unit increase in X. Equations 125 and 126 show the intercept and slope, respectively.

$$(125) \qquad \text{Intercept} \quad a \; = \; \overline{Y} - r_{XY}\!\left(\frac{s_Y}{s_X}\right)\!\overline{X}\,,$$

where $\overline{Y}$ = the mean of the Y values, $\overline{X}$ = the mean of the X values, s_Y = the standard deviation of the Y scores, and s_X = the standard deviation of the X scores.

$$(126) \qquad \text{Slope} \quad b \; = \; r_{XY}\!\left(\frac{s_Y}{s_X}\right)$$

In our example, the data for graduates of Alameda University yield the following:

$\overline{Y}$ = 3.333 offers, $\overline{X}$ = 2.333 courses, s_Y = 2.225 offers, and s_X = 2.024 courses.

We insert these values into equations 125 and 126 to derive the intercept and slope illustrated in Figure 9.3:

$$a = 3.333 - .767\!\left(\frac{2.225}{2.024}\right)\!2.333 = 3.333 - .767\,(1.099)(2.333) = 3.333 - 1.967 = 1.366 \text{ offers}$$

$$b = .767\!\left(\frac{2.225}{2.024}\right) = .767\,(1.099) = .843 \text{ offers}$$

The intercept (*a*) is 1.366 job offers. When X = 0 courses in Figure 9.3, the $\hat{Y}$ line appears to be at a Y value of about 1.3 or 1.4, consistent with a calculated intercept of 1.366. The slope (*b*) is .843 job offers. For every increment in X (one calculus or statistics course), the number of job offers increases by .843. The $\hat{Y}$ line's angle reflects this increase. From left to right, it goes up at a bit less than a 45° angle. With similarly scaled X and Y axes, a line at about 45° indicates a 1:1 relationship between courses and job offers, an increase of one job offer for every additional course taken. The angle of the line in Figure 9.3 is consistent with an increase of slightly less, .843 job offers.

To use Figure 9.3, we predict a Y value falling on the line for a given value of X. Since this is the one line that best represents the linear relationship between X and Y, $\hat{Y}$ is the Y value on the line intercepted by X. To get a $\hat{Y}$ value, we trace a line upward through the graph from any value of X on the lower axis. When we intercept the $\hat{Y}$ line we read directly across to the left and parallel to the lower axis. The point at which we intercept the Y axis is our $\hat{Y}$ value. (See the dotted lines in Figure 9.3.)

It is a tedious process, however, to draw a scatterplot and regression line just to get $\hat{Y}$. Because the line is mathematically determined by its intercept (a) and slope (b), a formula for $\hat{Y}$ incorporates their expressions (equations 125 and 126) and allows us to insert any value of X. Equation 127 shows the general regression equation, the $\hat{Y}$ equation. Equation 128 substitutes actual quantities for a and b and reduces the equation to its most usable terms. We use equation 128 to predict a Y score ($\hat{Y}$) for any value of X.

(127) $$\hat{Y} = a + b\,(X)$$

(128) $$\hat{Y} = \overline{Y} + r_{XY}\left(\frac{s_Y}{s_X}\right)(X - \overline{X})$$

3. Given the previous data for Alameda University, how many job offers would you predict for someone who had taken four statistics or calculus courses?

From our previous calculations and figures, we have:
$r_{XY} = .767$, $\overline{Y} = 3.333$ offers, $\overline{X} = 2.333$ courses, $s_Y = 2.225$ offers, and $s_X = 2.024$ courses. The given X value is 4 courses. Therefore,

$$\hat{Y} = \overline{Y} + r_{XY}\left(\frac{s_Y}{s_X}\right)(X - \overline{X}) = 3.333 + .767\left(\frac{2.225}{2.024}\right)(4 - 2.33)$$

$$= 3.333 + .767\,(1.099)\,(1.667) = 3.333 + 1.405 = 4.738 \ \text{job offers}$$

We predict that taking four calculus or statistics courses results in 4.738 recruitment offers. If one were to trace the line in Figure 9.3, it approximates this many job offers, but the formula is much more precise.

A Confidence Interval for $\hat{Y}$

We know an *XY* correlation is rarely, if ever, perfect. As a result, whenever we calculate $\hat{Y}$ based on *X*, there is predictive error involved. By using equation 128, we predict a *Y* value falling on the regression line. That line, however, represents an average of sorts. Actual *Y* values occur both above and below the *Y*-predicted line. This is apparent with just the 15 data points in Figure 9.3. When *X* = 0, for example, the *Y* values are 0, 2, and 3 and lie above and below the $\hat{Y}$ line. We see similar variations in *Y* scores for other values of *X*. For all *X* scores, people received more job offers than predicted, fewer offers, or both. As the regression line progresses from left to right, it runs through the middle or the average of these data points. It is the best single straight line that passes through the middle of the dots such that all variations or distances of data points from the line, both above and below it, are as small as possible. (Hence the term "line of best fit.") But lacking perfect correlations, there are always variations around the regression line, and we make an assumption about those variations.

With large data sets, we assume the actual *Y* values for any *X* score are normally distributed around the $\hat{Y}$ line. Figure 9.4 is drawn on a flat page and in two-dimensional space, of course, but imagine the small curves in a third dimension, coming straight at you off the page. They represent the normal distributions of actual *Y* scores around predicted $\hat{Y}$ scores. Any $\hat{Y}$ value (on the line) then becomes the central point estimate as we construct a confidence interval for our prediction.

To establish a confidence interval for $\hat{Y}$ we need a measure of the actual spread around the $\hat{Y}$ line. Do the data points (dots) deviate markedly from the line or do they cluster fairly close to it? The spread around the line is directly related to our predictive error: The more spread, the more the actual *Y* values differ from our predicted $\hat{Y}$ (on the line), and the more predictive error we have. The measure of this spread or predictive error is the standard error of estimate. A standard deviation of sorts, the **standard error of estimate** *is the standard error for the amount of vertical spread around the $\hat{Y}$ line*. It measures the variation in the distribution of the actual data points around the line. The less the variation, the smaller the standard error, and the more accurate our predictions.

The standard error of estimate is written as s_{YX} (read as "the standard error of *Y* on *X*"), and is the standard error when predicting *Y* from *X*. It can be calculated by using equation 129.

(129)
$$s_{Y.X} = s_Y \sqrt{1 - r_{XY}^2}$$

How far above and below the line (above and below $\hat{Y}$) does a given confidence level take us? How many standard errors of estimate? This depends upon our confidence level and on whether we use *z* or *t* to derive our confidence interval.

Figure 9.4 For Any Given Value of *X*, the Normal Variation of Possible *Y* Values Around the Line of Best Fit

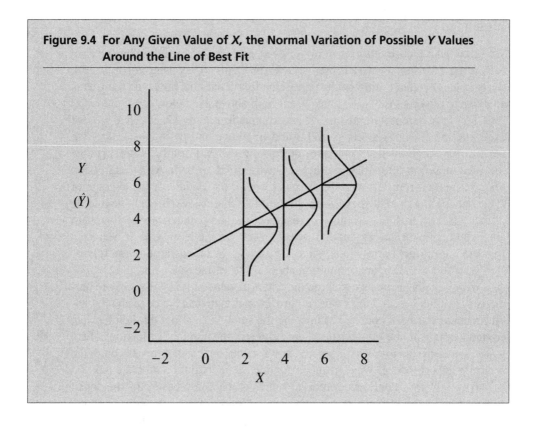

The now-familiar rule applies: if $n < 30$, use t; otherwise, use z. t has $(n - 2)$ degrees of freedom, just as for the earlier hypothesis test involving r_{XY}. Equation 130 is shown with t, but z can be substituted as appropriate.

(130) $$\text{C.I. Estimate for } \hat{Y} = \hat{Y} \pm t \, (s_{YX}),$$

where t has $(n - 2)$ degrees of freedom.

4. *Given four calculus or statistics courses, what is the 95% C.I. estimate for the number of job offers received?*

The point estimate, $\hat{Y} = 4.738$ offers, $s_Y = 2.225$, and with $n - 2 = 13$ degrees of freedom at the .95 confidence level, $t = \pm 2.160$.

$$s_{Y.X} = s_Y \sqrt{1 - r_{XY}^2} = 2.225\sqrt{1 - (.767)^2} = 2.225\sqrt{.412} = 2.225(.642) = 1.428$$

.95 C.I. Est. for $\hat{Y} = \hat{Y} \pm t \, (s_{YX}) = 4.738 \pm 2.160 \, (1.428) = 4.738 \pm 3.084$ offers

For the C.1.'s lower limit: 4.738 – 3.084 = 1.654 recruitment offers

For the C.I.'s upper limit: 4.738 + 3.084 = 7.822 recruitment offers

The distribution curve is shown in Figure 9.5.

Based on our study, we may be 95% certain that a student graduating from Alameda University with 4 calculus or statistics courses receives between 1.654 and 7.822 (roughly 2 to 8) job offers. Or, we estimate he or she receives 4.738 job offers, with a margin of error of ±3.084 offers.

Up to this point, we first derive a correlation coefficient as a measure of our *XY* association, and we subject this coefficient to a hypothesis test. If r_{XY} comes from a population in which ρ_{XY} differs significantly from zero, we may use it as a basis for predicting $\hat{Y}$ based upon any value of *X*. As a final step, we establish a projected confidence interval estimate around $\hat{Y}$. Beyond this, there is one final function for Pearson's r_{XY}.

Does *X* Explain *Y*? A Coefficient of Determination

Our focus here is the variation in the *Y* scores. Their **total variation** *reflects how much they vary around the mean* of *Y*, or $\overline{Y}$. This is measured by s_Y, the standard deviation for the *Y* scores (see Chapter 2). In a scatterplot, this is how much the actual data points deviate above and below a horizontal line drawn through the dots at $\overline{Y}$ (see Figure 9.6). In a formula measuring this variation, the numerator includes the term $\Sigma (Y - \overline{Y})$, the sum of variations from the *Y* average? However, we may partition this total variation into two parts.

Considering the influence of *X* on *Y*, the total variation in the dependent variable (*Y*) actually consists of both explained and unexplained variation. Explained

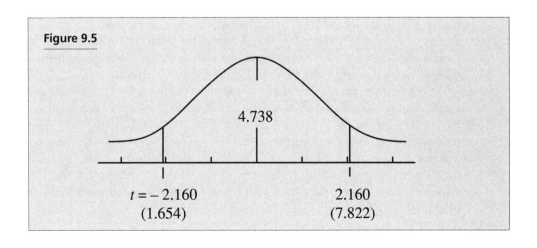

Figure 9.5

4.738

$t = -2.160$ 2.160
(1.654)   (7.822)

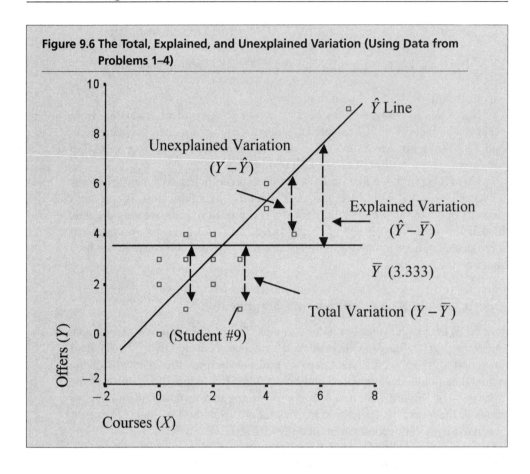

Figure 9.6 The Total, Explained, and Unexplained Variation (Using Data from Problems 1–4)

variation takes the *XY* correlation into account when making predictions rather than just using the mean of *Y* ($\overline{Y}$). **Explained variation** *is the amount of the total variation we are able to account for with our regression line*. It is how much less variation there is from the $\hat{Y}$ regression line than from the *Y* average. If there is any *XY* correlation whatever, the data points fall closer to the line of best fit than to the $\overline{Y}$ line. After all, by definition, the line of best fit minimizes the deviation of the *XY* data points from itself. The mean of *Y* does not; it takes into account only the *Y* values, not the *XY* bits of information. Thus, with an *XY* correlation, there is less variation around the regression line than around the mean of *Y*. Assuming *X* and *Y* *are* correlated, we gain some explanatory or predictive power by using the $\hat{Y}$ line rather than predicting everyone will get the same average $\overline{Y}$ score. This improvement is measured by $\Sigma\,(\hat{Y} - \overline{Y})$; for all values of *X*, it represents the sum of the differences between the improved predictive $\hat{Y}$ values and the $\overline{Y}$ average.

Implied in this is the fact that, if there is any XY correlation at all, s_{YX} is smaller than s_Y. s_{YX} reflects how much the data points deviate from the $\hat{Y}$ line, and this deviation should be less than the deviation around the mean of Y ($\overline{Y}$) as measured by s_Y. Only if there is absolutely no XY correlation—only if r_{XY} is virtually zero—are these two figures about equal. Under those circumstances, any regression line gives us about the same amount of explanatory power as does $\overline{Y}$. If there is no XY correlation, knowing X and using a $\hat{Y}$ line to make predictions offers no advantage. We have just as much accuracy predicting the mean score of Y regardless of X.

The second source of variation is that for which the XY correlation cannot account, the unexplained variation. The **unexplained variation** *is the deviation of the actual* Y *scores from the* $\hat{Y}$ *line (the* $\hat{Y}$ *scores)*. It is the random variation in the Y scores that even the regression line does not explain. In our scatterplot (Figure 9.6), this is depicted by the variation of the data points from the $\hat{Y}$ regression line. Even though, by definition, the $\hat{Y}$ line fits or represents the data best, it is still not perfect. Unless we have a perfect linear correlation and $r_{XY} = 1.00$ (or -1.00), some of the data points have to vary from the regression line. Even using our $\hat{Y}$ line to make predictions (an improvement over using $\overline{Y}$), there is still random or residual variation in the Y scores that we cannot explain or account for statistically. Some of them *still* vary from the $\hat{Y}$ line. This variation is expressed as $\Sigma\,(Y - \hat{Y})$: for all X values, it is the sum of how much the true Y scores vary from the predicted $\hat{Y}$ scores.

These sources of variation may be illustrated by examining one data point in Figure 9.6 more closely. Student #9 took 3 statistics or calculus courses ($X = 3$) and received 1 job offer ($Y = 1$). The total variation in his case is $Y - \overline{Y}$ or $1 - 3.333 = -2.333$. In Figure 9.6, student #9 appears to be slightly more than 2 job offers below the average. For the explained variation, the predicted number of job offers is 3.895 (without showing the math); that is, $\hat{Y} = 3.895$ when $X = 3$. As expected, the $\hat{Y}$-line shows a value just below 4 when $X = 3$. Finally, for the residual or unexplained variation, we compare the actual number of job offers to those predicted, or $Y - \hat{Y} = 1 - 3.895 = -2.895$ for student #9. He or she is almost 3 job offers below what we would expect, given our $\hat{Y}$ line. This is student #9's contribution to the sum of squares for the overall error factor, or by how much our predictions for this student might vary from his or her actual number of job offers.

Returning to Pearson's r_{XY}, the **coefficient of determination,** r_{XY}^2, *is the ratio of the explained variation to the total variation*. What proportion of the total variation does the $\hat{Y}$ regression line explain? Put another way, how much of our predictive error do we eliminate by using our knowledge of X and the XY correlation to estimate Y as opposed to using only $\overline{Y}$? r_{XY}^2 tells us this, and, for this reason, is described as a **proportional reduction in error** (or PRE) statistic (see equation 131).

(131) $$r_{XY}{}^2 = \frac{\text{Explained variation}}{\text{Total variation}} = \frac{\Sigma(\hat{Y}-\overline{Y})}{\Sigma(Y-\overline{Y})}$$

We square our correlation coefficient to determine how much of the variation in the Y scores is explained by or attributable to variations in the X scores. For our example, $r_{XY} = .767$, and $(.767)^2 = .588$. That means 58.8% of the variation in the number of job offers is statistically attributable to (or explained by) the number of calculus or statistics courses a graduate took.

To review the steps in correlation and regression, an additional problem puts them into one comprehensive question.

5. *A college student wondered whether the numbers of students' comments and questions asked were related to class size. In fact, she suspected that larger classes produced fewer student remarks. She videotaped randomly selected lecture classes for four class meetings each and then carefully reviewed each tape. Her data for 7 cases or classes are shown below.*

Class:	1	2	3	4	5	6	7
Class size:	30	80	24	50	60	44	18
Number of comments and questions:	55	42	50	32	33	34	64

    a. *Is there an* XY *correlation between class size and the number of student remarks?*
    b. *Is the* r_{XY} *correlation significant at the .05 alpha level?*
    c. *How many comments and questions would the researcher predict for a class of 30 students?*
    d. *What is the .90 confidence interval estimate for the student's prediction made in (c)?*

For part (a) of the problem, class size is assumed to be the predictor variable and is designated X. The Y variable, or dependent variable, becomes the number of students' remarks recorded. The parameters for this problem are presented in Table 9.2.

From equation 120,

$$r_{XY} = \frac{\Sigma XY - \dfrac{(\Sigma X)(\Sigma Y)}{n}}{\sqrt{\Sigma X^2 - \dfrac{(\Sigma X)^2}{n}}\sqrt{\Sigma Y^2 - \dfrac{(\Sigma Y)^2}{n}}} = \frac{12{,}438 - \dfrac{(306)(310)}{7}}{\sqrt{16{,}236 - \dfrac{(306)^2}{7}}\sqrt{14{,}654 - \dfrac{(310)^2}{7}}}$$

$$= \frac{-1113.429}{\sqrt{2859.429}\,\sqrt{925.429}} = \frac{-1113.429}{(53.474)(30.421)} = \frac{-1113.429}{1626.733} = -.684$$

Table 9.2

Class	Size (X)	Remarks (Y)	X^2	Y^2	XY
1	30	55	4900	3025	1650
2	80	42	6400	1764	3360
3	24	50	576	2500	1200
4	50	32	2500	1024	1600
5	60	33	3600	1089	1980
6	44	34	1936	1156	1496
7	18	64	324	4096	1152
Totals	$\Sigma X = 306$	$\Sigma Y = 310$	$\Sigma X^2 = 16,236$	$\Sigma Y^2 = 14,654$	$\Sigma XY = 12,438$

There is a reasonably strong negative correlation. As class size increases, there is an obvious tendency for the number of student questions and remarks to decrease. The researcher's suspicions appear to be correct.

For part (b) of problem 5, since the student has specific expectations for her correlation—that Y will decrease as X increases—she does a lower-tailed test. The question is: Is r_{XY} from a population in which ρ_{XY} is significantly negative and less than .00? For the t test, $t_c = -2.015$ with $(n - 2) = 5$ degrees of freedom at the .05 alpha level. For r_{XY} with $n = 7$, Table H shows a value of .669, which is $-.669$ in the lower tail. The hypotheses are: $H_A: \rho_{XY} < .00$ $H_0: \rho_{XY} \geq .00$. The null hypothesis curve is shown in Figure 9.7. Using equation 123, we get:

$$t_O = r_{XY}\sqrt{\frac{n-2}{1-r_{XY}^2}} = -.684\sqrt{\frac{7-2}{1-(-.684)^2}} = -.684\sqrt{\frac{5}{.532}}$$

$$= -.684\sqrt{9.398} = -.684(3.066) = -2.097$$

There is sufficient evidence to reject H_0 because $t_c = 2.015$, and $t_o = -2.097$. Confirming these results, $r_{XY_c} = -.669$, and $r_{XY} = -.684$. r_{XY} comes from a population in which ρ_{XY} must be significantly less than .00 at the .05 alpha level. With this result, the student is justified in using r_{XY} as a basis for predicting $\hat{Y}$ values.

In addition, $r_{XY}^2 = .468$. Without taking any other factors into account, about 47% of the variation in numbers of student comments is statistically attributable to class size.

To answer part (c) of the problem, even though an X value of 30 occurs as one of the data points in the sample, we cannot merely cite its corresponding Y score (55) as an answer. That represents just one observation. Our prediction must be based upon the XY correlation over all seven cases.

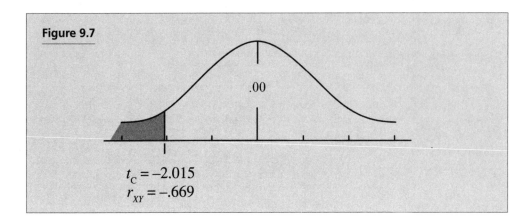

Figure 9.7

.00

$t_c = -2.015$
$r_{XY} = -.669$

We also need the averages and standard deviations. The averages are easy, of course, and earlier calculations for r_{XY} now allow us to get the standard deviations quickly. First, the averages:

$$\bar{X} = \frac{\sum X}{n} = \frac{306}{7} = 43.714$$

$$\bar{Y} = \frac{\sum Y}{n} = \frac{310}{7} = 44.286$$

The standard deviations use the *sums of squares* (Chapter 2), or the sum of the squared deviations from the mean. The sums of squares for X and for Y appear in the denominator of the r_{XY} formula (equation 121). Dividing those sums of squares by $n-1$ before taking the square roots gives us equation 8 and our standard deviations. For example, for the X variable in the Pearson's r_{XY} denominator:

$$\sqrt{\sum X^2 - \frac{(\sum X)^2}{n}}$$

and for the X variable in equation 8 for s_X

$$s_X = \sqrt{\frac{\sum X^2 - \frac{(\sum X)^2}{n}}{n-1}}$$

To get our standard deviations, first for X, the sum of squares was 2859.429, and

$$s_X = \sqrt{\frac{2859.429}{n-1}} = \sqrt{\frac{2859.429}{6}} = \sqrt{476.500} = 21.829$$

For the Y variable, the sum of squares was 925.429:

$$s_Y = \sqrt{\frac{925.429}{n-1}} = \sqrt{\frac{925.429}{6}} = \sqrt{154.238} = 12.419$$

We now have all the necessary bits of information to get $\hat{Y}$ from equation 128:

$$\hat{Y} = \overline{Y} + r_{XY} \ \frac{s_Y}{s_X} (X - \overline{X}) = 44.286 + [(-.684) \ \frac{12.419}{21.829} \ (30 - 43.714)]$$

$$= 44.286 + [-.684(.569)(-13.714)] = 44.286 + 5.337 = 49.623$$

For a class of 30 students and based upon r_{XY}, we expect 49.623 or about 50 comments and questions over four class meetings. To add to the estimate, we finally determine a confidence interval estimate for $\hat{Y}$.

Finally, in answer to part (d) of problem 5, the confidence interval requires the standard error of estimate for Y on X, $s_{Y.X}$, and the t value (not z, since $n < 30$) for the .90 confidence level with $(n - 2)$ degrees of freedom.

$$s_{Y.X} = s_Y \sqrt{1 - r_{XY}^2} = 12.419\sqrt{.532} = 12.419(.729) = 9.058$$

(Note that $1 - r_{XY}^2 = .532$ was already calculated for t_O in part (b).)

Then, $t = \pm 2.015$ with 5 d.f. at the .90 confidence level, according to Table C, and

.90 C.I. Est. for $\hat{Y} = \hat{Y} \pm t \ (s_{Y.X}) = 49.623 \pm 2.015 \ (9.058) = 49.623 \pm 18.252$

For the C.I.'s lower limit: $49.623 - 18.252 = 31.371$ comments/questions

For the C.I.'s upper limit: $49.623 + 18.252 = 67.875$ comments/questions

The distribution curve for $\hat{Y}$ is shown in Figure 9.8.

With 90% confidence, the researcher can estimate between 31.4 and 67.9 comments and questions in a class of 30 people over 4 class meetings. Her margin of error is a large ± 18.3 remarks around her earlier $\hat{Y}$ figure. The downside, of course, is not her confidence level but a low n of just 7 cases. Her predictive precision suffers somewhat due to a small sample size.

Finally, we have a correlation measure for ordinal or rankable data. This is a handy correlation tool when one or both quantitative variables do not meet the rigor of interval-ratio measurement. Spearman's r_s is less complicated than r_{XY}, but it does permit a measure of correlation and a test of its significance.

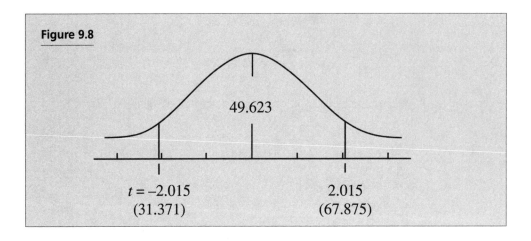

Figure 9.8

49.623

$t = -2.015$ 2.015
(31.371)   (67.875)

Ranks: Spearman's Correlation (r_s)

Spearman's rank-order correlation coefficient, as the name suggests, is suitable for ordinal or ranked data. **Spearman's r_s** *is appropriate for numerical measurements lacking the precision and reliability of interval-ratio data.* To calculate r_s, we separately rank the X and Y scores and then calculate a measure of association between respondents' X and Y ranks.

We do the same as we did earlier when we ranked ordinal scores in Chapters 6 and 7. We now ask: Do those ranked high on X also tend to rank high on Y? Or is the ranking high on X and low on Y? After ranking both the X and Y observations, we obtain a ***d* score** (d), the difference between the rank on X and the rank on Y, as before, for each pair of observations. We calculate r_s using the d scores. Consistent with Pearson's r_{XY}, r_s varies only between -1.00 and $+1.00$, and its interpretation is the same as that for r_{XY}.

(132)
$$r_S = 1 - \left(\frac{6\sum d^2}{n(n^2 - 1)} \right)$$

The significance test for r_s is also identical to that for r_{XY}.

Is r_s Significant?

The Spearman's correlation test has t and z versions and a table of significant values for r_s. The t and z tests use the same formulas as for r_{XY}, substituting r_s for r_{XY}. The t_O and z_O formulas are merely equations 123 and 124 rewritten to include r_s.

(133)
$$t_O = r_S \sqrt{\frac{n-2}{1 - r_S^2}}$$

where t_O has $(n-2)$ degrees of freedom, and

(134)
$$z_O = r_S\sqrt{n-1}$$

In using equations 133 and 134, we follow the now-familiar rules. If we have fewer than 30 cases or paired scores, we use t; with 30 cases or more, we use z. In lieu of the formulas, however, we have Table I and critical values for r_S. The r_{S_C} values are found in the table at the intersection of our sample sizes (n's) and alpha levels.

6. *A professor gave an examination consisting of objective and essay parts. Up to 20 points were possible on the objective segment and as many as 30 points on the essay component. For the 12 students taking the exam, the raw scores on the respective parts of the test are shown below.*

Student: *1   2   3   4   5   6   7   8   9 10 11 12*
Objective score (X): *18 15 12 19 10 14 17 18 13 20 11 16*
Essay score (Y):   *24 25 20 21 19 23 28 27 22 26 17 18*

a. *What is the correlation between the objective and essay scores, and how would you describe it?*
b. *At the .05 alpha level, is the professor correct in assuming there is a positive and significant association between the ranks?*

To answer part (a), recognizing that we may not have interval-ratio data for both the *X* and *Y* variables, we first convert our raw data to ranks in each distribution. As in earlier chapters, rules apply. First, the *X* and *Y* scores may be ranked from highest to lowest (as done in Table 9.3) or in the reverse order. Either way works, but you must be consistent and rank both variables the same way. Second, tied *X* scores or tied *Y* scores are each given the average of the ranks they would otherwise occupy, e.g., identical *X* scores at ranks 3 and 4 are each ranked 3.5. Finally, as common sense suggests, all the *d* scores are calculated in a consistent fashion (here, *X* rank minus *Y* rank). The magnitudes of the positive or negative differences (*d* scores) should cancel each other out and total to zero.

The Spearman's correlation r_S is found by using equation 132:

$$r_S = 1 - \left(\frac{6\Sigma d^2}{n(n^2-1)}\right) = 1 - \left(\frac{6(108.50)}{12(144-1)}\right) = 1 - \left(\frac{651}{1716}\right) = 1 - .379 = .621$$

This value for r_S indicates that there is a fairly strong correlation between the ranks of *X* and those of *Y*. There is an obvious tendency for someone ranking

Table 9.3

Student	Objective (X)	Essay (Y)	Rank on X	Rank on Y	d	d^2
1	18	24	3.5	5	−1.5	2.25
2	15	25	7	4	3	9.00
3	12	20	10	9	1	1.00
4	19	21	2	8	−6	36.00
5	10	19	12	10	2	4.00
6	14	23	8	6	2	4.00
7	17	28	5	1	4	16.00
8	18	27	3.5	2	1.5	2.25
9	13	22	9	7	2	4.00
10	20	26	1	3	−2	4.00
11	11	17	11	12	−1	1.00
12	16	18	6	11	−5	25.00
Totals					$\Sigma d = 0.0$	$\Sigma d^2 = 108.50$

high on the objective test to also rank high on the essay test. Students doing well on one test did well on the other.

To answer part (b) of problem 6, with $n = 12$, a t test and equation 133 work. The professor wishes to test for a positive association between students' objective and essay ranks. Table C shows that for an upper-tailed test at the .05 alpha level with $(n - 2) = 10$ degrees of freedom, $t_C = 1.812$. From Table I, the critical value for r_s is $r_{s_C} = .503$ when $n = 12$ and $\alpha = .05$.

Finally, to frame the professor's hypotheses, we again use rho (ρ) as the symbol for r_s's population parameter: ρ_s. We are asking whether $r_s = .621$ may come from a population in which ρ_s is significantly greater than .00. Our hypotheses are: $H_A: \rho_s > .00$; $H_0: \rho_s \leq .00$. The distribution curve is shown in Figure 9.9.

Figure 9.9

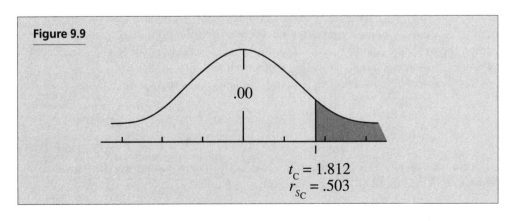

$$t_C = 1.812$$
$$r_{s_C} = .503$$

From equation 133, we get

$$t_O = r_S \sqrt{\frac{n-2}{1-r_S^2}} = .621 \sqrt{\frac{12-2}{1-(.621)^2}} = .621 \sqrt{\frac{10}{.614}} = .621\sqrt{16.287} = .621(4.036) = 2.506$$

The professor may reject her null hypothesis at the .05 alpha level because t_O exceeds t_C, and r_S is larger than r_{SC}. Her suspicions are correct. A positive and statistically significant association exists between students' objective and essay test ranks.

Remember that when we conduct the H_O test for Spearman's correlation, we compare t_O and t_C, *or* we evaluate r_S against r_{SC}. We never combine these two separate statistical tests and compare, for example, t_O to r_{SC} or r_S to t_C.

Also, note that the H_O test for ρ_S is as far as we may go. There is no regression equivalent for Spearman's r_S. We do not predict $\hat{Y}$ values given significant r_S results; the data do not justify that sort of extrapolation. Without interval-ratio data we cannot calculate the required means and standard deviations. We do not use our original X and Y values at all, other than to convert them to ranks. Spearman's r_S is for rank-order correlation only, not regression.

7. *For a series of reviews in their college newspaper, a team of students sampled campus-area restaurants serving reasonably priced hamburgers. On a 0 to 50 scale, they rated each for the burgers' quality and appeal. On a similar scale, they also rated the promptness, thoroughness, and friendliness of the service. Their data are given below.*

Restaurant:	*1*	*2*	*3*	*4*	*5*	*6*	*7*	*8*	*9*	*10*	*11*
Food rating (X):	27	42	35	40	33	38	45	31	43	30	48
Service rating (Y):	41	23	20	32	42	32	27	39	26	38	40

   a. *Is there a correlation between the students' food and service ratings, and how would you describe their association?*
   b. *Is the relationship between the restaurants' food and service ratings significant at the .05 alpha level?*

To answer part (a), rank the variables to determine a value for d^2, as shown in Table 9.4, in order to determine r_S from equation 132.

$$r_S = 1 - \left(\frac{6\Sigma d^2}{n(n^2-1)} \right) = 1 - \left(\frac{6(310.50)}{11(121-1)} \right) = 1 - \left(\frac{1863}{1320} \right) = 1 - 1.411 = -.411$$

The student team's evaluations reveal a negative association between the rankings of local eateries' hamburgers and service. When restaurants' burgers

Table 9.4

Restaurant	Food (*X*)	Service (*Y*)	Rank on *X*	Rank on *Y*	*d*	*d²*
1	27	41	11	2	9	81.00
2	42	23	4	10	−6	36.00
3	35	20	7	11	−4	16.00
4	40	32	5	6.5	−1.5	2.25
5	33	42	8	1	7	49.00
6	38	32	6	6.5	−.5	.25
7	45	27	2	8	−6	36.00
8	31	39	9	4	5	25.00
9	43	26	3	9	−6	36.00
10	30	38	10	5	5	25.00
11	48	40	1	3	−2	4.00
Totals					$\Sigma d = 0.0$	$\Sigma d^2 = 310.50$

rank higher, there is a moderate tendency for their service to rank lower and vice versa.

The question in part (b) of problem 6 asks about "the association," not about a positive correlation between the rankings. A two-tailed test is appropriate. With the students' *n* = 11 ratings, a *t* test works. The hypotheses are: H_A: $\rho_S \neq .00$; H_O: $\rho_S = .00$.

Let us first consult Table I to determine r_{S_C}. The critical value r_{S_C} (*n* = 11 and $\alpha = .05$) is ±.618. Our r_S value of −.411 is therefore not significant. We may not reject H_O. The association between restaurants' hamburgers and service is not significant. Despite the students' rankings, their r_S comes from a population in which ρ_S equals zero.

The *t* test, although technically unnecessary at this point, confirms these results. From equation 132:

$$t_O = r_S \sqrt{\frac{n-2}{1-r_S^2}} = .621 \sqrt{\frac{12-2}{1-(.621)^2}} = .621 \sqrt{\frac{10}{.614}} = .621\sqrt{16.287} = .621(4.036) = 2.506$$

At the .05 alpha level, with *n* − 2 = 9 degrees of freedom. Table C indicates that $t_C = \pm 2.262$. The distribution curve is shown in Figure 9.10.

Again, since $t_C = \pm 2.262$ and $t_O = -1.353$, we lack evidence to reject the null hypothesis. The burger and service rankings are unrelated at the .05 alpha level.

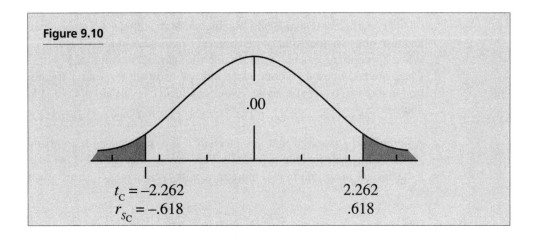

Figure 9.10

$t_C = -2.262$ 2.262
$r_{S_C} = -.618$ .618

Summary

The correlation coefficient, r_{XY}, and the rank-order correlation coefficient, r_S, are the two principal measures of correlation for quantitative data. With r_{XY} we may also predict $\hat{Y}$ values based upon statistically significant XY correlations. The rank-order correlation coefficient does not permit that step, but it does allow us to usefully work with quantitative ordinal data.

Exercises

1. A sample of 8 people was asked the numbers of people living in their households and the typical numbers of times the family or household dined out together per month. The results are:

Size of household: 3   2   5   7   4   3   4   4
Times dined out: 3   5   2   0   1   4   2   1

 a. What are the intercept and slope for this set of data?
 b. Please construct a scatterplot for these data and approximate the regression line using the Y intercept and slope.
 c. Given your line of best fit, what $\hat{Y}$ value would you predict for a 4-person household?

For Exercises 2 through 17, please:

 a. Calculate and describe the appropriate measure of correlation, including the coefficient of determination when using r_{XY}.
 b. State your hypotheses and conclusions in determining whether your cor-

relation coefficient is statistically significant. Start with the .05 alpha level, and also note whether your correlations are significant at the .01 level.
c. When appropriate, calculate the $\hat{Y}$ value for the median value of X.
d. Calculate and interpret a confidence interval estimate for your $\hat{Y}$ prediction in part (c), being sure to vary your confidence levels among .90, .95, and .99.

2. An employer suspected that absenteeism was associated with company tenure. He randomly selected the records of 7 employees and recorded the data below. Is he correct in assuming that more time with the company means fewer absences?

Years with company: 3   6   7   7   5   2   1
Days absent last quarter: 10   0   1   4   3   7   5

3. A movie studio convened a focus group to rate a new release. The studio compared each person's number of movies seen in theaters during the last 3 months with his or her rating (on a 0–15 point scale) of its new film. Are movie attendance and participants' ratings related?

Movies seen: 4   7   9   3   1   7   8   5   12   10   2   0
Rating for new release: 7   15   5   4   6   12   6   8   14   11   2   1

4. Below are data for a sample of people convicted of identical crimes committed under similar circumstances. Do older people receive lighter sentences?

Age at conviction: 23   27   18   34   28   33   21
Sentence (in years): 12   10   14   9   11   11   12

5. Does the percentage of children living in poverty correlate with an area's cost of living (COL)? For the 12 largest US cities, census data permit indices of children living in poverty or extreme poverty and cities' costs of living compared to the US average, scored at 1.00.

COL compared to
  US mean: 1.57 1.49 1.37 1.34 1.33 1.32 1.21 1.14 1.00 .97 .96 .92
Percentage minors
  in poverty: 14   39   44   34   32   48   60   31   43   22   40   47

6. A medical sociologist suspected an indirect association between personal income and time since last physical examination. She asked a small sample of people about their before-tax monthly incomes (recorded in thousands of dollars) and the approximate numbers of months since their last physical exams. Given this evidence, are her suspicions correct?

Income (in $1000s): 2.2 3.7 5.8 7.1 4.2 8.3 2.9 5.2 6.4 3.8 9.1 4.6 4.9
Months since
  last physical: 23   9   8   4   15   5   30   7   10   13   36   24   11

7. A psychologist asked office workers to rank the amenities and décor of their workplaces (personal spaces, plants, music, opening windows, and so on) and also measured their job satisfaction. His physical environment scale went from a low of 0 to a high of 20, and his job satisfaction measure from 10 to 50. Does décor appreciation correlate with greater job satisfaction?

Office rating: 18   10   5   12   15   16   9   17   20   6
Job satisfaction: 42   31   28   30   34   38   23   43   45   24

8. Looking at videotapes, the Midwestern Athletic Conference graded the performances of 11 officials refereeing its men's and women's basketball games. It gave officials letter grades, and it wishes to correlate those grades with officials' years of experience. Are conference administrators correct in assuming more experience correlates with higher grades (after quantifying grades on a 12-point scale: A+ = 12 points, A = 11, A– = 10, B+ = 9, and so on down to F = 0)?

Years experience: 4   1   7   3   5   12   9   11   6   2   8
Grade: B–   C   A+   C+   D+   A   B   B+ A–   C–   B

9. A company contacted randomly selected people who had recently called its customer service department. They were asked about how long they had been on hold before talking to a representative and how they rated the company's customer service response, with ratings scored on a scale of 7 (–) to 21 (+). Does more time on hold mean a lower rating?

Estimated minutes
  on hold: 8   2   3   5   7   1   10   6   4   5   9   6   13   11
Service rating: 9   18   15   10   9   20   8   14   16   13   11   12   10   17

10. Randomly selected commuting students were asked the times they normally arrived on campus and the number of minutes it usually took them to find parking and be on their way to class. The approximate arrival times were recorded as 7.0 for 7:00 a.m., 7.5 for 7:30 a.m., and so on. Do earlier arrivals find parking more quickly?

Arrival time:	7.0	7.5	8.0	8.5	9.0	9.5	10.0	10.5	11.0	11.5	12.0
Minutes to park:	5	6	10	17	24	21	20	22	18	15	20

11. Two chronic features of end-of-the-year celebrations are budgets and diets. Are they correlated? If people manage to maintain or even lose weight over the holidays (and feel better about their self-images), do they tend to spend more or less money? We asked 9 randomly selected adults how many pounds they had gained or lost and how much money they had spent that was probably frivolous or unnecessary.

Weight gain/loss:	7	–2	8	5	4	–3	6	0	1
Extra money spent:	200	110	160	150	0	100	240	50	80

12. A random sample of English-as-a-Second-Language (ESL) immigrants was asked what percentage of the time they spoke English and also to complete an assimilation index. The researcher suspected more frequent English speakers would be more highly assimilated (i.e., score more highly on their 20- to 80-point scale).

Percent time speaking English:	70	20	50	90	25	95	10	33
Assimilation index:	67	32	49	62	40	72	54	35

13. College students were asked how many units they were taking, and to judge, from a low of "very poorly" (1) to a high of "very well" (10), how adequately they were able to balance work, school, and family obligations. Do students taking more units find it more difficult to balance their varied obligations?

Units:	17	13	12	18	8	9	11	15	14	16
Balance rating:	2	6	8	5	3	7	10	9	4	8

14. Based upon promotional pictures, a realtor asked a random sample of her colleagues and a client of each to complete a questionnaire ranking the appearance and "curb appeal" of a particular property. Do the two samples agree? Given a maximum score of 60, is there a positive correlation between their ratings?

Realtor: 40 52 48 56 46 50 53 53 45
Client: 45 48 55 51 42 53 40 47 50

15. A child development professor conducted a study in which a sample of seventh graders was asked how many books or magazines each had read for pleasure in the last two weeks and about how many hours of television each watched per day (rounded to the nearest whole number). Does more TV correlate with less reading?

TV hours per day: 5   4   5   6   3   1   2   5
Books or magazines read: 2   2   0   0   4   3   3   2

16. A youth soccer coach, hovering between appreciation of and frustration with her players' parents and grandparents, wondered whether her team scored more goals with fewer family members at games. She kept records for the next 6 matches. Is she correct? Do fewer family members mean more goals?

Total family present: 27 36 19 24 31 20
Goals scored: 1   0   3   3   1   2

17. As part of his initial interviews, a bankruptcy counselor asked clients how many years of education they had and the peak percentages of their savings ever held in stocks and bonds. Do higher educational levels correlate with more savings ever held in stocks and bonds?

Years of education: 11 14 17 9 10 18 13 12 17 16 20 15
Percent in stocks
  and bonds: 20 26 11 5   0   0 12 18 15 32 24 10

18. Why does Spearman's r_s use data ranks rather than the X values themselves?

19. What are explained and unexplained variation, and how are they part of correlation and regression?

20. What information do Tables H and I provide? Why do they sometimes substitute for t or z tests in checking the statistical significance of r_{XY} and r_s, respectively?

21. What are ρ_{XY} and ρ_s? What is the rationale for using these terms in hypotheses tests of r_{XY} and r_s?

22. In your own words, what is the regression (or $\hat{Y}$) line, its intercept, and its slope?

23. Why do we calculate $\hat{Y}$ only if r_{XY} is statistically significant?

24. Why do we not calculate a $\hat{Y}$ value if r_s is statistically significant?

25. In your own words, define s_{YX} and explain why it is typically smaller than s_Y. Under what conditions would it not be smaller? (Review your answer to Exercise 19. Your answer here should include references to both the total and unexplained variation.)

APPENDIX 1

Glossary

Numbers in brackets at the end of each entry indicate the chapter in which the term was introduced.

Absolute value. The mathematical value of a number without regard to its sign, positive or negative. The absolute value of –3 is 3, for example. **[2]**

Additive rule. The probability of any of a set of mutually exclusive outcomes is found by adding their respective probabilities. **[3]**

Alpha error (Type I error). Rejecting a null hypothesis that is really true. The sample statistic falls into the alpha region due to chance, not the experimental treatment or independent variable's effect. **[5]**

Alpha level. The proportion of the null hypothesis curve designated as a rejection or alpha region, usually 5% or 1% of the curve's tail area(s). **[5]**

Alpha region (rejection region). The tail(s) of the null hypothesis curve. Reject the null hypothesis when the sample statistic falls into the alpha region. **[5]**

Alternative hypothesis (H_A). The complement of the null hypothesis. H_A states that the sample statistic *does* differ from the population parameter, from random chance, or from a designated value. It may or may not state a direction of that difference also. **[5]**

Analysis of variance (ANOVA). A statistical test for significant differences between three or more sample means. **[8]**

Area (of curve). Shows the relative frequency or proportion of cases falling into certain parts of the distribution. [**3**]

Average, mean. The arithmetic balance point in a set of interval-ratio numbers such that the total of the deviations of the data points above (larger than) the mean are the same magnitude as the total of the deviations of data points below (less than) the mean. [**2**]

Average product of the z scores. A measure of the XY correlation. [**9**]

Beta error (Type II error). The failure to reject a null hypothesis that is really false. Despite belonging to an alternative population, the sample statistic falls within the bounds of normal variation of the null hypothesis curve. [**5**]

Between-groups variation (Treatment variation). With ANOVA, the variation between the means of the treatment groups. [**8**]

Bimodal. Describing a distribution in which two scores or categories have the largest and identical frequencies. A distribution with two modes. [**2**]

Binomial (variable). A variable for which each outcome or trial is either a 1 or a 0, Yes or No, Agree or Disagree, or a Success or Failure type of measurement and in which X is the number of "successes" over n trials. [**3**]

Binomial probability distribution. A distribution showing the probabilities of getting X successes (including each possible value of X) from n trials. When graphed above a horizontal axis representing all values of X, the distribution may be discrete and block-like, or it may on occasion appear continuous and to approximate the normal distribution. [**3**]

Binomial probability experiments, variables. Situations in which there are only two possible results or outcomes (e.g., a coin flip resulting in heads or tails). In the contexts of particular problems, the two outcomes are typically designated as success or failure. [**3**]

Central tendency. The typical or most representative observation(s) in a set of data. Measures of central tendency include the mode, the median, and the mean or arithmetic average. [**2**]

Chi square "goodness of fit" test (χ^2 test of normality) A version of the one-way chi square procedure in which observed frequencies are compared to hypothetical frequencies that would typify a normal distribution, or possibly compared to some other idealized set of frequencies. [7]

Chi square test. A hypothesis test in which actual or observed categorical frequencies are systematically compared to hypothetical frequencies, typically those dictated by random chance. [7]

Chi square test of independence. The two-way chi square test for the categorical frequencies of two cross-tabulated variables. The cell frequencies of intersecting categories are compared to random chance frequencies. If actual frequencies deviate from random chance at given alpha levels, the two variables are not independent. [7]

Coefficient of determination (r_{XY}^2). The XY correlation coefficient, Pearson's r_{XY}, squared. It measures the proportion or percentage of variation in Y that is statistically accounted for by variation in X. *See also* Proportional reduction in error. [9]

Coefficient of Variation (CV). The ratio of the standard deviation to the mean times 100. A way of comparing the relative sizes of standard deviations. [2]

Combinations. A mathematical procedure for deriving the total number of *different* sets or combinations of X items that may be selected from a larger total of *n* items. [3]

Complement. The opposite or counterpart of an event. In a coin flip, heads is the complement of tails. Among *n* items, if X is the event, $n - X$ is the complement. [3]

Confidence interval estimation. Based upon sample data, the construction of a range or interval into which, with a certain level of surety or confidence, the unknown population parameter is expected to fall. [4]

Confidence level. Based upon statistical theory and with confidence interval estimation, the degree of surety or the probability that the unknown population parameter will indeed fall into the interval. Common confidence levels are .90, .95, and .99. [4]

Constant. Something that is true of *every* person or element; opposite of variable. [1]

Contingency tables (cross-classification tables). A way of presenting data in which the categories of two variables are cross-classified. One variable becomes the column variable and the other the row variable. The intersection of their respective categories forms cells in which are presented joint frequencies or probabilities. [7]

Continuous data, variables. Data or measurements made in increments of less than whole units (i.e., in fractions). When graphed, exceedingly small increments appear as unbroken and inseparable units along one X continuum of measurements and may be represented by one unbroken line. *See also* Discrete data, variables. [3]

Correction factor. An adjustment sometimes applied in statistics that typically makes a sample statistic more representative of a population parameter. The use of a correction factor is often made in response to sampling contingencies. [2]

Cramér's V. A measure of association based upon the two-way chi square. May be used with contingency tables of any size. [7]

Criteria for continuity. Used with binomial distributions. The criteria (np and nq) used to determine whether the binomial probability distribution approximates the continuous normal distribution. [3]

Critical value (of a statistic). In a null hypothesis curve and along the horizontal axis, the actual value of a statistic falling at the critical point of rejection/acceptance for the null hypothesis. It is the equivalent of z-critical or t-critical, but is expressed in the original units of the sample statistic in question (e.g., $\overline{X}_C$, p'_C, or r_{XY_C}). [5]

Cumulative percentage. In a frequency distribution for rankable measurements, the running total of the proportion or percentage of cases accounted for as one proceeds from the lowest to the highest observations, or vice versa. [2]

***d* scores.** The differences between pairs of matched scores. In the case of related or matched samples, statistical methods are often based upon the differences (d scores) between each pair of values. [6]

Degrees of freedom (d.f.). Ways of taking sample size (n) or the number of categories (k) into account when interpreting statistical results. The relevant degrees of freedom are often $n - 1$ or $k - 1$. [4]

Dependent variable. In a multivariate statistical relationship, the variable presumably influenced by at least one other factor, the independent variable(s). Statistically, the task is typically to explain variation in the dependent variable by reference to either a known population parameter's variation or to possible variation in another known figure. *See also* Independent/predictor variable. **[5]**

Direct relationship, correlation, association. A relationship, correlation, or association in which the values of one variable increase as do the values of the other variable (or decrease as do values of the second variable). Also known as a positive correlation or association. **[9]**

Discrete data, variables. Data or measurements recordable only in whole units, not fractions or parts of units, such as the number of successes over n binomial trials: 0, 1, 2, etc. *See also* Continuous data, variables. **[3]**

Element. A generic term for a population or sample member. Particularly relevant where the population may consist of inanimate units and not people. **[1]**

Empirical indicators. Specific and concrete measurements used as indices of more abstract concepts and variables. For example, the use of income and/or years of education as measures of general socioeconomic level. **[1]**

Error term. *See* Margin of error.

Error variation. *See* Within-group variation.

Estimation. Deriving the population average or the population proportion based upon sample data. **[4]**

Event. The defined outcome of interest in a probability question. The outcome whose probability is sought. *See also* Outcome. **[3]**

Expected frequencies (E). In chi square, the frequencies we expect merely by random chance—or, occasionally, the frequencies dictated by a specified criterion, such as normality. **[7]**

Expected value $E(X)$. On the average, given the probability of success, how many successes we should expect over a given number of trials. **[3]**

Explained variation. In regression, that part of the total variation (around $\overline{Y}$) we are able to improve upon or reduce by using the $\hat{Y}$ line to make predictions instead of using $\overline{Y}$. The reduction in variation made by our $\hat{Y}$ line over the $\overline{Y}$ line. **[9]**

***F* ratio, *F* test (ANOVA).** The F ratio is the final calculation or statistic representing the ratio of the between-groups (treatment) variation and the within-groups (error) variation in analyses of variance F tests. To determine statistical significance, F-observed is compared to F-critical. **[8]**

Factorials (!). A common mathematical procedure but used in statistics when calculating discrete binomial probabilities. A factorial is a number multiplied by all whole numbers smaller than itself down to 1 (e.g., $3! = 3 \cdot 2 \cdot 1 = 6$). **[3]**

Failure (q). In binomial probability, the absence of the outcome or event of interest, that is, its failure to occur. Also, the proportion of binomial trials resulting in failures. **[3]**

Finite population correction (fpc). The hypergeometric correction factor used in confidence interval estimation and hypothesis testing when sampling more than 10% of a population without replacement. Corrects for reduced population variation and sampling error when taking either comparatively large samples or when taking samples from small (finite) populations. **[4]**

Frequencies, counts. The tallies or totals of specific observations in a distribution. The number of times particular X or Y values, outcomes, or measurements occur. **[2]**

Frequency distributions. Displays (tables, graphs, etc.) showing how all cases are distributed across various categories or measurements. Ways of presenting sample or population data. **[1]**

Goodness-of-fit test. *See* One-way chi square test.

Grand mean. When dealing with interval-ratio data and multiple samples, as in ANOVA, the mean or average of *all* observations without regard to particular treatment groups. **[8]**

Harmonic mean. An "average," or a weighted mean, if we have unequal sample sizes, and no single value for n. **[8]**

HSD, Tukey's. *See* Tukey's HSD.

Hypergeometric correction factor. A correction factor used in confidence interval estimation and hypothesis testing to modify and reduce the standard error term. *See also* Finite population correction. **[4]**

Hypotheses. The researcher's/statistician's expected outcomes. **[5]**

Hypothesis testing. The procedure of comparing sample statistics or distributions to either known or assumed population characteristics or to other samples' statistics. The use of probability to determine whether the sample features are "statistically normal" or "differ significantly" from the population or random chance figures being used. [5]

Independent/predictor variable. In a bivariate relationship, assumed to be the antecedent or causal variable. Given knowledge of the independent variable (IV), one may statistically better predict variation in the dependent variable (DV). Given gender (the IV), may one better predict income (the DV) than when not knowing gender? Statistically, to what extent does gender predict income, if at all? *See also* Dependent variable. [5]

Independent samples. Samples drawn from different populations and by independent and random procedures. [6]

Independent trials. What happens on one trial (i.e., what outcome occurs) has no effect on what happens on any other trial. A key consideration in binomial probabilities. [3]

Index of dispersion, index of qualitative variation. Examines the actual number of paired differences in the data (comparing each respondent's answer to that of every other respondent) versus the total possible number of such differences, which is a function of sample size and the number of categories. [2]

Intercept, *a*. With regression analyses and when constructing an XY scatterplot (as on graph paper), the point at which the $\hat{Y}$ regression line crosses the vertical Y axis when $X = 0$ on the horizontal axis; the $\hat{Y}$ value when $X = 0$. [9]

Interquartile range, semi-interquartile range. A measure of variation for numerical ordinal data. The difference between the 75th and 25th percentile ranks, or (semi) one-half that difference. [2]

Interval or ratio levels of measurement. Data consisting of reliably, legitimately quantitative, and unambiguous measurements. Data for which we may use the actual X measurements in calculations. Intervals and ratios between measurement increments are constant, reliable, absolute, and verifiable. [1]

Joint probability. The simultaneous occurrence of two or more events. The probability that they will occur jointly or together. The probabilities of the respective events are multiplied to derive the probability of their joint occurrence. [3]

Level of measurement. The quantitative precision with which variables are measured. Levels vary from the nonquantitative, nominal/categorical differences (e.g., female/male) to the rankable (e.g., high-medium-low workout schedule), to the reliably numerical (e.g., miles driven from home to job). [1]

Line of best fit, regression line, $\hat{Y}$ line. A straight line running through the data points in a scatterplot that most closely represents those points, that is, it minimizes the variation between the line and the data values. May be used to approximate Y values ($\hat{Y}$) based upon given X values. [9]

Lower-tailed test. *See* One-tailed test.

Major mode. In a frequency distribution, the one X value or category that occurs with the greatest frequency but is accompanied by an X value or category with a slightly lower but still relatively high count, the minor mode. [2]

Mann-Whitney U test. A hypothesis test for the difference between ranked (ordinal) data from two independent samples. [7]

Margin of error, error term. The half-width of a confidence interval. The plus-or-minus variation around the point estimate. The distance or variation around the point estimate that produces the actual *interval* of estimation. [4]

Matched samples. *See* Related, paired samples.

Mean. *See* Average.

Mean absolute deviation (MAD). A rarely used measure of variation. The average absolute value of deviations from the mean. The use of absolute values renders this statistic unusable with more complex statistical procedures. [2]

Mean difference ($\bar{d}$). With matched or related samples, the average difference between all the matched pairs. [6]

Mean square (MS). In ANOVA, the average of the sum of squares values. [8]

Mean square between groups (MS_B). The typical variation between treatment groups. [8]

Mean square within groups (MS_W). Represents the average error variation within the respective treatment groups. [8]

Measures of association. Statistics describing the strengths of bivariate associations. Vary between 0 and 1 or between –1 and 1. [7]

Measures of central tendency. Measures that summarize a set of data by showing the typical response or observation. They comprise the mean, median, and mode. [2]

Measures of variation. Measures of the spread in a data set or distribution, the most commonly used measures being the standard deviation and the variance. [2]

Median. The middle-ranking observation or score in a set of data. By definition, the 50th percentile rank. Half the scores rank above the median and half below. [2]

Minor mode. In a frequency distribution, the *X* value or category with the second greatest frequency. Not as numerous as the *major* modal score or category, but having a higher count than remaining values or categories. [2]

Mode. In a frequency distribution, the most commonly occurring value or category. [2]

Multiplicative rule. *See* Joint probability.

Negative, indirect, inverse correlation or association. A correlation between two rankable or interval-ratio variables in which the high values of one variable are associated with the low values or ranks of the other. [9]

Nominal level of measurement. Data or measurements of nonquantitative differences, typological differences; nonrankable. [1]

Nonparametric tests. Hypothesis tests not using known population parameters. Sample data are compared instead to random chance outcomes or to other criteria, not to population parameters. [6]

Normal curve, distribution. A symmetrical, bell-shaped distribution of continuous data, which peaks at a central mode, median, and mean. The area of the curve indicates the relative frequency with which given *X* values occur. From a central peak, the distribution or curve tapers off to represent comparatively small frequencies at low and high extremes. [3]

Null hypothesis (H_O). A hypothesis stating that frequencies or the statistic under study, either a single sample statistic or the difference between two samples, does *not* differ from a particular criterion (i.e., from a population parameter, the assumed difference between two population parameters, random chance, or from an alternative criterion such as normality). The null hypothesis is always tested against this known or presumed criterion and either retained or rejected at a given alpha level. [5]

Observed frequencies (O). In chi square, the actual frequencies or counts observed in the sample data. [7]

One-tailed test (upper, lower) A hypothesis test determining whether the sample statistic differs significantly from the known population parameter or criterion in a specified direction. We may test whether the statistic is significantly *less* than the population criterion or whether it is significantly *greater*. Accordingly, the alpha region occupies just one tail, lower or upper, respectively, of the null hypothesis curve. [5]

One-way chi square test. A chi square with a single variable. The null hypothesis tested is that the observed frequency distribution does not differ significantly from that expected by random chance or by normal distribution guidelines. [7]

Operating characteristics/curve for an H_0 test. The probability of committing a beta or Type II error. The sample statistic actually belongs to an alternative population centered around μ_A or p_A but is an extreme case in that population. It falls into that part of the H_A curve that overlaps the bounds of normal variation in the H_O curve. That part of the H_A curve represents the operating characteristic of the test. [5]

Ordinal level of measurement. Rankable measurements that may be ordered from highest to lowest or vice versa. The data may consist of rankable (but not interval-ratio) quantitative measurements or of rankable, hierarchical categories. [1]

Outcomes. In probability, a term for possible results or events that may occur in a probability trial or experiment. [3]

Pearson's r_{XY}. The correlation coefficient, a popular measure of association for paired interval-ratio measurements. A measure of correlation for interval-ratio data and the basis for linear regression analyses involving $\hat{Y}$. [9]

Percentile rank. In a set of data, a rank determined by the percentage of cases falling at or below a given value. For example, by definition, 70% of all cases fall at or below an X value at the 70th percentile rank. **[2]**

Phi coefficient. For 2×2 tables only, a measure of association based upon the two-way chi square test. **[7]**

Point estimate. In confidence interval estimation, the sample statistic (e.g., $\overline{X}$, p', or $\hat{Y}$) around which an interval is constructed. **[4]**

Pooled variance. In the case of independent samples, a combined or pooled measure of the underlying population variance based upon the two sample variances. **[6]**

Population average, mean, μ. Average, or mean, of the entire population, often estimated by the mean of a sample, $\overline{X}$. **[2]**

Population standard deviation, σ. Standard deviation of the entire population, often estimated by the standard deviation of a sample, s. **[2]**

Population variance, σ². Variance of the entire population, often estimated by the standard deviation of a sample, s^2. **[2]**

Positive/direct relationship, correlation, or association. A bivariate correlation in which the high values of one variable are associated with the high values of the other, and the low values of one with the low values of the other. **[9]**

Power, power curve for an H₀ test. The ability of a hypothesis test to correctly reject a false null hypothesis. Ideally, and assuming our sample statistic *does* belong to the alternative (μ_A) population and curve, the power of the test is represented by that part of the μ_A curve *not* falling into the H_O curve's bounds of normal variation. **[5]**

Probability distribution. A probability distribution depicts all the possible outcomes in a particular situation along with their respective probabilities. **[3]**

Probability table. A form of probability distribution. In a probability situation, a way of showing each possible outcome along with its respective probability. All probabilities must total to 1.00 or 100%. Cross-classification probability tables including two variables are particularly useful for depicting joint frequencies and probabilities. **[3]**

Proportional reduction in error (PRE). Statistical measures of association based upon reductions in predictive error. In the case of Pearson's r_{XY}, what proportion of the errors in predicting $\hat{Y}$ are reduced by using X as the predictor rather than merely using the mean of Y, $\overline{Y}$? This is measured by $r_{XY}{}^2$. *See also* Coefficient of determination. **[9]**

Random sample. A representative sample selected such that each element in the population (or subpopulation) has an equal chance to be drawn. A key prerequisite for statistical inference. **[1]**

Random variables. In probability situations, variables having more than two possible outcomes. **[3]**

Range. In a set of quantitative measurements, the difference between the greatest or highest score and the least or lowest. **[1]**

Rankable scores, data. Numerical data or data in ordered categories which may be put in hierarchical order from the highest to the lowest score or category, or vice versa. **[1]**

Regression line. *See* Line of best fit.

Related, paired samples. Samples in which two observations are taken either from the same person or source (e.g., at Time 1 and again at Time 2), or are taken from separate but matched sources. **[6]**

Relative frequency. The height of the probability curve, representing the proportional frequency with which various X values occur. **[3]**

Rho, ρ (ρ_{XY} or ρ_S). The correlational population parameter. The population parameter for Pearson's r_{XY} (ρ_{XY}) and for Spearman's r_S (ρ_S). Used in testing null hypotheses regarding the statistical significance of r_{XY} and r_S. **[9]**

Sampling distribution of the difference. With two samples, the hypothetical distribution of the differences between two statistics recorded over many such differences. For example, the distribution of differences ($\overline{P}$) found between many pairs of sample proportions, or the distribution of differences ($\overline{X}_1 - X_2$ or $\overline{d}$ for related samples) found between many pairs of sample averages. **[6]**

Sampling distribution of the mean. The hypothetical distribution of averages taken from a large number of random samples. Used in confidence interval estimation for μ and as the H_O curve in hypothesis testing. **[4]**

Sampling distribution of the proportion. The hypothetical distribution of proportions taken from a large number of random samples. Used in confidence interval estimation for p and as the H_O curve in hypothesis testing. [4]

Sampling error. Unavoidable error associate with sampling, even random sampling. The inaccuracies or degree of error when using samples to make inferences about populations. Measured and taken into account by the standard error. [4]

Scatterplot. Especially with Pearson's r_{XY}, the depiction of XY intersections or data points plotted along X (horizontal) and Y (vertical) axes. Also used to show the regression line or $\hat{Y}$ line of best fit. [9]

Semi-interquartile range. *See* Interquartile range.

Shift. In hypothesis testing and along the horizontal $\overline{X}$ or p' axis, the difference between central values of the H_O and presumed H_A curves. The shift or difference from μ_O to μ_A or from p_O to p_A. Important in calculating the P(Beta error), the operating characteristic of an H_O test, and the power of a test. [5]

Sign test. A hypothesis test for related samples with ordinal-level data. Based upon the signed (+/–) differences in pairs of scores. [7]

Skew. A non-normal or asymmetrical distribution that includes extreme scores at one end of the continuum or range. [2]

Slope, *b*. In an XY scatterplot, the angle of the regression line as it proceeds from left to right above the horizontal (X) axis. Formally defined as the line's change in Y for every increment in X. If the line angles upward, it indicates a positive change in Y and a positive slope; if it angles downward, it indicates a negative change and slope. [9]

Spearman's r_s. A measure of correlation for numerical ordinal/rankable data. A substitute for Pearson's r_{XY} with ordinal data. Also testable for statistical significance. [9]

Standard deviation. The basic measure of variation. Measures variation from the sample mean (s) or from the population mean (σ). The typical or usual deviation from the average. [2]

Standard deviation of the difference (s_d). With related samples and in situations involving averages, the standard deviation of the d scores (differences) between the paired scores. [6]

Standard error of the difference between two population means. A measure of variation for a hypothetical sampling distribution of differences between averages from two independent samples. [6]

Standard error of the difference between two population proportions. A measure of variation for a hypothetical sampling distribution of differences between proportions from two independent samples. [6]

Standard error of estimate ($s_{Y.X}$). In correlation and regression, a measure of the unexplained variation, the variation of *actual Y* values from those predicted by the $\hat{Y}$ line. [9]

Standard error of the mean. The measure of variation in the sampling distribution of the mean. A measure of sampling error and how much a typical sample mean would vary from the population mean. A function of variation in the population and of sample size. [4]

Standard error of the mean difference ($s_{\bar{d}}$). A measure of variation in a sampling distribution of the mean differences between related sample averages. A measure of variation around $\bar{D}$, the difference between related population averages. [6]

Standard error of the proportion. The measure of variation in the sampling distribution of the proportion. A measure of sampling error and how much the typical sample proportion varies from the population proportion. A function of variation in the population and of sample size. [4]

Standardized scores (*z* scores). Standardized scores (e.g., percentages and *z* scores) that transform dissimilar measurement scales into comparable (or standardized) units. The *z* distribution represents such a standardized scale with a mean of 0.00 and a standard deviation of 1.00. [3]

Statistical independence/nonindependence. Whether two outcomes or variables are statistically related to each other. Whether or not one outcome affects the probability of another outcome. [9]

Statistical significance. A term describing a statistically rare event. An event or outcome so unlikely to happen by random chance that its occurrence is regarded as significant, as when a sample statistic falls into an H_O test's alpha or rejection region. [7]

Success. The outcome of interest in a binomial probability situation and also a reference to the probability or proportion of trials resulting in successes. [3]

Sum of squares (SS) and SS$_T$, SS$_B$, and SS$_W$. SS is an abbreviated term for the sum of the squared deviations from an average. In analysis of variance (ANOVA), SS$_T$ refers to the sum of the squared deviations of all cases from the grand mean, SS$_B$ to the sum of the squared deviations of the treatment group means from the grand mean, and SS$_W$ to the sum of the squared deviations of individual scores from their respective treatment group means. **[2]**

t-critical (t_C). The t scores cutting off an alpha region or critical percentage of the curve, readily available in t tables. **[5]**

t curves, distributions, t test. Sampling distributions (of the mean, the difference between two means, ρ_{XY}, and ρ_S) based upon small samples and having more spread and variation than the normal distribution. Hypothesis tests using these sampling distributions are referred to as t tests. **[4]**

Total variation. The variation of all cases from the mean of a distribution, especially in ANOVA: the variation of all observations from the grand mean. **[8]**

Treatment groups. In ANOVA, a term for the different samples, that is, groups that have presumably been subjected to different experimental treatments. **[8]**

Treatment variation. *See* Between-group variation.

True limits, upper and lower The limits of discrete or whole numbers when considered part of continuous distributions. For example, the whole number 26 might actually begin at 25.50, its true lower limit, and extend to 26.49, its true upper limit. Sometimes used in the calculation of continuous binomial probabilities. **[3]**

Tukey's HSD. Tukey's test for honestly significant differences (HSD) when one has a statistically significant F ratio in ANOVA. By revealing the minimum difference between means necessary for statistical significance, Tukey's HSD sorts out which among several means differ significantly and which do not, following statistically significant ANOVA results. **[8]**

Two-way chi square test. *See* Chi square test of independence.

Two-tailed test. A nondirectional hypothesis test in which the sample statistic may differ significantly from the population parameter (or criterion such as random chance) in either direction, either above or below the specified criterion. **[5]**

Two-way chi square test. A nonparametric hypothesis test for a statistical relationship between two categorical variables. If the distribution of cases does *not* differ from random chance, the two variables are said to be statistically independent. *See also* Chi square test of independence. **[7]**

Type I error. *See* Alpha error.

Type II error. *See* Beta error.

Unbiased estimator. Allowing for sampling error, a sample statistic claimed to be a reliable and accurate representation of a population parameter. Derived from measurements on a random sample of the population in question. **[4]**

Unexplained variation. In correlation and regression, the remaining variation of the Y scores around the $\hat{Y}$ regression line. The variation in Y scores not explained or accounted for by the XY regression line. **[9]**

Upper-tailed test. *See* One-tailed test.

Valid percentage. In a frequency distribution, a percentage of the actual number of respondents, not necessarily the total sample size. A percentage of those actually responding to an item or question and discounting the "missing values," i.e., those who omitted the item. **[1]**

Variables. As opposed to constants, measurable characteristics, behaviors, or opinions that may vary from one case to the next. **[1]**

Variance. A measure of variation from an average. The square of the standard deviation. **[2]**

Variation, measures of. *See* Measures of variation.

Wilcoxon rank-sum test. A hypothesis test for independent samples and quantitative ordinal/rankable data. Looks at the relative rankings of scores from the respective samples. **[7]**

Within-group variation (error/random variation). In ANOVA, the variation of individual scores from their respective treatment group means. Represents the error or random variation due to individual differences and not to the experimental treatment or independent variable. **[8]**

$\hat{Y}$, Y-predicted. In correlation and regression, a Y value predicted to result from a particular value of X. Based upon an XY correlation and may be determined by formula or estimated by the $\hat{Y}$ regression line. **[9]**

Yates' correction for continuity. A correction factor used with chi square in cases of (1) small (<5) expected frequencies, and/or (2) small (2 × 1 or 2 × 2) tables having just one degree of freedom. **[7]**

z **scores,** *z* **test.** As measurements used with the normal distribution, *z* scores indicate the number of standard deviations given *X* values lie from the mean. A *z* test is a hypothesis test using the normal distribution as the H_O curve and conducted as a comparison of *z*-observed and *z*-critical. **[3]**

APPENDIX 2

Reference Tables

Table A Areas of the Normal Distribution

A z	B Area from z to mean	C Tail area beyond z	A z	B Area from z to mean	C Tail area beyond z	A z	B Area from z to mean	C Tail area beyond z
0.00	.0000	.5000	0.30	.1179	.3821	0.60	.2257	.2743
0.01	.0040	.4960	0.31	.1217	.3783	0.61	.2291	.2709
0.02	.0080	.4920	0.32	.1255	.3745	0.62	.2324	.2676
0.03	.0120	.4880	0.33	.1293	.3707	0.63	.2357	.2643
0.04	.0160	.4840	0.34	.1331	.3669	0.64	.2389	.2611
0.05	.0199	.4801	0.35	.1368	.3632	0.65	.2422	.2578
0.06	.0239	.4761	0.36	.1406	.3594	0.66	.2454	.2546
0.07	.0279	.4721	0.37	.1443	.3557	0.67	.2486	.2514
0.08	.0319	.4681	0.38	.1480	.3520	0.68	.2517	.2483
0.09	.0359	.4641	0.39	.1517	.3483	0.69	.2549	.2451
0.10	.0398	.4602	0.40	.1554	.3446	0.70	.2580	.2420
0.11	.0438	.4562	0.41	.1591	.3409	0.71	.2611	.2389
0.12	.0478	.4522	0.42	.1628	.3372	0.72	.2642	.2358
0.13	.0517	.4483	0.43	.1664	.3336	0.73	.2673	.2327
0.14	.0557	.4443	0.44	.1700	.3300	0.74	.2704	.2296
0.15	.0596	.4404	0.45	.1736	.3264	0.75	.2734	.2266
0.16	.0636	.4364	0.46	.1772	.3228	0.76	.2764	.2236
0.17	.0675	.4325	0.47	.1808	.3192	0.77	.2794	.2206
0.18	.0714	.4286	0.48	.1884	.3156	0.78	.2823	.2177
0.19	.0753	.4247	0.49	.1879	.3121	0.79	.2852	.2148
0.20	.0793	.4207	0.50	.1915	.3085	0.80	.2881	.2119
0.21	.0832	.4168	0.51	.1950	.3050	0.81	.2910	.2090
0.22	.0871	.4129	0.52	.1985	.3015	0.82	.2939	.2061
0.23	.0910	.4090	0.53	.2019	.2981	0.83	.2967	.2033
0.24	.0948	.4052	0.54	.2054	.2946	0.84	.2995	.2005
0.25	.0987	.4013	0.55	.2088	.2912	0.85	.3023	.1977
0.26	.1026	.3974	0.56	.2123	.2877	0.86	.3051	.1949
0.27	.1064	.3936	0.57	.2157	.2843	0.87	.3078	.1922
0.28	.1103	.3897	0.58	.2190	.2810	0.88	.3106	.1894
0.29	.1141	.3859	0.59	.2224	.2776	0.89	.3133	.1867

(continues)

Table A continued

A	B	C	A	B	C	A	B	C
z	Area from z to mean	Tail area beyond z	z	Area from z to mean	Tail area beyond z	z	Area from z to mean	Tail area beyond z
0.90	.3159	.1841	1.20	.3849	.1151	1.50	.4332	.0668
0.91	.3186	.1814	1.21	.3869	.1131	1.51	.4345	.0655
0.92	.3212	.1788	1.22	.3888	.1112	1.52	.4357	.0643
0.93	.3238	.1762	1.23	.3907	.1093	1.53	.4370	.0630
0.94	.3264	.1736	1.24	.3925	.1075	1.54	.4382	.0618
0.95	.3289	.1711	1.25	.3944	.1056	1.55	.4394	.0606
0.96	.3315	.1685	1.26	.3962	.1038	1.56	.4406	.0594
0.97	.3340	.1660	1.27	.3980	.1020	1.57	.4418	.0582
0.98	.3365	.1635	1.28	.3997	.1003	1.58	.4429	.0571
0.99	.3389	.1611	1.29	.4015	.0985	1.59	.4441	.0559
1.00	.3413	.1587	1.30	.4032	.0968	1.60	.4452	.0548
1.01	.3438	.1562	1.31	.4049	.0951	1.61	.4463	.0537
1.02	.3461	.1539	1.32	.4066	.0934	1.62	.4474	.0526
1.03	.3485	.1515	1.33	.4082	.0918	1.63	.4484	.0516
1.04	.3508	.1492	1.34	.4099	.0901	1.64	.4495	.0505
1.05	.3531	.1469	1.35	.4115	.0885	1.65	.4505	.0495
1.06	.3554	.1446	1.36	.4131	.0869	1.66	.4515	.0485
1.07	.3577	.1423	1.37	.4147	.0853	1.67	.4525	.0475
1.08	.3599	.1401	1.38	.4162	.0838	1.68	.4535	.0465
1.09	.3621	.1379	1.39	.4177	.0823	1.69	.4545	.0455
1.10	.3643	.1357	1.40	.4192	.0808	1.70	.4554	.0409
1.11	.3665	.1335	1.41	.4207	.0793	1.71	.4564	.0436
1.12	.3686	.1314	1.42	.4222	.0778	1.72	.4573	.0427
1.13	.3708	.1292	1.43	.4236	.0764	1.73	.4582	.0418
1.14	.3729	.1271	1.44	.4251	.0749	1.74	.4591	.0409
1.15	.3749	.1251	1.45	.4265	.0735	1.75	.4599	.0401
1.16	.3770	.1230	1.46	.4279	.0721	1.76	.4608	.0392
1.17	.3790	.1210	1.47	.4292	.0708	1.77	.4616	.0384
1.18	.3810	.1190	1.48	.4306	.0694	1.78	.4625	.0375
1.19	.3830	.1170	1.49	.4319	.0681	1.79	.4633	.0367

(continues)

Table A continued

A	B	C	A	B	C	A	B	C
z	Area from z to mean	Tail area beyond z	z	Area from z to mean	Tail area beyond z	z	Area from z to mean	Tail area beyond z
1.80	.4641	.0359	2.10	.4821	.0179	2.40	.4918	.0082
1.81	.4649	.0351	2.11	.4826	.0174	2.41	.4920	.0080
1.82	.4656	.0344	2.12	.4830	.0170	2.42	.4922	.0078
1.83	.4664	.0336	2.13	.4834	.0166	2.43	.4925	.0075
1.84	.4671	.0329	2.14	.4838	.0162	2.44	.4927	.0073
1.85	.4678	.0322	2.15	.4842	.0158	2.45	.4929	.0071
1.86	.4686	.0314	2.16	.4846	.0154	2.46	.4931	.0069
1.87	.4693	.0307	2.17	.4850	.0150	2.47	.4932	.0068
1.88	.4699	.0301	2.18	.4854	.0146	2.48	.4934	.0066
1.89	.4706	.0294	2.19	.4857	.0143	2.49	.4936	.0064
1.90	.4713	.0287	2.20	.4681	.0139	2.50	.4938	.0062
1.91	.4719	.0281	2.21	.4864	.0136	2.51	.4940	.0060
1.92	.4726	.0274	2.22	.4868	.0132	2.52	.4941	.0059
1.93	.4732	.0268	2.23	.4871	.0129	2.53	.4943	.0057
1.94	.4738	.0262	2.24	.4875	.0125	2.54	.4945	.0055
1.95	.4744	.0256	2.25	.4878	.0122	2.55	.4946	.0054
1.96	.4750	.0250	2.26	.4881	.0119	2.56	.4948	.0052
1.97	.4756	.0244	2.27	.4884	.0116	2.57	.4949	.0051
1.98	.4761	.0239	2.28	.4887	.0113	2.58	.4951	.0049
1.99	.4767	.0233	2.29	.4890	.0110	2.59	.4952	.0048
2.00	.4772	.0228	2.30	.4893	.0107	2.60	.4953	.0047
2.01	.4778	.0222	2.31	.4896	.010	2.61	.4955	.0045
2.02	.4783	.0217	2.32	.4898	.0102	2.62	.4956	.0044
2.03	.4788	.0212	2.33	.4901	.0099	2.63	.4957	.0043
2.04	.4793	.0207	2.34	.4904	.0096	2.64	.4959	.0041
2.05	.4798	.0202	2.35	.4906	.0094	2.65	.4960	.0040
2.06	.4803	.0197	2.36	.4909	.0091	2.66	.4961	.0039
2.07	.4808	.0192	2.37	.4911	.0089	2.67	.4962	.0038
2.08	.4812	.0188	2.38	.4913	.0087	2.68	.4963	.0037
2.09	.4817	.0183	2.39	.4916	.0084	2.69	.4964	.0036

(continues)

Table A continued

A	B	C	A	B	C	A	B	C
z	Area from z to mean	Tail area beyond z	z	Area from z to mean	Tail area beyond z	z	Area from z to mean	Tail area beyond z
2.70	.4965	.0035	2.95	.4984	.0016	3.20	.4993	.0007
2.71	.4966	.0034	2.96	.4985	.0015	3.21	.4993	.0007
2.72	.4967	.0033	2.97	.4985	.0015	3.22	.4994	.0006
2.73	.4968	.0032	2.98	.4986	.0014	3.23	.4994	.0006
2.74	.4969	.0031	2.99	.4986	.0014	3.24	.4994	.0006
2.75	.4970	.0030	3.00	.4987	.0013	3.25	.4994	.0006
2.76	.4971	.0029	3.01	.4987	.0013	3.30	.4995	.0005
2.77	.4972	.0028	3.02	.4987	.0013	3.35	.4996	.0004
2.78	.4973	.0027	3.03	.4988	.0012	3.40	.4997	.0003
2.79	.4974	.0026	3.04	.4988	.0012	3.45	.4997	.0003
2.80	.4974	.0026	3.05	.4989	.0011	3.50	.4998	.0002
2.81	.4975	.0025	3.06	.4989	.0011	3.55	.4998	.0002
2.82	.4976	.0024	3.07	.4989	.0011	3.60	.4998	.0002
2.83	.4977	.0023	3.08	.4990	.0010	3.65	.4999	.0001
2.84	.4977	.0023	3.09	.4990	.0010	3.70	.4999	.0001
2.85	.4978	.0022	3.10	.4990	.0010	3.75	.4999	.0001
2.86	.4979	.0021	3.11	.4991	.0009	3.80	.4999	.0001
2.87	.4979	.0021	3.12	.4991	.0009	3.85	.4999	.0001
2.88	.4980	.0020	3.13	.4991	.0009	3.90	.49995	.00005
2.89	.4981	.0019	3.14	.4992	.0008	3.95	.49996	.00004
2.90	.4981	.0019	3.15	.4992	.0008	4.00	.49997	.00003
2.91	.4982	.0018	3.16	.4992	.0008	4.05	.49997	.00003
2.92	.4982	.0018	3.17	.4992	.0008	4.10	.49998	.00002
2.93	.4983	.0017	3.18	.4993	.0007	4.15	.49998	.00002
2.94	.4984	.0016	3.19	.4993	.0007	4.20	.49999	.00001

Note: Adapted from Table II, page 44, in R.A. Fisher and F. Yates: *Statistical Tables for Biological, Agricultural, and Medical Research.* 6th edition. London: Longman Group Limited. 1963. Reproduced with permission of Pearson Education Ltd.

Table B Selected Binomial Probabilities up to *n* = 15

For any given set of *n*, *X*, and *p* values (e.g., *n* = 4, *X* = 0 through 4, and *p* = .20), all the probabilities should total to 1.000, allowing for rounding. Also, instead of writing ".000" many times, probabilities too low to register at the third decimal place are omitted.

To solve for binomial probabilities where *p* > .5, merely turn the problem inside out, as it were. Solve for the probability of *n* – *X* failures and use *q* as if it were the *p* value. *X* successes and *n* – *X* failures are actually the same event or outcome, but we must use the *q* value when looking up the *P*(*n* – *X*) failures.

For example: To get the *P*(*X* = 4 | *n* = 6 and *p* = .70), use *n* – *X* and *q*. *n* – *X* = 6 – 2 = 2, and *q* = 1 – *p* = .30. Simply look up the *P*(*X* = 2 | *n* = 6 and *p* = .30). You should get .324. This is the *P* of *X* = 4 successes with *p* = .70 *and also* the *P* of *n* – *X* failures when letting *"p"* (i.e., *q*) equal .30.

Table B Selected Binomial Probabilities

							p				
n	*X*	.10	.15	.20	.25	.30	.33	.35	.40	.45	.50
2	0	.810	.723	.640	.563	.490	.449	.423	.360	.303	.250
	1	.180	.255	.320	.375	.420	.442	.455	.480	.495	.500
	2	.010	.023	.040	.063	.090	.109	.123	.160	.203	.250
3	0	.729	.617	.512	.422	.343	.301	.275	.216	.166	.125
	1	.243	.325	.384	.422	.441	.444	.444	.432	.408	.375
	2	.027	.057	.096	.141	.189	.219	.239	.288	.334	.375
	3	.001	.003	.008	.016	.027	.036	.043	.064	.091	.125
4	0	.656	.522	.410	.316	.240	.202	.179	.130	.092	.063
	1	.292	.368	.410	.422	.412	.397	.384	.346	.299	.250
	2	.049	.098	.154	.211	.265	.293	.311	.346	.368	.375
	3	.004	.011	.026	.047	.076	.096	.111	.154	.200	.250
	4		.001	.002	.004	.008	.012	.015	.025	.041	.063

(continues)

Table B   continued

n	X	.10	.15	.20	.25	.30	.33	.35	.40	.45	.50
5	0	.590	.444	.328	.237	.168	.135	.116	.078	.050	.031
	1	.328	.392	.410	.396	.360	.332	.312	.259	.206	.156
	2	.073	.138	.205	.264	.309	.328	.336	.346	.337	.313
	3	.008	.024	.051	.088	.132	.161	.181	.230	.276	.313
	4		.002	.006	.015	.028	.040	.049	.077	.113	.156
	5				.001	.002	.004	.005	.010	.018	.031
6	0	.531	.377	.262	.178	.118	.090	.075	.047	.028	.016
	1	.354	.399	.393	.356	.303	.267	.244	.187	.136	.094
	2	.098	.176	.246	.297	.324	.329	.328	.311	.278	.234
	3	.015	.041	.082	.132	.185	.216	.235	.276	.303	.313
	4	.001	.005	.015	.033	.060	.080	.095	.138	.186	.234
	5		.001	.002	.004	.010	.016	.020	.037	.061	.094
	6					.001	.001	.002	.004	.008	.016
7	0	.478	.321	.210	.133	.082	.061	.049	.028	.015	.008
	1	.372	.396	.367	.311	.247	.209	.185	.131	.087	.055
	2	.124	.210	.275	.311	.318	.309	.298	.261	.214	.164
	3	.023	.062	.115	.173	.227	.253	.268	.290	.292	.273
	4	.003	.011	.029	.058	.097	.125	.144	.194	.239	.273
	5		.001	.004	.012	.025	.037	.047	.077	.117	.164
	6				.001	.004	.006	.008	.017	.032	.055
	7							.001	.002	.004	.008
8	0	.430	.272	.168	.100	.058	.041	.032	.017	.008	.004
	1	.383	.385	.336	.267	.198	.160	.137	.090	.055	.031
	2	.149	.238	.294	.311	.296	.276	.259	.209	.157	.109
	3	.033	.084	.147	.208	.254	.272	.279	.279	.257	.219
	4	.005	.018	.046	.087	.136	.167	.188	.232	.263	.271
	5		.003	.009	.023	.047	.066	.081	.124	.172	.219
	6			.001	.004	.010	.016	.022	.041	.070	.109
	7					.001	.002	.003	.008	.016	.031
	8								.001	.002	.004

(continues)

Table B   continued

n	X	p									
		.10	.15	.20	.25	.30	.33	.35	.40	.45	.50
9	0	.387	.232	.134	.075	.040	.027	.021	.010	.005	.002
	1	.387	.368	.302	.225	.156	.121	.100	.060	.034	.018
	2	.172	.260	.302	.300	.267	.238	.216	.161	.111	.070
	3	.045	.107	.176	.234	.267	.273	.272	.251	.212	.164
	4	.007	.028	.066	.117	.172	.202	.219	.251	.260	.246
	5	.001	.005	.017	.039	.074	.099	.118	.167	.213	.246
	6		.001	.003	.009	.021	.033	.042	.074	.116	.164
	7				.001	.004	.007	.010	.021	.041	.070
	8						.001	.002	.004	.008	.018
	9									.001	.002
10	0	.349	.197	.107	.056	.028	.018	.013	.006	.003	.001
	1	.387	.347	.268	.188	.121	.090	.072	.040	.021	.010
	2	.194	.276	.302	.282	.233	.199	.176	.121	.076	.044
	3	.057	.130	.201	.250	.267	.261	.252	.215	.166	.117
	4	.011	.040	.088	.146	.200	.225	.238	.251	.238	.205
	5	.001	.008	.026	.058	.103	.133	.154	.201	.234	.246
	6		.001	.006	.016	.037	.055	.069	.111	.160	.205
	7			.001	.003	.009	.015	.021	.042	.075	.117
	8					.001	.003	.004	.011	.023	.044
	9							.001	.002	.004	.010
	10										.001
11	0	.314	.167	.086	.042	.020	.012	.009	.004	.001	
	1	.384	.325	.236	.155	.093	.066	.052	.027	.013	.005
	2	.213	.287	.295	.258	.200	.163	.140	.089	.051	.027
	3	.071	.152	.221	.258	.257	.241	.225	.177	.126	.081
	4	.016	.054	.111	.172	.220	.237	.243	.236	.206	.161
	5	.002	.013	.039	.080	.132	.164	.183	.221	.236	.226
	6		.002	.010	.027	.057	.081	.099	.147	.193	.226
	7			.002	.006	.017	.028	.038	.070	.113	.161
	8				.001	.004	.007	.010	.023	.046	.081
	9					.001	.001	.002	.005	.013	.027
	10								.001	.002	.005

(continues)

Table B continued

n	X	.10	.15	.20	.25	.30	.33	.35	.40	.45	.50
12	0	.282	.142	.069	.032	.014	.008	.006	.002	.001	
	1	.377	.301	.206	.127	.071	.048	.037	.017	.008	.003
	2	.230	.292	.283	.232	.168	.131	.109	.064	.034	.016
	3	.085	.172	.236	.258	.240	.215	.195	.142	.092	.054
	4	.021	.068	.133	.194	.231	.238	.237	.213	.170	.121
	5	.004	.019	.053	.103	.158	.188	.204	.227	.222	.193
	6		.004	.016	.040	.079	.108	.128	.177	.212	.226
	7		.001	.003	.011	.029	.046	.059	.101	.149	.193
	8			.001	.002	.008	.014	.020	.042	.076	.121
	9					.001	.003	.005	.012	.028	.054
	10							.001	.002	.007	.016
	11									.001	.003
	12										
13	0	.254	.121	.055	.024	.010	.005	.004	.001		
	1	.367	.277	.179	.103	.054	.035	.026	.011	.004	.002
	2	.245	.294	.268	.206	.139	.104	.084	.045	.022	.010
	3	.100	.190	.246	.252	.218	.187	.165	.111	.066	.035
	4	.028	.084	.154	.210	.234	.231	.222	.184	.135	.087
	5	.006	.027	.069	.126	.180	.205	.215	.221	.199	.157
	6	.001	.006	.023	.056	.103	.134	.155	.197	.217	.209
	7		.001	.006	.019	.044	.066	.083	.131	.177	.209
	8			.001	.005	.014	.024	.034	.066	.109	.157
	9				.001	.003	.007	.010	.024	.050	.087
	10					.001	.001	.002	.006	.016	.035
	11								.001	.004	.010
	12										.002
	13										

(continues)

Table B   continued

						p					
n	X	.10	.15	.20	.25	.30	.33	.35	.40	.45	.50
14	0	.229	.103	.044	.018	.007	.004	.002	.001		
	1	.356	.254	.154	.083	.041	.025	.018	.007	.003	.001
	2	.257	.291	.250	.180	.113	.081	.063	.032	.014	.006
	3	.114	.206	.250	.240	.194	.160	.137	.084	.046	.022
	4	.035	.100	.172	.220	.229	.216	.202	.155	.104	.061
	5	.008	.035	.086	.147	.196	.213	.218	.207	.170	.122
	6	.011	.009	.032	.073	.126	.157	.176	.207	.209	.183
	7		.002	.009	.028	.062	.089	.108	.157	.195	.209
	8			.002	.008	.023	.038	.051	.092	.140	.183
	9				.002	.007	.013	.018	.041	.076	.122
	10					.001	.003	.005	.014	.031	.061
	11						.001	.002	.003	.009	.022
	12								.001	.002	.006
	13										.001
	14										
15	0	.206	.087	.035	.013	.005	.002	.002			
	1	.343	.231	.132	.067	.031	.018	.013	.005	.002	
	2	.267	.286	.231	.156	.092	.063	.048	.022	.009	.003
	3	.129	.218	.250	.225	.170	.134	.111	.063	.032	.014
	4	.043	.116	.188	.225	.219	.198	.179	.127	.078	.042
	5	.010	.045	.103	.165	.206	.214	.212	.186	.140	.092
	6	.002	.013	.043	.092	.147	.176	.191	.207	.191	.153
	7		.003	.014	.039	.081	.111	.132	.177	.201	.196
	8		.001	.003	.013	.035	.055	.071	.118	.065	.196
	9			.001	.003	.012	.021	.030	.061	.105	.153
	10				.001	.003	.006	.010	.024	.051	.092
	11					.001	.001	.002	.007	.019	.042
	12								.002	.005	.014
	13									.001	.003
	14										
	15										

Table C Critical Values for the *t* Distributions

Confidence Level:	.90	.95	.98	.99
Two-Tailed Test α:	.10	.05	.02	.01
One-Tailed Test α:	.05	.025	.01	.005

d.f.				
1	6.314	12.706	31.821	63.657
2	2.920	4.303	6.965	9.925
3	2.353	3.182	4.541	5.841
4	2.132	2.776	3.747	4.604
5	2.015	2.571	3.365	4.032
6	1.943	2.447	3.143	3.707
7	1.895	2.365	2.998	3.499
8	1.860	2.306	2.896	3.355
9	1.833	2.262	2.821	3.250
10	1.812	2.228	2.764	3.169
11	1.796	2.201	2.718	3.106
12	1.782	2.179	2.681	3.055
13	1.771	2.160	2.650	3.012
14	1.761	2.145	2.624	2.977
15	1.753	2.131	2.602	2.947
16	1.746	2.120	2.583	2.921
17	1.740	2.110	2.567	2.898
18	1.734	2.101	2.552	2.878
19	1.729	2.093	2.539	2.861
20	1.725	2.086	2.528	2.845
21	1.721	2.080	2.518	2.831
22	1.717	2.074	2.508	2.819
23	1.714	2.069	2.500	2.807
24	1.711	2.064	2.492	2.797
25	1.708	2.060	2.485	2.787
26	1.706	2.056	2.479	2.779
27	1.703	2.052	2.473	2.771
28	1.701	2.048	2.467	2.763
29	1.699	2.045	2.462	2.756
30	1.697	2.042	2.457	2.750
40	1.684	2.021	2.423	2.704
60	1.671	2.000	2.390	2.660
120	1.658	1.980	2.358	2.617
∞ (z)	1.645	1.960	2.326	2.575

Note: Adapted from Table III, page 46, in R.A. Fisher and F. Yates: *Statistical Tables for Biological, Agricultural, and Medical Research.* 6th edition. London: Longman Group Limited. 1963. Reproduced with permission of Pearson Education Ltd.

Table D-1 Critical Values for the Mann-Whitney U Test: α = .05

n_1	2	3	4	5	6	7	8	9	10	11	12	13	14	15	16	17	18	19	20
n_2																			
2				0	0	0	1	1	1	1	2	2	2	3	3	3	4	4	4
							0	**0**	**0**	**0**	**1**	**1**	**1**	**1**	**1**	**2**	**2**	**2**	**2**
3		0	0	1	2	2	3	3	4	5	5	6	7	7	8	9	9	10	11
				0	**1**	**1**	**2**	**2**	**3**	**3**	**4**	**4**	**5**	**5**	**6**	**6**	**7**	**7**	**8**
4		0	1	2	3	4	5	6	7	8	9	10	11	12	14	15	16	17	18
			0	**1**	**2**	**3**	**4**	**4**	**5**	**6**	**7**	**8**	**9**	**10**	**11**	**11**	**12**	**13**	**13**
5	0	1	2	4	5	6	8	9	11	12	13	15	16	18	19	20	22	23	25
		0	**1**	**2**	**3**	**5**	**6**	**7**	**8**	**9**	**11**	**12**	**13**	**14**	**15**	**17**	**18**	**19**	**20**
6	0	2	3	5	7	8	10	12	14	16	17	19	21	23	25	26	28	30	32
		1	**2**	**3**	**5**	**6**	**8**	**10**	**11**	**13**	**14**	**16**	**17**	**19**	**21**	**22**	**24**	**25**	**27**
7	0	2	4	6	8	11	13	15	17	19	21	24	26	28	30	33	35	37	39
		1	**3**	**5**	**6**	**8**	**10**	**12**	**14**	**16**	**18**	**20**	**22**	**24**	**26**	**28**	**30**	**32**	**34**
8	1	3	5	8	10	13	15	18	20	23	26	28	31	33	36	39	41	44	47
	0	**2**	**4**	**6**	**8**	**10**	**13**	**15**	**17**	**19**	**22**	**24**	**26**	**29**	**31**	**34**	**36**	**38**	**41**
9	1	3	6	9	12	15	18	21	24	27	30	33	36	39	42	45	48	51	54
	0	**2**	**4**	**7**	**10**	**12**	**15**	**17**	**20**	**23**	**26**	**28**	**31**	**34**	**37**	**39**	**42**	**45**	**48**
10	1	4	7	11	14	17	20	24	27	31	34	37	41	44	48	51	55	58	62
	0	**3**	**5**	**8**	**11**	**14**	**17**	**20**	**23**	**26**	**29**	**33**	**36**	**39**	**42**	**45**	**48**	**52**	**55**
11	1	5	8	12	16	19	23	27	31	34	38	42	46	50	54	57	61	65	69
	0	**3**	**6**	**9**	**13**	**16**	**19**	**23**	**26**	**30**	**33**	**37**	**40**	**44**	**47**	**51**	**55**	**58**	**62**
12	2	5	9	13	17	21	26	30	34	38	42	47	51	55	60	64	68	72	77
	1	**4**	**7**	**11**	**14**	**18**	**22**	**26**	**29**	**33**	**37**	**41**	**45**	**49**	**53**	**57**	**61**	**65**	**69**
13	2	6	10	15	19	24	28	33	37	42	47	51	56	61	65	70	75	80	84
	1	**4**	**8**	**12**	**16**	**20**	**24**	**28**	**33**	**37**	**41**	**45**	**50**	**54**	**59**	**63**	**67**	**72**	**76**
14	2	7	11	16	21	26	31	36	41	46	51	56	61	66	71	77	82	87	92
	1	**5**	**9**	**13**	**17**	**22**	**26**	**31**	**36**	**40**	**45**	**50**	**55**	**59**	**64**	**67**	**74**	**78**	**83**
15	3	7	12	18	23	28	33	39	44	50	55	61	66	72	77	83	88	94	100
	1	**5**	**10**	**14**	**19**	**24**	**29**	**34**	**39**	**44**	**49**	**54**	**59**	**64**	**70**	**75**	**80**	**85**	**90**
16	3	8	14	19	25	30	36	42	48	54	60	65	71	77	83	89	95	101	107
	1	**6**	**11**	**15**	**21**	**26**	**31**	**37**	**42**	**47**	**53**	**59**	**64**	**70**	**75**	**81**	**86**	**92**	**98**
17	3	9	15	20	26	33	39	45	51	57	64	70	77	83	89	96	102	109	115
	2	**6**	**11**	**17**	**22**	**28**	**34**	**39**	**45**	**51**	**57**	**63**	**67**	**75**	**81**	**87**	**93**	**99**	**105**
18	4	9	16	22	28	35	41	48	55	61	68	75	82	88	95	102	109	116	123
	2	**7**	**12**	**18**	**24**	**30**	**36**	**42**	**48**	**55**	**61**	**67**	**74**	**80**	**86**	**93**	**99**	**106**	**112**
19	4	10	17	23	30	37	44	51	58	65	72	80	87	94	101	109	116	123	130
	2	**7**	**13**	**19**	**25**	**32**	**38**	**45**	**52**	**58**	**65**	**72**	**78**	**85**	**62**	**99**	**106**	**113**	**119**
20	4	11	18	25	32	39	47	54	62	69	77	84	92	100	107	115	123	130	138
	2	**8**	**13**	**20**	**27**	**34**	**41**	**48**	**55**	**62**	**69**	**76**	**83**	**90**	**98**	**105**	**112**	**119**	**127**

Notes: Critical values for one-tailed tests are in Roman type, followed by those for two-tailed tests in **boldface** type. Blank spaces indicate that no decisions are possible at the .05 alpha level with such small sample sizes.

Adapted from Table D.12 in Roger E. Kirk, *Statistics: An Introduction.* 4th edition. Pages 728–729. 1999. Reproduced with permission of Thomson Learning: www.thomsonrights.com.

Table D-2 Critical Values for the Mann-Whitney U Test: $\alpha = .01$

$n_2 \backslash n_1$	2	3	4	5	6	7	8	9	10	11	12	13	14	15	16	17	18	19	20
2												0	0	0	0	0	0	1	1
																		0	**0**
3					0	0	1	1	1	2	2	2	3	3	4	4	4	5	
								0	**0**	**0**	**1**	**1**	**1**	**2**	**2**	**2**	**2**	**3**	**3**
4			0	1	1	2	3	3	4	5	5	6	7	7	8	9	9	10	
				0	**0**	**1**	**1**	**2**	**2**	**3**	**3**	**4**	**5**	**5**	**6**	**6**	**7**	**8**	
5		0	1	2	3	4	5	6	7	8	9	10	11	12	13	14	15	16	
			0	**1**	**1**	**2**	**3**	**4**	**5**	**6**	**7**	**7**	**8**	**9**	**10**	**11**	**12**	**13**	
6		1	2	3	4	6	7	8	9	11	12	13	15	16	18	19	20	22	
		0	**1**	**2**	**3**	**4**	**5**	**6**	**7**	**9**	**10**	**11**	**12**	**13**	**15**	**16**	**17**	**18**	
7	0	1	3	4	6	7	9	11	12	14	16	17	19	21	23	24	26	28	
		0	**1**	**3**	**4**	**6**	**7**	**9**	**10**	**12**	**13**	**15**	**16**	**18**	**19**	**21**	**22**	**24**	
8	0	2	4	6	7	9	11	13	15	17	20	22	24	26	28	30	32	34	
		1	**2**	**4**	**6**	**7**	**9**	**11**	**13**	**15**	**17**	**18**	**20**	**22**	**24**	**26**	**28**	**30**	
9	1	3	5	7	9	11	14	16	18	21	23	26	28	31	33	36	38	40	
	0	**1**	**3**	**5**	**7**	**9**	**11**	**13**	**16**	**18**	**20**	**22**	**24**	**27**	**29**	**31**	**33**	**36**	
10	1	3	6	8	11	13	16	19	22	24	27	30	33	36	38	41	44	47	
	0	**2**	**4**	**6**	**9**	**11**	**13**	**16**	**18**	**21**	**24**	**26**	**29**	**31**	**34**	**37**	**39**	**42**	
11	1	4	7	9	12	15	18	22	25	28	31	34	37	41	44	47	50	53	
	0	**2**	**5**	**7**	**10**	**13**	**16**	**18**	**21**	**24**	**27**	**30**	**33**	**36**	**39**	**42**	**45**	**48**	
12	2	5	8	11	14	17	21	24	28	31	35	38	42	46	49	53	56	60	
	1	**3**	**6**	**9**	**13**	**15**	**18**	**21**	**24**	**27**	**31**	**34**	**37**	**41**	**44**	**47**	**51**	**54**	
13	0	2	5	9	12	16	20	23	27	31	35	39	43	47	51	55	59	63	67
		1	**3**	**7**	**10**	**13**	**17**	**20**	**24**	**27**	**31**	**34**	**38**	**42**	**45**	**49**	**53**	**56**	**60**
14	0	2	6	10	13	17	22	26	30	34	38	43	47	51	56	60	65	69	73
		1	**4**	**7**	**11**	**15**	**18**	**22**	**26**	**30**	**34**	**38**	**42**	**46**	**50**	**54**	**58**	**63**	**67**
15	0	3	7	11	15	19	24	28	33	37	42	47	51	56	61	66	70	75	80
		2	**5**	**8**	**12**	**16**	**20**	**24**	**29**	**33**	**37**	**42**	**46**	**51**	**55**	**60**	**64**	**69**	**73**
16	0	3	7	12	16	21	26	31	36	41	46	51	56	61	66	71	76	82	87
		2	**5**	**9**	**13**	**18**	**22**	**27**	**31**	**36**	**41**	**45**	**50**	**55**	**60**	**65**	**70**	**74**	**79**
17	0	4	8	13	18	23	28	33	38	44	49	55	60	66	71	77	82	88	93
		2	**6**	**10**	**15**	**19**	**24**	**29**	**34**	**39**	**44**	**49**	**54**	**60**	**65**	**70**	**75**	**81**	**86**
18	0	4	9	14	19	24	30	36	41	47	53	59	65	70	76	82	88	94	100
		2	**6**	**11**	**16**	**21**	**26**	**31**	**37**	**42**	**47**	**53**	**58**	**64**	**70**	**75**	**81**	**87**	**92**
19	1	4	9	15	20	26	32	38	44	50	56	63	69	75	82	88	94	101	107
	0	**3**	**7**	**12**	**17**	**22**	**28**	**33**	**39**	**45**	**51**	**56**	**63**	**69**	**74**	**81**	**87**	**93**	**99**
20	1	5	10	16	22	28	34	40	47	53	60	67	73	80	87	93	100	107	114
	0	**3**	**8**	**13**	**18**	**24**	**30**	**36**	**42**	**48**	**54**	**60**	**67**	**73**	**79**	**86**	**92**	**99**	**105**

Notes: Critical values for one-tailed tests are in Roman type, followed by those for two-tailed tests in **boldface** type. Blank spaces indicate that no decisions are possible at the .05 alpha level with such small sample sizes.

Adapted from Table D.12 in Roger E. Kirk, *Statistics: An Introduction*. 4th edition. Pages 728–729. 1999. Reproduced with permission of Thomson Learning: www.thomsonrights.com.

Table E Critical Values for the Chi Square Tests

			Significance Level		
*d.f.**	.10	.05	.02	.01	.001
1	2.706	3.841	5.412	6.635	10.827
2	4.605	5.991	7.824	9.210	13.815
3	6.251	7.815	9.837	11.345	16.266
4	7.790	9.488	11.668	13.277	18.467
5	9.236	11.070	13.388	15.086	20.515
6	10.645	12.592	15.033	16.812	22.457
7	12.017	14.067	16.622	18.475	24.322
8	13.362	15.507	18.168	20.090	26.125
9	14.684	16.919	19.679	21.666	27.877
10	15.987	18.307	21.161	23.209	29.588
11	17.275	19.675	22.618	24.725	31.264
12	18.549	21.026	24.054	26.217	32.909
13	19.812	23.362	25.472	27.688	34.528
14	21.064	23.685	26.873	29.141	36.123
15	22.307	24.996	28.259	30.578	37.697
16	23.542	26.296	29.633	32.000	39.252
17	24.769	27.587	30.995	33.409	40.790
18	25.989	28.869	32.346	34.805	42.312
19	27.204	30.144	33.687	36.191	43.820
20	28.412	31.410	35.020	37.566	45.315
21	29.615	32.671	36.343	38.932	46.797
22	30.813	33.924	37.659	40.289	48.268
23	32.007	35.172	38.968	41.638	49.728
24	33.196	36.415	40.270	42.980	51.179
25	34.382	37.652	41.566	44.314	52.620
26	35.563	38.885	42.856	45.642	54.052
27	36.741	40.113	44.140	46.963	55.476
28	37.916	41.337	45.419	48.278	56.893
29	39.087	42.557	46.693	49.588	58.302
30	40.256	43.773	47.962	50.892	59.703
32	42.585	46.194	50.487	53.486	62.487
34	44.903	48.602	52.995	56.061	65.247
36	47.212	50.999	55.489	58.619	67.985
38	49.513	53.384	57.969	61.162	70.703

(continues)

Table E continued

d.f.*	Significance Level				
	.10	.05	.02	.01	.001
40	51.805	55.759	60.436	63.691	73.402
42	54.090	58.124	62.892	66.206	76.084
44	56.369	60.481	65.337	68.710	78.750
46	58.641	62.830	67.771	71.201	81.400
48	60.907	65.171	70.197	73.683	84.037
50	63.167	67.505	72.613	76.154	86.661
52	65.422	69.832	75.021	78.616	89.272
54	67.673	72.153	77.422	81.069	91.872
56	69.919	74.468	79.815	83.513	94.461
58	72.160	76.778	82.201	85.950	97.039
60	74.397	79.082	84.580	88.379	99.607
62	76.630	81.381	86.953	90.802	102.166
64	78.860	83.675	89.320	93.217	104.716
66	81.085	85.965	91.681	95.626	107.258
68	83.308	88.250	94.037	98.028	109.791
70	85.527	90.531	96.388	100.425	112.317

Notes: *For odd d.f. values between 30 and 70, the mean of the tabular values for d.f. − 1 and d.f. + 1 may be taken.

Adapted from Table IV, page 47, in R.A. Fisher and F. Yates: *Statistical Tables for Biological, Agricultural, and Medical Research.* 6th edition. London: Longman Group Limited. 1963. Reproduced with permission of Pearson Education Ltd.

Table F ANOVA: Critical Values for the *F* Ratio

Degrees of Freedom for Numerator

Denom. d.f.	1	2	3	4	5	6	7	8	9	10	11	12	14	16	20	24	30
2	18.51	19.00	19.16	19.25	19.30	19.33	19.36	19.37	19.38	19.39	19.40	19.41	19.42	19.43	19.44	19.45	19.46
	98.49	**99.00**	**99.17**	**99.25**	**99.30**	**99.33**	**99.34**	**99.36**	**99.38**	**99.40**	**99.41**	**99.42**	**99.43**	**99.45**	**99.46**	**99.47**	**99.48**
3	10.13	9.55	9.28	9.12	9.01	8.94	8.88	8.84	8.81	8.78	8.76	8.74	8.71	8.69	8.66	8.64	8.62
	34.12	**30.82**	**29.46**	**28.71**	**28.24**	**27.91**	**27.67**	**27.49**	**27.34**	**27.23**	**27.13**	**27.05**	**26.92**	**26.83**	**26.69**	**26.60**	**26.50**
4	7.71	6.94	6.59	6.39	6.26	6.16	6.09	6.04	6.00	5.96	5.93	5.91	5.87	5.84	5.80	5.77	5.74
	21.20	**18.00**	**16.69**	**15.98**	**15.52**	**15.21**	**14.98**	**14.80**	**14.66**	**14.54**	**14.45**	**14.37**	**14.24**	**14.15**	**14.02**	**13.93**	**13.74**
5	6.61	5.79	5.41	5.19	5.05	4.95	4.88	4.82	4.78	4.74	4.70	4.68	4.64	4.60	4.56	4.53	4.50
	16.26	**13.27**	**12.06**	**11.39**	**10.97**	**10.67**	**10.45**	**10.27**	**10.15**	**10.05**	**9.96**	**9.89**	**9.77**	**9.68**	**9.55**	**9.47**	**9.38**
6	5.99	5.14	4.76	4.53	4.39	4.28	4.21	4.15	4.10	4.06	4.03	4.00	3.96	3.92	3.87	3.84	3.81
	13.74	**10.92**	**9.78**	**9.15**	**8.75**	**8.47**	**8.26**	**8.10**	**7.98**	**7.87**	**7.79**	**7.72**	**7.60**	**7.52**	**7.39**	**7.31**	**7.23**
7	5.59	4.47	4.35	4.12	3.97	3.87	3.79	3.73	3.68	3.63	3.60	3.57	3.52	3.49	3.44	3.41	3.38
	12.25	**9.55**	**8.45**	**7.85**	**7.46**	**7.19**	**7.00**	**6.84**	**6.71**	**6.62**	**6.54**	**6.47**	**6.35**	**6.27**	**6.15**	**6.07**	**5.98**
8	5.32	4.46	4.07	3.84	3.69	3.58	3.50	3.44	3.39	3.34	3.31	3.28	3.23	3.20	3.15	3.12	3.08
	11.26	**8.65**	**7.59**	**7.01**	**6.63**	**6.37**	**6.19**	**6.03**	**5.91**	**5.82**	**5.74**	**5.67**	**5.56**	**5.48**	**5.36**	**5.28**	**5.20**
9	5.12	4.26	3.86	3.63	3.48	3.37	3.29	3.23	3.18	3.13	3.10	3.07	3.02	2.98	2.93	2.90	2.86
	10.56	**8.02**	**6.99**	**6.42**	**6.06**	**5.80**	**5.62**	**5.47**	**5.35**	**5.26**	**5.18**	**5.11**	**5.00**	**4.92**	**4.80**	**4.73**	**4.64**
10	4.96	4.10	3.71	3.48	3.33	3.22	3.14	3.07	3.02	2.97	2.94	2.91	2.86	2.82	2.77	2.74	2.70
	10.04	**7.56**	**6.55**	**5.99**	**5.64**	**5.39**	**5.21**	**5.06**	**4.95**	**4.85**	**4.78**	**4.71**	**4.60**	**4.52**	**4.41**	**4.33**	**4.25**
11	4.84	3.98	3.59	3.36	3.20	3.09	3.01	2.95	2.90	2.86	2.82	2.79	2.74	2.70	2.65	2.61	2.57
	9.65	**7.20**	**6.22**	**5.67**	**5.32**	**5.07**	**4.88**	**4.74**	**4.63**	**4.54**	**4.46**	**4.40**	**4.29**	**4.21**	**4.10**	**4.02**	**3.94**
12	4.75	3.88	3.49	3.26	3.11	3.00	2.92	2.85	2.80	2.76	2.72	2.69	2.64	2.60	2.54	2.50	2.46
	9.33	**6.93**	**5.95**	**5.41**	**5.06**	**4.82**	**4.65**	**4.50**	**4.39**	**4.30**	**4.22**	**4.16**	**4.05**	**3.98**	**3.86**	**3.78**	**3.70**
13	4.67	3.80	3.41	3.18	3.02	2.92	2.84	2.77	2.72	2.67	2.63	2.60	2.55	2.51	2.46	2.42	2.38
	9.07	**6.70**	**5.74**	**5.20**	**4.86**	**4.62**	**4.44**	**4.30**	**4.19**	**4.10**	**4.02**	**3.96**	**3.85**	**3.78**	**3.67**	**3.59**	**3.51**
14	4.60	3.74	3.34	3.11	2.96	2.85	2.77	2.70	2.65	2.60	2.56	2.53	2.48	2.44	2.39	2.35	2.31
	8.86	**6.51**	**5.56**	**5.03**	**4.69**	**4.46**	**4.28**	**4.14**	**4.03**	**3.94**	**3.86**	**3.80**	**3.70**	**3.62**	**3.51**	**3.43**	**3.34**
15	4.54	3.68	3.29	3.06	2.90	2.79	2.70	2.64	2.59	2.55	2.51	2.48	2.43	2.39	2.33	2.29	2.25
	8.68	**6.36**	**5.42**	**4.89**	**4.56**	**4.32**	**4.14**	**4.00**	**3.89**	**3.80**	**3.73**	**3.67**	**3.56**	**3.48**	**3.36**	**3.29**	**3.20**
16	4.49	3.63	3.24	3.01	2.85	2.74	2.66	2.59	2.54	2.49	2.45	2.42	2.37	2.33	2.28	2.24	2.20
	8.53	**6.23**	**5.29**	**4.77**	**4.44**	**4.20**	**4.03**	**3.89**	**3.78**	**3.69**	**3.61**	**3.55**	**3.45**	**3.37**	**3.25**	**3.18**	**3.10**
17	4.45	3.59	3.20	2.96	2.81	2.70	2.62	2.55	2.50	2.45	2.41	2.38	2.33	2.29	2.23	2.19	2.15
	8.40	**6.11**	**5.18**	**4.67**	**4.34**	**4.10**	**3.93**	**3.79**	**3.68**	**3.59**	**3.52**	**3.45**	**3.35**	**3.27**	**3.16**	**3.08**	**3.00**
18	4.41	3.55	3.16	2.93	2.77	2.66	2.58	2.51	2.46	2.41	2.37	2.34	2.29	2.25	2.19	2.15	2.11
	8.28	**6.01**	**5.09**	**4.58**	**4.25**	**4.01**	**3.85**	**3.71**	**3.60**	**3.51**	**3.44**	**3.37**	**3.27**	**3.19**	**3.07**	**3.00**	**2.91**

(continues)

Table F continued

Degrees of Freedom for Numerator

Denom. d.f.	1	2	3	4	5	6	7	8	9	10	11	12	14	16	20	24	30
19	4.38	3.52	3.13	2.90	2.74	2.63	2.55	2.48	2.43	2.38	2.34	2.31	2.26	2.21	2.15	2.11	2.07
	8.18	**5.93**	**5.01**	**4.50**	**4.17**	**3.94**	**3.77**	**3.63**	**3.52**	**3.43**	**3.36**	**3.30**	**3.19**	**3.12**	**3.00**	**2.92**	**2.84**
20	4.35	3.49	3.10	2.87	2.71	2.60	2.52	2.45	2.40	2.35	2.31	2.28	2.23	2.18	2.12	2.08	2.04
	8.10	**5.85**	**4.94**	**4.43**	**4.10**	**3.87**	**3.56**	**3.45**	**3.37**	**3.30**	**3.23**	**3.28**	**3.13**	**3.05**	**2.94**	**2.83**	**2.77**
21	4.32	3.47	3.07	2.84	2.68	2.57	2.49	2.42	2.37	2.32	2.28	2.25	2.20	2.15	2.09	2.05	2.00
	8.02	**5.78**	**4.87**	**4.37**	**4.04**	**3.81**	**3.65**	**3.51**	**3.40**	**3.31**	**3.24**	**3.17**	**3.07**	**2.99**	**2.88**	**2.80**	**2.72**
22	4.30	3.44	3.05	2.82	2.66	2.55	2.47	2.40	2.35	2.30	2.26	2.23	2.18	2.13	2.07	2.03	1.98
	7.94	**5.72**	**4.82**	**4.31**	**3.99**	**3.76**	**3.59**	**3.45**	**3.35**	**3.26**	**3.18**	**3.12**	**3.02**	**2.94**	**2.83**	**2.75**	**2.67**
23	4.28	3.42	3.03	2.80	2.64	2.53	2.45	2.38	2.32	2.28	2.24	2.20	2.14	2.10	2.04	2.00	1.96
	7.88	**5.66**	**4.76**	**4.26**	**3.94**	**3.71**	**3.54**	**3.41**	**3.30**	**3.21**	**3.14**	**3.07**	**2.97**	**2.89**	**2.78**	**2.70**	**2.62**
24	4.26	3.40	3.01	2.78	2.62	2.51	2.43	2.36	2.30	2.26	2.22	2.18	2.13	2.09	2.02	1.98	1.94
	7.82	**5.61**	**4.72**	**4.22**	**3.90**	**3.67**	**3.50**	**3.36**	**3.25**	**3.17**	**3.09**	**3.03**	**2.93**	**2.85**	**2.74**	**2.66**	**2.58**
25	4.24	3.38	2.99	2.76	2.60	2.49	2.41	2.34	2.28	2.24	2.20	2.16	2.11	2.06	2.00	1.96	1.92
	7.77	**5.57**	**4.68**	**4.18**	**3.86**	**3.63**	**3.46**	**3.32**	**3.21**	**3.13**	**3.05**	**2.99**	**2.89**	**2.81**	**2.70**	**2.62**	**2.54**
26	4.22	3.37	2.98	2.74	2.59	2.47	2.39	2.32	2.27	2.22	2.18	2.15	2.10	2.05	1.99	1.95	1.90
	7.72	**5.53**	**4.64**	**4.14**	**3.82**	**3.59**	**3.42**	**3.29**	**3.17**	**3.09**	**3.02**	**2.96**	**2.86**	**2.77**	**2.66**	**2.58**	**2.50**
27	4.21	3.35	2.96	2.73	2.57	2.46	2.37	2.30	2.25	2.20	2.16	2.13	2.08	2.03	1.97	1.93	1.88
	7.68	**5.49**	**4.60**	**4.11**	**3.79**	**3.56**	**3.39**	**3.26**	**3.14**	**3.06**	**2.98**	**2.93**	**2.83**	**2.74**	**2.63**	**2.55**	**3.47**
28	4.20	3.34	2.95	2.71	2.56	2.44	2.36	2.29	2.24	2.19	2.15	2.12	2.06	2.02	1.96	1.91	1.87
	7.64	**5.45**	**4.57**	**4.07**	**3.76**	**3.53**	**3.36**	**3.23**	**3.11**	**3.03**	**2.95**	**2.90**	**2.80**	**2.71**	**2.60**	**2.52**	**2.44**
29	4.18	3.33	2.93	2.70	2.54	2.43	2.35	2.28	2.22	2.18	2.14	2.10	2.05	2.00	1.94	1.90	1.85
	7.60	**5.42**	**4.54**	**4.04**	**3.73**	**3.50**	**3.33**	**3.20**	**3.08**	**3.00**	**2.92**	**2.87**	**2.77**	**2.68**	**2.57**	**2.49**	**2.41**
30	4.17	3.32	2.92	2.69	2.53	2.42	2.34	2.27	2.21	2.16	2.12	2.09	2.04	1.99	1.93	1.89	1.84
	7.56	**5.39**	**4.51**	**4.02**	**3.70**	**3.47**	**3.30**	**3.17**	**3.06**	**2.98**	**2.90**	**2.84**	**2.74**	**2.66**	**2.55**	**2.47**	**2.38**
32	4.15	3.30	2.90	2.67	2.51	2.40	2.32	2.25	2.19	2.14	2.10	2.07	2.02	1.97	1.91	1.86	1.82
	7.50	**5.34**	**4.46**	**3.97**	**3.66**	**3.42**	**3.25**	**3.12**	**3.01**	**2.94**	**2.86**	**2.80**	**2.70**	**2.62**	**2.51**	**2.42**	**2.34**
34	4.13	3.28	2.88	2.65	2.49	2.38	2.30	2.23	2.17	2.13	2.08	2.05	2.00	1.95	1.89	1.84	1.80
	7.44	**5.29**	**4.42**	**3.93**	**3.61**	**3.38**	**3.21**	**3.08**	**2.97**	**2.89**	**2.82**	**2.76**	**2.66**	**2.58**	**2.47**	**2.38**	**2.30**
36	4.11	3.26	2.86	2.63	2.48	2.36	2.28	2.21	2.15	2.10	2.06	2.03	1.98	1.93	1.87	1.82	1.78
	7.39	**5.25**	**4.38**	**3.89**	**3.58**	**3.35**	**3.18**	**3.04**	**2.94**	**2.86**	**2.78**	**2.72**	**2.62**	**2.54**	**2.43**	**2.35**	**2.26**
38	4.10	3.25	2.85	2.62	2.46	2.35	2.26	2.19	2.14	2.09	2.05	2.02	1.96	1.92	1.85	1.80	1.76
	7.35	**5.21**	**4.34**	**3.86**	**3.54**	**3.32**	**3.15**	**3.02**	**2.91**	**2.82**	**2.75**	**2.69**	**2.59**	**2.51**	**2.40**	**2.32**	**2.22**
40	4.08	3.23	2.84	2.61	2.45	2.34	2.25	2.18	2.12	2.07	2.04	2.00	1.95	1.90	1.84	1.79	1.74
	7.31	**5.18**	**4.31**	**3.83**	**3.51**	**3.29**	**3.12**	**2.99**	**2.88**	**2.80**	**2.73**	**2.66**	**2.56**	**2.49**	**2.37**	**2.29**	**2.20**

(continues)

Table F continued

Degrees of Freedom for Numerator

Denom. d.f.	1	2	3	4	5	6	7	8	9	10	11	12	14	16	20	24	30
42	4.07	3.22	2.83	2.59	2.44	2.32	2.24	2.17	2.11	2.06	2.02	1.99	1.94	1.89	1.82	1.78	1.73
	7.27	**5.15**	**4.29**	**3.80**	**3.49**	**3.26**	**3.10**	**2.96**	**2.86**	**2.77**	**2.70**	**2.64**	**2.54**	**2.46**	**2.35**	**2.26**	**2.17**
44	4.06	3.21	2.82	2.58	2.43	2.31	2.23	2.16	2.10	2.05	2.01	1.98	1.92	1.88	1.81	1.76	1.72
	7.24	**5.12**	**4.26**	**3.78**	**3.46**	**3.24**	**3.07**	**2.94**	**2.84**	**2.75**	**2.68**	**2.62**	**2.52**	**2.44**	**2.32**	**2.24**	**2.15**
46	4.05	3.20	2.81	2.57	2.42	2.30	2.22	2.14	2.09	2.04	2.00	1.97	1.91	1.87	1.80	1.75	1.71
	7.21	**5.10**	**4.24**	**3.76**	**3.44**	**3.22**	**3.05**	**2.92**	**2.82**	**2.73**	**2.66**	**2.60**	**2.50**	**2.42**	**2.30**	**2.22**	**2.13**
48	4.04	3.19	2.80	2.56	2.41	2.30	2.21	2.14	2.08	2.03	1.99	1.96	1.90	1.86	1.79	1.74	1.70
	7.19	**5.08**	**4.22**	**3.74**	**3.42**	**3.20**	**3.04**	**2.90**	**2.80**	**2.71**	**2.64**	**2.58**	**2.48**	**2.40**	**2.28**	**2.20**	**2.11**
50	4.03	3.18	2.79	2.56	2.40	2.29	2.20	2.13	2.07	2.02	1.98	1.95	1.90	1.85	1.78	1.74	1.69
	7.17	**5.06**	**4.20**	**3.72**	**3.41**	**3.18**	**3.02**	**2.88**	**2.78**	**2.70**	**2.62**	**2.56**	**2.46**	**2.39**	**2.26**	**2.18**	**2.10**
55	4.02	3.17	2.78	2.54	2.38	2.27	2.18	2.11	2.05	2.00	1.97	1.93	1.88	1.83	1.76	1.72	1.67
	7.12	**5.01**	**4.16**	**3.68**	**3.37**	**3.15**	**2.98**	**2.85**	**2.75**	**2.66**	**2.59**	**2.53**	**2.43**	**2.35**	**2.23**	**2.15**	**2.06**
60	4.00	3.15	2.76	2.52	2.37	2.25	2.17	2.10	2.04	1.99	1.95	1.92	1.86	1.81	1.75	1.70	1.65
	7.08	**4.98**	**4.13**	**3.65**	**3.34**	**3.12**	**2.95**	**2.82**	**2.72**	**2.63**	**2.56**	**2.50**	**2.40**	**2.32**	**2.20**	**2.12**	**2.03**
65	3.99	3.14	2.75	2.51	2.36	2.24	2.15	2.08	2.02	1.98	1.94	1.90	1.85	1.80	1.73	1.68	1.63
	7.04	**4.95**	**4.10**	**3.62**	**3.31**	**3.09**	**2.93**	**2.79**	**2.70**	**2.61**	**2.54**	**2.47**	**2.37**	**2.30**	**2.18**	**2.09**	**2.00**
70	3.98	3.13	2.74	2.50	2.35	2.23	2.14	2.07	2.01	1.97	1.93	1.89	1.84	1.79	1.72	1.67	1.62
	7.01	**4.92**	**4.08**	**3.60**	**3.29**	**3.07**	**2.91**	**2.77**	**2.67**	**2.59**	**2.51**	**2.45**	**2.35**	**2.28**	**2.15**	**2.07**	**1.98**
80	3.96	3.11	2.72	2.48	2.33	2.21	2.12	2.05	1.99	1.95	1.91	1.88	1.82	1.77	1.70	1.65	1.60
	6.96	**4.88**	**4.04**	**3.56**	**3.25**	**3.04**	**2.87**	**2.74**	**2.64**	**2.55**	**2.48**	**2.41**	**2.32**	**2.24**	**2.11**	**2.03**	**1.94**
100	3.94	3.09	2.70	2.46	2.30	2.19	2.10	2.03	1.97	1.92	1.88	1.85	1.79	1.75	1.68	1.63	1.57
	6.90	**4.82**	**3.98**	**3.51**	**3.20**	**2.99**	**2.82**	**2.69**	**2.59**	**2.51**	**2.43**	**2.36**	**2.26**	**2.19**	**2.06**	**1.98**	**1.89**
125	3.92	3.07	2.68	2.44	2.29	2.17	2.08	2.01	1.95	1.90	1.86	1.83	1.77	1.72	1.65	1.60	1.55
	6.84	**4.78**	**3.94**	**3.47**	**3.17**	**2.95**	**2.79**	**2.65**	**2.56**	**2.47**	**2.40**	**2.33**	**2.23**	**2.15**	**2.03**	**1.94**	**1.85**
150	3.91	3.06	2.67	2.43	2.27	2.16	2.07	2.00	1.94	1.89	1.85	1.82	1.76	1.71	1.64	1.59	1.54
	6.81	**4.75**	**3.91**	**3.44**	**3.14**	**2.92**	**2.76**	**2.62**	**2.53**	**2.44**	**2.37**	**2.30**	**2.20**	**2.12**	**2.00**	**1.91**	**1.83**
200	3.89	3.04	2.65	2.41	2.26	2.14	2.05	1.98	1.92	1.87	1.83	1.80	1.74	1.69	1.62	1.57	1.52
	6.76	**4.71**	**3.88**	**3.41**	**3.11**	**2.90**	**2.73**	**2.60**	**2.50**	**2.41**	**2.34**	**2.28**	**2.17**	**2.09**	**1.97**	**1.88**	**1.79**
400	3.86	3.02	2.62	2.39	2.23	2.12	2.03	1.96	1.90	1.85	1.81	1.78	1.72	1.67	1.60	1.54	1.49
	6.70	**4.66**	**3.83**	**3.36**	**3.06**	**2.85**	**2.69**	**2.55**	**2.46**	**2.37**	**2.29**	**2.23**	**2.12**	**2.04**	**1.92**	**1.84**	**1.74**
1000	3.85	3.00	2.61	2.38	2.22	2.10	2.02	1.95	1.89	1.84	1.80	1.76	1.70	1.65	1.58	1.53	1.47
	6.66	**4.62**	**3.80**	**3.34**	**3.04**	**2.82**	**2.66**	**2.53**	**2.43**	**2.34**	**2.26**	**2.20**	**2.09**	**2.01**	**1.89**	**1.81**	**1.71**
∞	3.84	2.99	2.60	2.37	2.21	2.09	2.01	1.94	1.88	1.83	1.79	1.75	1.69	1.64	1.57	1.52	1.46
	6.64	**4.60**	**3.78**	**3.32**	**3.02**	**2.80**	**2.64**	**2.51**	**2.41**	**2.32**	**2.24**	**2.18**	**2.07**	**1.99**	**1.87**	**1.79**	**1.69**

Notes: Adapted from Table A14 in G.W. Snedecor and W.G. Cochran: *Statistical Methods.* 7th edition. Ames, Iowa: The Iowa State University Press. 1980. Reproduced with permission of The Iowa State University Press.

Critical values for the .05 alpha level are in Roman type, followed by those for the .01 level in **boldface type.**

Table G Tukey's HSD: Critical Values for the Studentized Range (Q) Statistic

SS_W (error) d.f.	2	3	4	5	6	7	8	9	10	12	14	16	18	20
5	3.64	4.60	5.22	5.67	6.03	6.33	6.58	6.80	6.99	7.32	7.60	7.83	8.03	8.21
	5.70	**6.98**	**7.80**	**8.42**	**8.91**	**9.32**	**9.67**	**9.97**	**10.24**	**10.70**	**11.08**	**11.40**	**11.68**	**11.93**
6	3.46	4.34	4.90	5.30	5.63	5.90	6.12	6.32	6.49	6.79	7.03	7.24	7.43	7.59
	5.24	**6.33**	**7.03**	**7.56**	**7.97**	**8.32**	**8.61**	**8.87**	**9.10**	**9.48**	**9.81**	**10.08**	**10.32**	**10.54**
7	3.34	4.16	4.68	5.06	5.36	5.61	5.82	6.00	6.16	6.43	6.66	6.85	7.02	7.17
	4.95	**5.92**	**6.54**	**7.01**	**7.37**	**7.68**	**7.94**	**8.17**	**8.37**	**8.71**	**9.00**	**9.24**	**9.46**	**9.65**
8	3.26	4.04	4.53	4.89	5.17	5.40	5.60	5.77	5.92	6.18	6.39	6.57	6.73	6.87
	4.75	**5.64**	**6.20**	**6.62**	**6.96**	**7.24**	**7.47**	**7.68**	**7.86**	**8.18**	**8.44**	**8.66**	**8.85**	**9.03**
9	3.20	3.95	4.41	4.76	5.02	5.24	5.43	5.59	5.74	5.98	6.19	6.36	6.51	6.64
	4.60	**5.43**	**5.96**	**6.35**	**6.66**	**6.91**	**7.13**	**7.33**	**7.49**	**7.78**	**8.03**	**8.23**	**8.41**	**8.57**
10	3.15	3.88	4.33	4.65	4.91	5.12	5.30	5.46	5.60	5.83	6.03	6.19	6.34	6.47
	4.48	**5.27**	**5.77**	**6.14**	**6.43**	**6.67**	**6.87**	**7.05**	**7.21**	**7.49**	**7.71**	**7.91**	**8.08**	**8.23**
11	3.11	3.82	4.26	4.57	4.82	5.03	5.20	5.35	5.49	5.71	5.90	6.06	6.20	6.33
	4.39	**5.15**	**5.62**	**5.97**	**6.25**	**6.48**	**6.67**	**6.84**	**6.99**	**7.25**	**7.46**	**7.65**	**7.81**	**7.95**
12	3.08	3.77	4.20	4.51	4.75	4.95	5.12	5.27	5.39	5.61	5.80	5.95	6.09	6.21
	4.32	**5.05**	**5.50**	**5.84**	**6.10**	**6.32**	**6.51**	**6.67**	**6.81**	**7.06**	**7.26**	**7.44**	**7.59**	**7.73**
13	3.06	3.73	4.15	4.45	4.69	4.88	5.05	5.19	5.32	5.53	5.71	5.86	5.99	6.11
	4.26	**4.96**	**5.40**	**5.73**	**5.98**	**6.19**	**6.37**	**6.53**	**6.67**	**6.90**	**7.10**	**7.27**	**7.42**	**7.55**
14	3.03	3.70	4.11	4.41	4.64	4.83	4.99	5.13	5.25	5.46	5.64	5.79	5.91	6.03
	4.21	**4.89**	**5.32**	**5.63**	**5.88**	**6.08**	**6.26**	**6.41**	**6.54**	**6.77**	**6.96**	**7.13**	**7.27**	**7.39**
15	3.01	3.67	4.08	4.37	4.59	4.78	4.94	5.08	5.20	5.40	5.57	5.72	5.85	5.96
	4.17	**4.84**	**5.25**	**5.56**	**5.80**	**5.99**	**6.16**	**6.31**	**6.44**	**6.66**	**6.84**	**7.00**	**7.14**	**7.26**
16	3.00	3.65	4.05	4.33	4.56	4.74	4.90	5.03	5.15	5.35	5.52	5.66	5.79	5.90
	4.13	**4.79**	**5.19**	**5.49**	**5.72**	**5.92**	**6.08**	**6.22**	**6.35**	**6.56**	**6.74**	**6.90**	**7.03**	**7.15**

Number of means, treatment groups (k)

(continues)

Table G continued

SS$_W$ (error) d.f.	Number of means, treatment groups (k)													
	2	3	4	5	6	7	8	9	10	12	14	16	18	20
17	2.98	3.63	4.02	4.30	4.52	4.70	4.86	4.99	5.11	5.31	5.47	5.61	5.73	5.84
	4.10	**4.74**	**5.14**	**5.43**	**5.66**	**5.85**	**6.01**	**6.15**	**6.27**	**6.48**	**6.66**	**6.81**	**6.94**	**7.05**
18	2.97	3.61	4.00	4.28	4.49	4.67	4.82	4.96	5.07	5.27	5.43	5.57	5.69	5.79
	4.07	**4.70**	**5.09**	**5.38**	**5.60**	**5.79**	**5.94**	**6.08**	**6.20**	**6.41**	**6.58**	**6.73**	**6.85**	**6.97**
19	2.96	3.59	3.98	4.25	4.47	4.65	4.79	4.92	5.04	5.23	5.39	5.53	5.65	5.75
	4.05	**4.67**	**5.05**	**5.33**	**5.55**	**5.73**	**5.89**	**6.02**	**6.14**	**6.34**	**6.51**	**6.65**	**6.78**	**6.89**
20	2.95	3.58	3.96	4.23	4.45	4.62	4.77	4.90	5.01	5.20	5.36	5.49	5.61	5.71
	4.02	**4.64**	**5.02**	**5.29**	**5.51**	**5.69**	**5.84**	**5.97**	**6.09**	**6.28**	**6.45**	**6.59**	**6.71**	**6.82**
24	2.92	3.53	3.90	4.17	4.37	4.54	4.68	4.81	4.92	5.10	5.25	5.38	5.49	5.59
	3.96	**4.55**	**4.91**	**5.17**	**5.37**	**5.54**	**5.69**	**5.81**	**5.92**	**6.11**	**6.26**	**6.39**	**6.51**	**6.61**
30	2.89	3.49	3.85	4.10	4.30	4.46	4.60	4.72	4.82	5.00	5.15	5.27	5.38	5.47
	3.89	**4.45**	**4.80**	**5.05**	**5.24**	**5.40**	**5.54**	**5.65**	**5.76**	**5.93**	**6.08**	**6.20**	**6.31**	**6.41**
40	2.86	3.44	3.79	4.04	4.23	4.39	4.52	4.63	4.73	4.90	5.04	5.16	5.27	5.36
	3.82	**4.37**	**4.70**	**4.93**	**5.11**	**5.26**	**5.39**	**5.50**	**5.60**	**5.76**	**5.90**	**6.02**	**6.12**	**6.21**
60	2.83	3.40	3.74	3.98	4.16	4.31	4.44	4.55	4.65	4.81	4.94	5.06	5.15	5.24
	3.76	**4.28**	**4.59**	**4.82**	**4.99**	**5.13**	**5.25**	**5.36**	**5.45**	**5.60**	**5.73**	**5.84**	**5.93**	**6.01**
120	2.80	3.36	3.68	3.92	4.10	4.24	4.36	4.47	4.56	4.71	4.84	4.95	5.04	5.13
	3.70	**4.20**	**4.50**	**4.71**	**4.87**	**5.01**	**5.12**	**5.21**	**5.30**	**5.44**	**5.56**	**5.66**	**5.75**	**5.83**
∞	2.77	3.31	3.63	3.86	4.03	4.17	4.29	4.39	4.47	4.62	4.74	4.85	4.93	5.01
	3.64	**4.12**	**4.40**	**4.60**	**4.76**	**4.88**	**4.99**	**5.08**	**5.16**	**5.29**	**5.40**	**5.49**	**5.57**	**5.65**

Notes: Adapted from Table 29 in E. Pearson and H. Hartley, "Critical Values for the Studentized Range (Q) Distribution." *Biometrika Tables for Statisticians.* Vol. 1, 3rd edition. 1966. Reproduced with permission of Oxford University Press.
Critical values for the .05 alpha level are in Roman type, followed by those for the .01 level in **boldface** type.

Table H Critical Values for Pearson's r_{XY}

	Significance				
Two-Tailed Test	.10	.05	.02	.01	.001
One-Tailed Test	.05	.025	.01	.005	.0005
$n*$					
3	.988	.997	.999	.999	.999
4	.900	.950	.980	.990	.999
5	.805	.878	.934	.958	.991
6	.729	.811	.882	.917	.974
7	.669	.755	.833	.875	.951
8	.622	.707	.789	.843	.925
9	.582	.666	.750	.798	.898
10	.549	.632	.716	.765	.872
11	.521	.602	.685	.735	.847
12	.497	.579	.658	.708	.823
13	.476	.553	.634	.684	.801
14	.458	.532	.612	.661	.780
15	.441	.514	.592	.641	.760
16	.426	.497	.574	.623	.742
17	.412	.482	.558	.606	.725
18	.400	.468	.543	.590	.708
19	.389	.456	.529	.575	.693
20	.378	.444	.516	.561	.679
21	.369	.433	.503	.549	.665
22	.360	.423	.492	.537	.652
27	.323	.381	.445	.487	.597
32	.296	.349	.409	.449	.554
37	.275	.325	.381	.418	.519
42	.257	.304	.358	.393	.490
47	.243	.288	.338	.372	.465
52	.231	.273	.322	.354	.443
62	.211	.250	.295	.325	.408
72	.195	.232	.274	.302	.380
82	.183	.217	.257	.283	.357
92	.173	.205	.242	.267	.338
102	.164	.195	.230	.254	.321

Notes: *n* is the number of pairs, *not* the number of degrees of freedom, which is $n - 2$.
Adapted from Table VII, page 63, in R.A. Fisher and F. Yates: *Statistical Tables for Biological, Agricultural, and Medical Research.* 6th edition. London: Longman Group Limited. 1963. Reproduced with permission of Pearson Education Ltd.

Table I Critical Values for Spearman's r_s

	Significance			
Two-Tailed Test Test	.10	.05	.02	.01
One-Tailed Test Test	.05	.025	.01	.005
n				
6	.829	.886	.943	1.000
7	.714	.786	.893	.929
8	.643	.738	.833	.881
9	.600	.700	.783	.833
10	.564	.648	.745	.794
11	.536	.618	.709	.755
12	.503	.587	.671	.727
13	.484	.560	.648	.703
14	.464	.538	.622	.675
15	.443	.521	.604	.654
16	.429	.503	.582	.635
17	.414	.485	.566	.615
18	.401	.472	.550	.600
19	.391	.460	.535	.584
20	.380	.447	.520	.570
21	.370	.435	.508	.556
22	.361	.425	.496	.544
23	.353	.415	.486	.532
24	.344	.406	.476	.521
25	.337	.398	.466	.511
26	.331	.390	.457	.501
27	.324	.382	.448	.491
28	.317	.375	.440	.483
29	.312	.368	.433	.475
30	.306	.362	.425	.467
32	.296	.350	.412	.452
34	.287	.340	.399	.439
36	.279	.330	.288	.427
38	.271	.321	.378	.415
40	.264	.313	.368	.405
50	.235	.279	.329	.363
60	.214	.255	.300	.331
70	.198	.235	.278	.307
80	.185	.220	.260	.287
90	.174	.207	.245	.271
100	.163	.197	.233	.257

Notes: Adapted from J.H. Zar, "Significance Testing of the Spearman Rank Correlation Coefficient," *The Journal of the American Statistical Association.* Vol. 67. Pages 578–580. American Statistical Association. 1972. Reprinted with permission from *The Journal of the American Statistical Association.* All rights reserved.

Answers and Hints for Selected Exercises

This appendix presents answers to selected exercises for each chapter, but a few introductory qualifications are appropriate. First, the appendix does not include a math or algebra review. Many excellent reviews are freely available online, and it makes more sense to recommend those to you than to repeat that content here. Second, to give you practice in diagnosing statistical situations, questions requiring various procedures are more-or-less randomly ordered at the end of each chapter. Consequently, unlike some texts, the answers below do not necessarily cover all odd- or all even-numbered items. The answers given here are for at least half the questions of each chapter; they cover all procedures and types of problems. Third, for the exercises involving calculations, the math is done to three decimal places. Your answers may vary slightly if you are working to more or to fewer places. Finally, very brief hints—a word, a phrase, or sentence—are provided for each essay or discussion question. Answers will vary, of course, but the hints are key words, concepts, or issues that should pop to mind as you read those questions. They are included to jog your memory and to prompt you to formulate answers in your own words. Your answers should be complete and fully explanatory, however, and do more than just repeat these words or clues.

Chapter 1. Beginning Concepts

   1. Quantitative precision. Correct statistical tools.
   2, 3. Categories/counts, ranks, legitimate X values.
   4. Ambiguity. Precision.
   5. Sample and population.
   6. Representativeness.

Chapter 2. Getting Started: Descriptive Statistics

	Mode (Minor Mode)	Median	ΣX	ΣX^2	Mean $\overline{X}$	Standard Deviation s_X	(σ_X)
1.	13	13.5	137	1993	13.70	3.59	(3.41)
2.	0, 4	3.5	37	235	3.70	3.30	(3.13)
4.	none	14	145	2493	13.18	7.63	(7.27)
5.	4 (2)	4	76	698	5.43	4.69	(4.52)
7.	4 (1)	3.5	47	207	3.36	1.95	(1.87)
9.	20 (0)	15	192	4570	14.77	12.02	(11.55)

11. Degree of quantitative precision, accuracy. Frequencies, ranks, X values.
12. Spread. Average.
13. X scores comparatively spread out or clustered?
14. A distribution "pulled" in one direction or the other? Outliers? Extreme cases?
15. What does a negative standard deviation mean?
16. May X values be negative?
17. a. Frequencies.   d. Typical variation.
    b. Percentiles, ranks.   e. Highest and lowest.
    c. Mathematical weights. f. Squared deviations.
18. Levels of measurement. Suitable measures of central tendency.

Chapter 3. Probability: A Foundation for Statistical Decisions

Continuous Random Probabilities

1. a. $z = .44$, column C = .3330
   b. $z = -1.48$, column C = .0694
   c. $z = \pm.38$, column B = .1480 + .1480 = .2960
   d. $z_{30} = .44$, column C = .3300 ⎫
      $z_{40} = 1.71$, column C = .0436 ⎬ .2864
   e. $z = -.25$, $X = 24.62$ hours
   f. $z = 1.44$ for the top 7.5%, $X = 37.86$ hours or more
      $z = -1.44$ for the lowest 7.5%, $X = 15.30$ hours or less
   g. $P(X = 35)$, a single X value, cannot be done.
2. z for .0800 in column C = 1.405, $\mu = 134$ semester units, $\sigma = 3.56$ semester units
5. $z = .52$ or .525 for the top 30%, $\mu = 50$ months, $\sigma = 15.24$ months
7. a. $z_0 = 1.45$, column C = .0735
   b. $z_{-200} = -1.18$, column C = .1190
   c. $z_{20} = 1.71$, column B = .4564 ⎫
      $z_{50} = 2.11$, column B = .4826 ⎬ .0262

    d. z for the top $10\% = 1.28$, $X = -\$12.72$; lose $12.72 or less or even win

    e. z for .1500 in column B = .39 or .385, $X = -\$80.74$; lose $80.74

8. a. X.

    b. Proportion.

9. Standardization.

10. Directionality of deviation from the mean.

11. Equality of units. Equal units?

12. Proportions.

13. Proportion above? Below?

14. (Be more specific and descriptive in your own answers.)

    a. Average

    b. Downright scary.

    c. A rating we all give ourselves.

    d. Unpredictable.

15. How many standard deviations from the mean?

16. Spread, range.

17. Location. X axis.

18. Level of measurement.

Binomial Probabilities

19. Use binomial table: $P = .686$

22. Combinations: 156,800 different ways

23. a. Use binomial table. $P = .677$

    b. $z = -.20$, column B + .5000 = .0793 + .5000 = .5793

24. a. $z = -1.29$, column B + .5000 = .4015 + .5000 = .9015

    b. Use binomial table: $P = .828$

26. 2,598,960 different hands possible, 4 of which are royal flushes: $P = .00000154$

27. $z = 1.25$, column B + .5000 = .3944 + .5000 = .8944

28. Complement = $P(X \le 2) = .300 + .282 + .123 = .705$; $1 - .705 = .295$

32. a. $z = .85$, column C = .1977

    b. $P(X = 0) = .00012$

33. Use binomial table: $P = .421$

35. a. $P(X \ge 6) = .230 + .107 + .022 = .359$

    b. $z = -.96$, column C = .1685

Hints for Essay Questions

36. Values of p and q constant?

37. Independent trials. Sample size, number of trials.

38. Discrete versus continuous probability distributions.
39. a, b. Continuity, symmetry.
40. Continuity. Discrete versus continuous probability distributions.
41. Probability of failure. Ease of computation.

Chapter 4. Describing a Population: Estimation with a Single Sample

1. $z = 1.645$, use the fpc: 90 C.I. estimate for $p = .420 \pm .026$
2. $t = 2.861$, use the fpc: .99 C.I. estimate for $\mu = 6.94 \pm 1.38$ arrests
4. $z = 1.645$, use the fpc formula for the final answer: $n = 158$ students
7. $z = 1.96$: .95 C.I. estimate for $p = .34 \pm .047$
8. $z = 2.575$: $n = 509$ passengers
9. $z = 1.96$: $n = 2401$ policyholders
10. $t = 1.703$: .90 C.I. estimate for $\mu = \$10,430 \pm \$1054.34$
11. $z = 2.575$: .99 C.I. estimate for $p = .31 \pm .054$
13. $z = 2.575$, use the fpc formula for the final answer: $n = 815$ business majors
15. $z = 1.645$, $\sigma =$ range/4, use the fpc formula for the final answer: $n = 710$ seniors
17. $t = 1.714$, use the fpc: .90 C.I. estimate for $\mu = \$2987 \pm \$59.79$
19. a. $z = 1.645$: $n = 107$ ticket holders
    b. $z = 1.96$: $n = 152$ ticket holders
    c. $z = 2.575$: $n = 261$ ticket holders
22. $z = 1.96$, use the fpc: .95 C.I. estimate for $p = .23 \pm .035$
24. $z = 2.575$: .99 C.I. estimate for $\mu = 19.86 \pm .849$ hours per week
25. Sample statistics. Many samples.
26. Statistics and parameters. Sampling.
27. Known distribution of a statistic. Point estimate.
28. a. Normal variation.
    b. Known variation.
    c. Variation. Sampling.
29. Known variation/probabilities of a statistic.
30. Tails of the curve.
31, 32. Shape of the sampling distribution curve. Variation.
33. Sampling error. Sample size.
34. Spread in the sampling distribution of the mean.
35. Continuity. Continuous distributions?
36. Sampling distributions and the standard sampling error.
37. (One might sort the eight items into three groups: factors that would *increase* the margin of error, those that would *decrease* it, and those that would have *no effect*. Look for common or similar reasons why the factors in each group would have like effects on the margin of error.)
38. Sample size and variation.

39. Respective confidence intervals coincide?
40. Maximum variation.

Chapter 5. Testing a Hypothesis: Is Your Sample a Rare Case?

The answers below are sorted by type of problem rather than being strictly in order. Answers for selected hypothesis tests are presented first, followed by those for questions involving beta or Type II errors.

Hypothesis Tests

Means/ Propns.	Test (Tail) Direction	z or t	fpc	α	Crit. Value	Obs. Value	H_0 Decision	Conclusion
1. M	Lower	z	N	.05	−1.645	−2.103	Reject	County average is less
				.01	−2.33		Not reject	County average is not less
2. P	Upper	z	Y	.05	1.645	1.538	Not reject	% black and white in office is not higher
				.01	2.33		(The same, automatically.)	
3. P	2-tailed	z	N	.05	±1.96	2.051	Reject	Fraternity's percentage differs
4. M	Upper	t	Y	.05	1.711	1.606	Not reject	Students' cars are not older
				.01	2.492		(The same, automatically.)	
9. P	Lower	z	N	.05	−1.645	−2.100	Reject	Church members are less likely
11. P	Lower	z	Y	.05	−1.645	−1.763	Reject	County percentage is less
				.01	−2.33		Not reject	County percentage is not less
12. M	Upper	t	Y	.05	2.093	2.392	Reject	Temperature is warmer
				.01	2.861		Not reject	Temperature is not warmer
15. P	(Table 4.1 indicates n is too small when $q_0 = .30$. Cannot be done.)							
17. M	Lower	t	Y	.01	−2.473	−2.636	Reject	Average age has declined
18. P	2-tailed	z	N	.01	±2.575	2.857	Reject	County Latino percentage differs
19. M	Upper	z	Y	.05	1.645	1.583	Not reject	Commercials are not longer
22. M	2-tailed	t	N	.05	±2.365	−2.080	Not reject	Height does not differ from *Homo sapiens*
28. M	2-tailed	z	N	.05	±1.96	1.775	Not reject	Sample average score does not differ
29. P	Upper	z	Y	.05	1.645	1.833	Reject	Panel session has higher percentage

Determine P(Beta Error)

7. To pick out details, the problem involves averages, we may not use the fpc, and the question describes a two-tailed test. $\mu_O = \$37.20$, and the two possible H_A curves have μ_A values of $\pm\$6$ from that figure: $\$31.20$ and $\$43.20$, respectively. Use either H_A curve to determine P(Beta error), and then *double* that figure to take into account the other H_A curve's identical overlap with the central H_O curve. There are three steps; assume we are using the upper H_A curve (with $\mu_A = \$43.20$):

    a. Calculate the H_O curve's upper-tail $\overline{X}_C$ value: $\overline{X}_C = \$40.17$.
    b. Calculate the $\overline{X}_C$ value's z score in the upper H_A curve: $z_{\overline{X}_C} = -2.01$.
    c. Locate the area of the H_A curve falling beyond $z = -2.01$: Area $= .0222$.

The upper H_A curve has a beta area of $.0222$. In this case, for a final answer, *double* the $.0222$ figure to also include the beta part of the lower H_A curve. P(Beta error) $= .0222 + .0222 = .0444$ or 4.40%.

10. The problem involves proportions, a lower-tailed test, and, for the standard error, the $n{:}N$ ratio allows use of the fpc. $p_O = 58\%$ or $.58$, and the presumed p_A value for the H_A curve is 50% or $.50$. The three steps start with calculating the p'_C figure in the H_O curve, and then using that value to calculate P(Beta error) as part of the H_A curve.

    a. Calculate the H_O curve's lower-tail p'_C value: $p'_C = .54$.
    b. Calculate the p'_C value's z score in the H_A curve: $z_{p'_C} = 1.48$.
    c. Locate the area of the H_A curve falling beyond $z = 1.48$: Area $= .0694$.

The probability of committing a Type II error under the conditions described is $.0694$, or 6.94%.

Hints for Essay Questions

30, 31. Expected sampling error. Expected sampling outcomes. Probability of certain outcomes. Unusual or unexpected events.
    32. Unusual events. Low probability versus normality.
    33. Known population parameter.
    34. Required alpha region. Open tails of the H_O curve.
    35. Probability. Areas of the H_O curve. Location (overlap) of the H_A curve.
    36. Specific decision regarding H_O. Alpha and beta curve areas.
    37. Areas (tails) of the H_O curve. Probability. Rationale for H_A.
    38. Size, extremity of the alpha region
    39. Direction, if any, of H_A.
    40. Probability of a correct decision.
    41. How much might the H_O and H_A curves overlap?
    42. Shape of a sampling distribution.

43. a. Even smaller curve area.
    b, d. Where does z-observed or t-observed fall?
    c. Obvious decision?

Chapter 6. Estimates and Tests with Two Samples: Identifying Differences

The answers below are sorted by type of problem rather than being strictly in order. Problems involving confidence interval estimation are presented first, followed by answers to selected hypothesis tests.

Confidence Interval Estimation

Means/ Propns.	Samples Rel./ Ind.	Point Estimate	Std. Error	Conf. Level	z or t	z/t Coeff.	Confidence Int.
2. M	I	2.080 minutes	.599	.95	t	2.021	2.080 ± 1.211 minutes
6. P	R	.03	.024	.99	z	2.575	.03 ± .062
7. P	I	.09	.032	.90	z	1.645	.09 ± .053
8. M	R	$10.33	3.100	.90	t	1.740	$10.33 ± $5.39 per month
18. M	R	.967 moves	.273	.95	z	1.96	.967 ± .535 moves
21. M	I	5.530 hours	.595	.99	z	2.575	5.530 ± 1.532 hours
23. P	R	.09	.030	.95	z	1.96	.09 ± .059

Hypothesis Tests

Means/ Propns.	Test (Tail) Direct'n.	Samples Ind./Rel.	Difference (Numerator)	Std. Error	z or t	α	Crit. Value	Obs. Value	H_0 Decision
1. P	2-tailed	I	.772 – .690	.082	z	.05	±1.96	2.216	Reject
						.01	±2.575		Not reject
10. M	Upper	R	1.286	.699	z	.05	1.645	1.840	Reject
						.01	2.33		Not reject
16. P	Lower	R	.550 – .650	.045	z	.01	–2.33	–2.222	Not reject
17. P	Upper	I	.390 – .260	.045	z	.01	2.33	2.889	Reject
22. M	Lower	I	5.22 – 6.03	.338	t	.05	–1.697	–2.396	Reject
						.01	–2.457		Not reject
23. P	Upper	R	.700 – .610	.030	z	.01	2.33	3.000	Reject
24. M	2-tailed	R	.467	.456	t	.05	±2.145	1.024	Not reject

25. Comparisons of many paired samples. Comparison of independent samples.
26. Population estimates. Independent samples.
27. Population variances and sampling distributions.

Chapter 7. Exploring Ranks and Categories

1. One-way chi square. With 2 degrees of freedom, α^2-critical $= 5.99$ ($\alpha = .05$) and 9.21 ($\alpha = .01$). χ^2-observed $= 7.39$. Reject H_O at the .05 alpha level but not at the .01 alpha level. At the .05 alpha level, a disproportionate number of students (significantly more than expected) pay more than half of their personal expenses.

2. Wilcoxon test, 2-tailed. Renters are the smallest sample: $W_R = 181.500$. For the distribution of ranks, the mean (μ_W) is 168, and the standard deviation (σ_W) is 20.494. Then z_O equals .659. At the .01 alpha level, $z_C = \pm2.575$. We cannot reject H_O at the .01 alpha level (nor could we at the .05 level). The emergency preparedness scores of renters and homeowners do not differ significantly.

3. Two-way χ^2. Use the formula for 2×2 tables. $\chi^2_C = 3.85$ ($\alpha = .05$) and $\chi^2_C = 6.64$ ($\alpha = .01$). $\chi^2_O = 8.918$. Reject H_O at both levels. Cells b, c have more cases than expected: females are more likely to say No and males to say Yes. Phi $= .091$, indicating a weak association for a sample of 1000+ cases.

6. Mann-Whitney U test, 2-tailed. The sum of the ranks for $TA_1 = W_1 = 75.50$, and based upon TA_1's ranks, $U_O = 32.50$. For TA_2, we get a sum of the ranks $= W_2 = 39.50$, and $U_O = 39.50$. Comparing the smaller U_O value (32.50) to the critical value at the .05 alpha level, $U_C = 9$, we are unable to reject the null hypothesis. We conclude that the test scores given by the two TAs do not differ significantly at the .05 alpha level.

9. One-way chi square for a 2×1 table and using Yates' correction. With 1 degree of freedom, χ^2_C values $= 3.84$ ($\alpha = .05$) and 6.64 ($\alpha = .01$). With $\chi^2_O = 10.057$, we may reject H_O at both alpha levels. Compared to random chance, respondents were significantly more likely to answer Yes, they did watch sexually explicit TV programs.

12. Mann-Whitney U test, 2-tailed. The sum of the ranks for female managers is $W_F = 32.50$. Based upon the female ranks, $U_{OF} = 17.50$. For male managers, $W_M = 45.50$, and $U_{OM} = 17.50$. At the .05 alpha level, $U_C = 5$. Therefore, we may not reject the null hypothesis. The results of evaluations given to the female versus male managers do not differ significantly at the .05 alpha level.

13. A one-way χ^2 goodness-of-fit test. First, calculate the mean and standard deviation for the sample data given: $\overline{X} = 15.10$ representatives, and $s = 6.41$ reps. The distribution of available service representatives may be re-arranged into

categories based upon deviations from the mean. Then, for χ^2 and using Yates' correction, the numbers of actual cases in those categories may be compared to the number expected in a normal distribution, as shown below.

Distribution	Number of Representatives	Normal Curve Area	$E(f)$ $(n = 20)$	$O(f)$	$(O - E - .5)$	$\frac{(O - E - .5)^2}{E}$
$+2\sigma$ and greater	27.93 and greater	.0228	.456	1	.044	.004
$+1\sigma - 2\sigma$	21.52 – 27.92	.1359	2.718	4	.782	.225
$+0 - 1\sigma$	15.10 – 21.51	.3413	6.826	5	–2.326	.793
$-0 - 1\sigma$	8.69 – 15.10	.3413	6.826	7	–.326	.016
$-1\sigma - 2\sigma$	2.28 – 8.68	.1359	2.718	3	–.218	.018
-2σ and less	less than 2.28	.0228	.456	0	–.956	2.004
Totals		1.0000	$\Sigma E(f) =$ 20.000	$\Sigma O(f) = 20$		3.060

Thus, $\chi^2_0 = 3.060$. With $k - 1 = 5$ degrees of freedom, and using the .05 alpha level, $\chi^2_c = 11.07$. Therefore, there is insufficient evidence to reject the null hypothesis. The corporation's claim is correct. The distribution of available service representatives does not differ from an approximately normal distribution at the .05 alpha level.

14. Two-way χ^2 test of independence. $\chi^2_0 = 13.672$, and $\chi^2_c = 9.49$ ($\alpha = .05$) and 13.28 ($\alpha = .01$). There is sufficient evidence to reject H_0 at both alpha levels. The two variables are related and not independent. Disproportionately, liberal students tend to be humanities and liberal arts majors (cell *b*), moderates to be applied arts, business, or education majors (cell *d*), and conservative students are somewhat more likely to be science or engineering majors (cell *i*). Nevertheless, the association between liberalism-conservatism and major is not especially strong because $V = .099$, indicating a weak association for a sample of close to 700 respondents.

15. A two-way χ^2 problem using the 2×2 formula and Yates' correction. $\chi^2_0 = 5.951$. That is sufficient to reject H_0 at the .05 α level ($\chi^2_c = 3.84$) but not at the .01 level ($\chi^2_c = 6.64$). At the higher alpha level, citizens are disproportionately likely to say Yes to immigration restrictions and noncitizens to say No. The tendency is not sufficiently pronounced to be significant at the .01 level, however, and the phi coefficient is .092, suggesting a weak association for the sample of 698 students.

16. A 2-tailed Wilcoxon test. Patients under age 60 constitute the smaller sample, and their ranks sum to 347.000. The mean of the summed ranks' distribution is 304.000, and the standard deviation is 32.619. $z_0 = 1.318$ for these data,

but $z_C = \pm 1.96$. There is insufficient justification to reject the null hypothesis. The regimen compliance scores of patients under age 60 do not differ from those older than 60.

17. A 2-way χ^2 test for a 5×3 table, using Yates' correction due to a low expected frequency (for cell c). With 8 degrees of freedom, $\chi^2_C = 20.09$ at the .01 alpha level. The null hypothesis may be rejected based upon $\chi^2_O = 52.243$. African American, Asian, and Latino students are more likely to say Yes to increased faculty/staff diversity. White students are disproportionately likely to be unsure or say No. Students of other ethnic backgrounds generally respond according to random chance. Cramér's V = .223, suggesting a weak to moderate association between the two variables.

18. A 1-way χ^2 problem for a distribution with 9 categories. The observed and expected frequencies differ, and $\chi^2_O = 154.879$. With $k - 1 = 8$ degrees of freedom, $\chi^2_C = 26.12$ at the .001 alpha level. There is sufficient evidence to reject H_O. Students in the survey have distinct preferences for comedies or comedy/dramas and for noncrime dramas/adventure shows. Other program preferences approximate or fall below random chance probabilities.

Hints for Essay Questions

20. Known population parameter?

21. Use of data or data ranks/signs?

22. Rare events?

23. Ranks differ?

24. Given alpha, does random chance prevail?

25. Absolute value of $O - E$ differences versus possible differences from a known parameter.

26. Hypotheses? Expected frequencies? Degrees of freedom? Interpretation?

27. Random chance.

28. Unusual frequencies? Rare event?

29. The chi square distribution's continuity.

Chapter 8. Analysis of Variance: Do Multiple Samples Differ?

1.

Group	n	ΣX	ΣX^2	$\overline{X}$	Source	SS	d.f.	MS	F_O
High-tech	10	516	27,708	51.600	Between	8384.730	2	4192.365	33.763
Music	8	311	13,253	38.875	Within	3600.989	29	124.172	
Academia	14	1080	84,670	77.143					
Totals	32	1907	125,631						

With 2 and 29 degrees of freedom, $F_C = 5.42$ at the .01 alpha level. The sociologist rejects her null hypothesis with $\alpha = .01$. The average percentages of top administrators age 50 or older differ among the three organizational types. Next,

Tukey's HSD requires a weighted average for n and also q_α for the .01 level: $\bar{n}$ = 10.135, and q_α = 4.55 with 3 and 29 degrees of freedom. For Tukey's HSD, the minimum difference for significance at the .01 alpha level, we get 15.926. A difference of ±15.926% or more is statistically significant.

Groups	Means	Significant Difference
Academia vs. High-tech	77.143 – 51.600 = 25.543	Significant, α = .01
Academia vs. Music	77.143 – 38.875 = 38.268	Significant, α = .01
High-tech vs. Music	51.600 – 38.875 = 12.725	Not significant

The typical percentage of older top administrators found in academia is significantly greater than those found in either the high tech or music fields. However, the average figures for high tech and music do not differ significantly among themselves at the .01 alpha level.

4. Group	n	ΣX	ΣX^2	$\overline{X}$	Source	SS	d.f.	MS	F_o
Teen	12	853	61,901	71.083	Between	546.660	2	273.330	4.845
Adult	10	611	37,545	61.100	Within	1805.520	32	56.420	
Senior	13	873	58,951	67.154					
Totals	35	2337	158,397						

With 2 and 32 degrees of freedom, F_c = 3.30 at the .05 alpha level and 5.34 at the .01 level. The research team may reject its H_0 at the .05 α level but not when α = .01. Considering the .05 α level, Tukey's test reveals some support for the teacher's suspicions: $\bar{n}$ = 11.538, $q_{\alpha = .05}$ = 3.49 with 3 and 32 d.f., and then Tukey's HSD = ±7.718 decibels.

Groups	Means	Significant Difference
Teen vs. Adult	71.083 – 61.100 = 9.983	Significant, α = .05
Teen vs. Senior	71.083 – 67.154 = 3.929	Not significant
Senior vs. Adult	67.154 – 61.100 = 6.054	Not significant

The sample consisting mainly of teenage drivers averages a significantly higher dB count than do adult drivers, but it is not higher than that for older drivers. Similarly, adult and senior driver groups do not differ significantly, just those for teens and adults. The teacher's expectations are only partially confirmed.

7. Group	n	ΣX	ΣX^2	$\overline{X}$	Source	SS	d.f.	MS	F_o
Dorms	10	19	61	1.900	Between	5.600	2	2.800	1.372
Greek	10	27	89	2.700	Within	55.100	27	2.041	
Other	10	17	43	1.700					
Totals	30	63	193						

$F_O = 1.372$, but $F_C = 3.35$ at the .05 alpha level and with 2 and 27 degrees of freedom. There is insufficient evidence to reject the null hypothesis. At the .05 alpha level, the average number of times students have at least two drinks in a two-week period does not differ by housing group, and with no significant differences between groups for which to account, the Tukey test is not needed.

8.

Group	n	ΣX	ΣX^2	$\overline{X}$	Source	SS	d.f.	MS	F_O
Upper	8	180	4092	22.500	Between	771.287	2	385.644	37.412
Middle	9	251	7063	27.889	Within	247.380	24	10.308	
Lower	10	355	12,745	35.500					
Totals	27	786	23,900						

At the .01 alpha level, $F_C = 5.61$. The economist rejects her null hypothesis. The average percentages devoted to housing differ significantly among her three income levels. For her Tukey test: $\tilde{n} = 8.929$, $q_{\alpha = .01} = 4.54$ with 3 and 24 d.f., and Tukey's HSD = a minimum of 4.887%.

Groups	Means	Significant Difference
Lower vs. Upper	$35.500 - 22.500 = 13.000$	Significant, $\alpha = .01$
Lower vs. Middle	$35.500 - 27.889 = 7.611$	Significant, $\alpha = .01$
Middle vs. Upper	$27.889 - 22.500 = 5.389$	Significant, $\alpha = .01$

All group means differ significantly. The middle income sample, on the average, devoted a significantly higher percentage of its income to housing than did the upper income sample. Similarly, the average lower income household paid a significantly higher percentage of its income for housing than did either its upper or middle income counterparts.

Hints for Essay Questions

9. Expected, treatment variation versus error variation.
10. Samples used. Independence? Sources of variation.
11. Average variation.
12. Variation between samples exceed expected variation?
13. Differences between respective groups' averages meaningful or generally random and normal?

Chapter 9. *X* and *Y* Together: Correlation and Prediction

1. a. Intercept = 6.00 times per month. Slope = −.938 times per month
   b.

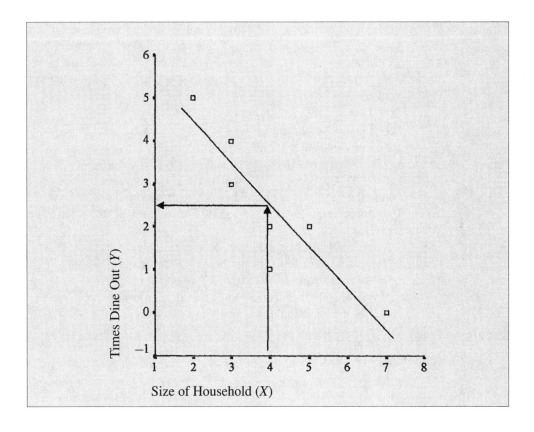

c. The scatterplot suggests a Y value of about 2.500 times dining out per month for a household of 4 people. The value for $\hat{Y} = 2.250$ times per month.

The answers below are sorted by type. Those for selected Pearson's r_{XY} problems appear first, followed by Spearman's r_s solutions.

Pearson's r_{XY}

2. a. $r_{XY} = -.670$, a strong and indirect XY association.

   $r_{XY}^2 = .449$. Variations in the length of time with the company explain about 45% of the variation in the number of days absent in the last quarter.

   b. Lower-tailed test. $\alpha = .05$: $= r_{XY_C} -.669$. Reject H_0. Indirect correlation is significant.

   $\alpha = .01$: $r_{XY_C} = -.833$. Fail to reject H_0. Correlation not significant.

c. $\Sigma X = 31, \Sigma X^2 = 173, s_X = 2.440, \overline{X} = 4.429$

 $\Sigma Y = 30, \Sigma Y^2 = 200, s_Y = 3.450, \overline{Y} = 4.286$

 $\Sigma XY = 99$

For a median value of 4 years, $Y = 4.692$ days absent.

d. $s_{Y \cdot X} = 2.561$. For t with 5 d.f.: .90 level = ±2.015, .95 level = ±2.571, .99 level = ±4.032

 .90 CI est. = 4.692 ± 5.160 days

 .95 CI est. = 4.692 ± 6.584 days

 .99 CI est. = 4.692 ± 10.326 days

5. a. $r_{XY} = -.248$, a weak to moderate and negative XY association.

 $r_{XY}^2 = .061$. Cities' costs of living explain only about 6% of the variation in the percentages of children living in poverty.

 b. Two-tailed test. $\alpha = .05$: $r_{XY_C} = \pm.579$. Fail to reject H_0. Correlation not significant.

  $\alpha = .01$: Same conclusion, automatically.

 c, d. Not justified due to a nonsignificant XY correlation.

16. a. $r_{XY} = -.873$, a strong and negative XY association.

 $r_{XY}^2 = .763$. The number of adult family members present explains 76.3% of the variation in the number of goals scored.

 b. Lower-tailed test. $\alpha = .05$: $r_{XY_C} = -.729$. Reject H_0. Indirect correlation is significant.

  $\alpha = .01$: $r_{XY_C} = -.882$. Fail to reject H_0. Correlation not significant.

 c. $\Sigma X = 157, \Sigma X^2 = 4323, s_X = 6.555, \overline{X} = 26.167$

  $\Sigma Y = 10, \Sigma Y^2 = 24, s_Y = 1.211, \overline{Y} = 1.667$

  $\Sigma XY = 227$

For a median value of 25.5 people present, $Y = 1.774$ goals.

 d. $s_{Y \cdot X} = .590$. For t with 4 d.f.: .90 level = ±2.132, .95 level = ±2.776, .99 level = ±4.604

  .90 CI est. = 1.774 ± 1.258 goals

  .95 CI est. = 1.774 ± 1.638 goals

  .99 CI est. = 1.774 ± 2.716 goals

17. a. $r_{XY} = .277$, a weak to moderate positive XY association.

 $r_{XY}^2 = .077$. Only 7.7% of the variation in percentages of savings invested in stocks and bonds may be explained by educational level.

 b. Upper-tailed test. $\alpha = .05$: $r_{XY_C} = .497$. Fail to reject H_0. Correlation not significant.

 $\alpha = .01$: Same conclusion, automatically.

 c, d. Not justified due to a nonsignificant XY correlation.

Spearman's r_s

3. a. $r_s = .643$. There is a strong positive association between the number of movies seen recently and the rating given the new film.

    b. Two-tailed test. $\alpha = .05$: $r_{s_C} = \pm.587$. Reject H_0. Positive correlation is significant.

    $\alpha = .01$: $r_{s_C} = \pm.727$. Fail to reject H_0. Correlation not significant.

8. a. $r_s = .725$. r_s indicates a strong positive association between number of years officiating and one's evaluation: more years officiating means a higher grade.

    b. Upper-tailed test. $\alpha = r_{s_C}$ .05: = .536. Reject H_0. Positive correlation is significant.

    $\alpha = .01$: $r_{s_C} = .709$. Reject H_0. Positive correlation is significant.

12. a. $r_s = .714$. There is a strong positive association between the percentage of time English is spoken and one's score/rank on the assimilation index.

    b. Upper-tailed test. $\alpha = .05$: $r_{s_C} = .643$. Reject H_0. Positive correlation is significant.

    $\alpha = .01$: $r_{s_C} = .833$. Fail to reject H_0. Correlation not significant.

13. a. $r_s = -.379$. Moderately strong negative association between the number of units being taken and how well one balances one's family, work, and school obligations: the more units, the lower the rank on the balance index.

    b. Lower-tailed test. $\alpha = .05$: $r_{s_C} = -.564$. Fail to reject H_0. Correlation not significant.

    $\alpha = .01$: Same conclusion, automatically.

Hints for Essay Questions

18. Level of measurement.
19. Variation. Regression line ($\hat{Y}$) versus the mean of Y ($\bar{Y}$).
20. Alpha regions and the horizontal axis of the H_0 curve.
21. Sampling distribution curves.
22. Reduction of variation and error.
23. Explained variation and predictability versus random error.
24. Level of measurement.
25. Variation. Regression line ($\hat{Y}$) versus the mean of Y ($\bar{Y}$).

Index

a, intercept, 292, 293, 294

Absolute value, 32

Additive rule for probabilities, 76–77

Alpha error (Type I error), 136–137, 142–144

Alpha level, region (rejection region), 136–137, 142–144, 216, 219, 237

Alternative hypothesis (H_A), 138–140, 141, 145, 147, 230, 233, 235, 239, 249, 269, 288, 289, 307, 308

    And one- vs. two-tailed tests, 138–140, 288, 289

Analysis of variance (ANOVA), 255, 265–276

    And degrees of freedom, 268, 269, 270

    And sums of squares. 267–68, 271, 272

    Calculating the *F* Ratio, 266, 268–269, 272, 274, 276

    Critical values of the *F* Ratio, Table F, 266, 269, 272, 274, 276, 348

    Critical values for the studentized range (Q) distribution, Table G, 269, 351

    Hypotheses for, 269

    Interpretation of significant *F* ratio, 266, 269, 270

    Means squares (MS), 268–269, 270, 272, 275

    Theory, logic, concepts of the test , 265–267

    With Tukey's HSD test, 267, 269–270, 272–274, 276

Area of normal curve, 46, 48, 54–61, 67–70, 136–137, 144–45, 160–161, 163–166

Association, measures of, 118–22, 228–229

Average, mean, 2, 3, 4, 25–29, 31, 64, 118–122, 180–185, 190–195, 228–229, 269, 273, 293, 297, 299, 302

    Compared to median, 27–29

    Confidence interval estimate for, 105–112, 118–122, 180–185, 200–203, 203–207

b, slope, 292, 293, 294

Beta error (Type II error), 144–148, 159–68

    And sample size (*n*), 159, 166–168

    Calculating *P*(Beta error), 159–166

    For H_O test with averages, 159–161

    For H_O test with proportions, 160–162

    For two-tailed test, 166–167

Between groups variation, treatment variation (*See also* Analysis of Variance), 266–269

Bimodal distributions, 22

Binomial distribution, probabilities, 3, 46, 60–82

    Continuous, 60–71

    Discrete, 62–63, 70–82

Binomial probabilities table, features and use of, 79–82

Binomial probabilities table, Table B, 338

Categorical data, 2, 4, 6–9, 10, 22–23, 25, 235–255
Central limit theorem, 94
Central tendency, location, 2, 20–29
Chi square "goodness of fit" test (χ^2 test of normality), 240–243
Chi square table of critical values, Table E, 237, 244–245, 346
Chi square test, 227, 234–253
  And degrees of freedom, 237, 244, 245
  And expected frequencies, 234–236, 241, 243–244
  Critical values for chi square: χ^2_C, 237, 244–245
  Goodness of fit test, Test of normality, 240–243, 235
  Interpretation of significant χ^2, 235, 237–238, 243, 246
  Null hypothesis (H_O) curve for χ^2, 237, 239, 244
  Measures of association based upon two-way χ^2, 247, 253–255
    Interpretation of, 253
    Cramér's V, 254–255
    Phi (2x2 tables), 253–254
  One-way χ^2, 235–243
  Theory, logic, and concepts, 234–238, 243–244
  Test of independence: two-way χ^2 test, 235, 243–249
  Variations on basic χ^2 test, 247–253
  Yates' correction for continuity (*See also* Yate's correction), 247–253
Coefficient of determination, r_{xy}^2, 283, 297–300, 301
Coefficient of Variation, 40–41
Combinations, 73–76
Complement, 76, 78
  Solving for, 78–79
Confidence interval estimation, 90–126, 177–189, 200–212, 283, 291, 295–297, 303
  Checklist for (one-sample estimate), 126

For the difference between two population means, 180–185, 203–207
For the difference between two population proportions, 185–189, 207–212
For the mean, 105–112
For the proportion, 112–116
For $\hat{Y}$, 283, 291, 295–297, 303
Sample size required for (one sample), 118–125
With an unknown σ, 98–99
Confidence level, 92, 100, 295–296
  And margin of error, 92, 100, 295–296
  And probability, 92
  z coefficients for, 100–101, 107, 113–114, 117
Constant, 5
Contingency tables (cross-classification tables), 243–246
Continuity correction, 66–67
Continuous data, variables, 3, 46, 47, 48–71
Correction factor, 35, 66–68, 103–105, 120–122, 124–125
Correlation and regression, 4–5, 280–308
  Pearson's r_{xy}, 282, 283–292, 300, 301
    Degrees of freedom for H_O test, 288, 296, 301
    H_O test for, 282, 287–292, 301
    Null hypothesis (H_O), 287–289
    Interpretation of, 281, 282, 284–285, 287, 301
    Limitations of, 282, 286–287, 292–293, 295
    Raw score formula, 284, 286, 300
    r_{xy}-critical values, features of Table H, 290–291, 301
    r_{xy} table of critical values, Table H, 290–291, 351
    r_{xy}^2, the coefficient of determination, 283, 297–300, 301
    t, z tests for, 288–291, 301
    Theory, logic, and concepts, 280–285
    z score formula for r_{xy}, 283–284
  Predicting a $\hat{Y}$ value, 283, 290–303

Theory, logic, and concepts, 292–294, 295

Confidence interval estimate for $\hat{Y}$, 295–297, 298, 299, 303

Degrees of freedom for t, 296

Spearman's r_s, 304–308

Degrees of freedom for t test, 304–305, 307, 308

H_O tests for, 304–308

Interpretation of, 305–308

r_s-critical and use of Table I, 305, 306

Table I: r_s critical values, 354

Use of ranks, d scores, 304, 305

Correlation ratio, eta, 287

Count, frequencies, 7–9, 228, 234–249

Cramér's V, 254–255

Criteria for continuity (for binomial distributions), 62–64

Critical value (of a statistic), 150–152, 160–161, 165–165, 290–291, 301

Cumulative percent, 10–12, 24–25

d scores, 200–201, 212, 304, 305

Data, Level of measurement, 1, 5–18

Degrees of freedom (d.f.), 100, 180–182, 203–204, 212, 237, 244, 245, 268–270, 288, 296, 301, 305, 306, 308

Dependent variable, 135, 244

Descriptive statistical analyses, 2, 18–19

Deviation procedure for the standard deviation, 30–36, 285

Direct correlation, association , 282, 285, 287

Discrete data, variables, 3, 46, 47, 62–63, 70–82, 234–253

Diversity, variation, 29–41, 95–99, 295–300

Element, 5

Empirical indicator, 6

Error term (*See also* Margin of error), 106–116, 118–125

When estimating n, 118–125

Error (within group) variation, 266–268

Estimation (*See* Confidence interval estimation)

Event, 72, 76, 78, 81

Expected frequencies (*See also* Chi square), 234–235, 236, 240

Expected value, average, 64

Explained variation, 283, 297–299

fpc (Finite population correction), 103–105, 120–122, 124–125, 153–154

When estimating n, 120–121, 124–125

F Ratio, F test for ANOVA (*See also* Analysis of variance), 266–269

Interpretation of significant F Ratio, 266

F test (ANOVA), 266–269

Factorial(s), 74–75

Failure, 46, 80–82

Frequencies, counts, 7–9, 228, 234–249

Frequency distributions, 8–17

Finite population correction, fpc, (Hypergeometric correction factor), 72, 103–105, 120–122

Glossary, 315

Goodness-of-fit test (chi square), 235, 240–243

Grand mean, 266

Harmonic mean, 270

HSD, Tukey's test for honestly significant differences (*See also* Analysis of variance), 269–270, 272–274, 276

Critical values for the Studentized Range (Q) distribution: Table G, 351

Hypergeometric correction factor (*See* the Finite Population Correction, or fpc)

Hypothesis(es), 194, 195, 237, 239, 245, 269, 270, 274, 306

Hypothesis testing, 3–4, 133–168, 190–200, 304–308

One sample: 3, 133–168

Checklist, 152, 155

One-tailed test, 139–140, 152–155

Procedures, 148–152, 304–308

Theory, logic, and concepts, 133–148, 159, 228, 234, 304–305

Two-tailed test, 136–137, 155–159, 308

Two-tailed vs. one-tailed tests, 139–140, 306, 308

Two sample, 3, 190–200, 212–220
  For averages, 190–195, 212–216
  For proportions, 177–181, 196–199, 216–219
  Generic formulas, 190–191, 203, 212
  Theory, logic, and concepts, 133–148, 177–181, 190–191, 200–203, 212, 228–229

Independent samples, 178–200, 228
  And pooled variances:
    For averages, means, 180, 182, 183
    For proportions, 181, 186
  Degrees of freedom with use of t, 180–182, 191
  H_O testing procedures:
    For averages, 190–195
    For proportions, 196–200
  Logic, rationale of H_O test for difference between two populations, 190–191
  Mann-Whitney U test, 228, 230–234
  p'_1, p'_2 and Table 4.1, 196, 198
  Sampling distribution of the difference, 177–181
  Standard error of the difference, 178–181, 182, 183
  Use of z vs. t for averages, 180–182, 191, 201, 207, 212, 213
  Wilcoxon rank-sum test, 228–231
Independent binomial trials, 62
Independent variable (*See also* Predictor variable), 135, 244, 266, 282, 300
Index of Dispersion (Qualitative variation), 41
Indirect correlation, association, 282, 285, 301
Inferential statistical analyses, 2, 18–19
Intercept, a, 292–294
Interquartile range, 41
Interval-ratio levels of measurement, 6, 14–18, 31, 282, 286

Inverse correlation, association, 282, 285, 301

Joint probability, 71, 237

k, number of categories in one-way chi square, 237
k, number of treatment groups in ANOVA, 268, 270

Level of measurement, data, 1, 5–18
Line of best fit, 292–294, 295, 297, 299
Lower-tailed test, 139–140, 148–152, 152–155, 198, 213, 217, 230

Major mode, 22, 23
Mann-Whitney U table of critical values, Table D, 344
Mann-Whitney U test, 227, 230–324
Margin of error (*See also* Error term), 106–116, 118–125, 230–234, 303
  And confidence level, 106–107, 111, 295
Matched samples (*See* Related samples)
Mean (*See* Average)
Mean absolute deviation (MAD), 32
Mean difference ($\bar{d}$), 200, 203
Mean square (MS), 268–269, 270, 272, 275
  MS_B, mean square between groups, 268–269, 270, 272, 275
  MS_W, mean square within groups, 268–269, 290, 272, 275
Measurement, level of, 1, 5–18
Measures of association based upon two-way chi square, 4, 253–255
  Interpretation of, 253
  Cramér's V, 254–255
  Phi (2x2 tables), 253–254
Measures of variation, 2, 21, 30–40, 50, 57, 94–97, 99, 100–101, 120, 122, 180, 293, 295, 296, 297, 299, 302, 303
Median, 23–25
  Compared to mean, 27–29
Midpoint, 23–24
Minor mode, 22, 23
Missing values, 8, 17
Mode, 22–23

Mu (population average, μ), 26–27
Multiplicative rule for joint probabilities, 71

n (sample size), 7, 61, 64, 79, 97–100,
109, 118–126, 207, 267, 268, 281,
282, 284, 290, 303
N (population size), 27, 103–104, 120–122,
124–125
Negative correlation, association, 282,
285, 301
Nominal (qualitative, categorical) level
of measurement, 6–9, 22–23, 25,
234–253
Non-parametric tests, 134, 227–128,
234–253
Normal curve, distribution, 45, 48–71, 92,
94, 117–118, 149–150, 181, 240–243,
288, 295
Normal distribution table, Table A, 52–54,
334
Normal variation, bounds of, 134–138,
149, 158, 190, 219, 235, 237, 269,
287–288
Null hypothesis (H_O) (*See also* Hypothesis
testing), 138–142, 228, 231, 233, 239,
282, 287–292, 301
Why always test H_O, 140–142, 212,
282, 287–288
Null hypothesis (H_O) curve, 138, 140–141,
145, 147, 160–167

Observed frequencies (O), 234–235, 236,
240
One-way chi square test, 235–240
One tailed test (*See also* Hypothesis
testing), 137, 152–55, 158, 159, 216,
230, 233, 287–288
Operating characteristics/curve for an H_O
test, 146–147, 159
Ordinal level of measurement, 9–14, 25,
31, 304, 305
Outcomes, 47, 70–76
Overview of text, 2–5

p, probability of success, proportion of
successes, 60–62, 70–73, 96

When estimating n, 123–126
p', probability of success in a sample,
96–97
Paired samples (*See* Related samples)
Pearson's r_{xy} (*See also* Correlation and
regression), 280–308
Pearson's r_{xy} critical values, Table H,
290–291, 301, 353
Percent, percentage, 3, 8, 10–12, 24–25
Percentile rank, 14–16, 24
Phi coefficient, 253–254
Pilot study, 120, 123
Point estimate, 106–107, 112, 181, 185,
187, 188, 189, 203, 296
Pooled variance, 35–36, 180–181, 182,
186–187
Population, 5, 26, 103–105, 287–288
Population parameter, 27, 34–35, 148
Positive correlation, association, 282, 285,
287
Power/power curve for an H_O test,
146–148, 159
Power of a statistic, 30–31, 162–165
Predictor variable (*See also* Independent
variable), 135, 244, 282, 300
Probability, 3, 45–82, 92
Probability distributions, Table B, 3, 45,
48–50, 71, 79–80, 338
Probability of exactly X successes over n
trials, 70–73, 76–82, 338
Proportion, 3, 123–125
Proportional reduction in error (PRE),
299–300

q, probability of failure, proportion of
failures, 60–62, 70–73
q', probability of failure in a sample, 97
q_a, critical value for Studentized Q Range,
Tukey's HSD, 269

Random chance, 4, 136–137, 144,
234–235, 237–238, 243, 244, 246
Random samples, 18, 124
Random variables, 46, 47
Range, 17
Use in σ estimation, 120, 120–122

Rankable scores, data, 4, 25, 228–234, 304–308

Ranks, 9–14, 23–24, 228–234, 304–305
    Tied ranks, 228–229, 305

Ratio level of measurement, 14–17

Raw score procedure for s_x, 36–40, 302–303

Regression (*See also* Correlation and regression), 292–303

Regression line (*See also* Line of best fit, $\hat{Y}$ line), 292–294, 295, 297, 299

Related (matched, paired) samples, 200–219
    Confidence interval estimates for the difference, 203–207
        Degrees of freedom with t, 203–204
        For averages, means, 203–207
        For proportions, 207–212
    Hypothesis test for the difference between two related populations, 200–203, 212–220
        For averages, means, 212–216
        For proportions, 216–219
    Sampling distributions of the difference, concepts, 200–203

Relative frequency, 48
    And probability, 48

Rho, ρ ($\rho_{xy} \cdot \rho_s$), 282, 287–288, 289, 306

R_s, Spearman's correlation, 3–4

r_{xy}-critical , 290–291, 301

r_{xy}, Pearson's correlation coefficient (*See also* Correlation and regression), 282, 283–287, 300

r_{xy}^2, coefficient of determination, 283, 297–300, 301

Sample, random, 18, 124, 228

Sample size (*See also* n), 7, 61, 64, 79, 94–95, 97–102, 109, 118–126, 202, 207, 267, 268, 281, 282, 284, 290, 303
    For estimating p, 123–125
    For estimating μ, 118–122

Sample vs. population, 26, 150

Sampling correction factor (*See* Finite population correction)

Sampling distributions, 3, 90–99, 105–106, 134–138, 178–181, 200–203, 287–288

And single sample H_O testing, 134–138

Sampling distribution of the difference (*See also* Independent samples, Related samples), 177–181, 200–201

Sampling distribution of the mean, 91, 93–95

Sampling distribution of the proportion, 91, 95–97, 100–102

Sampling error, 94–99, 103, 134
    Sampling error and standard error, 94–95, 97–99, 104–108

Scale of measurement, 1, 5–18

Scatterplot, 292, 293, 294, 298

Semi-interquartile range, 41

Shift, 146

Sigma, σ, 34–36, 40, 99

Significant difference (*See* Statistical significance)

Skew, skewed data, 28–29, 240–243

Slope, b, 292, 293, 294

Spearman's correlation (r_s), 304–309

SPSS, 9

Squaring deviation scores, 32

Standard deviation, 21, 30–40, 50, 57, 99, 95, 99, 120, 122, 180,
    Sample s_x as a substitute for population σ_x, 95, 99

Standard deviation of the difference (s_d), 200, 203, 204

Standard error, 94–97, 100–101, 103–104, 109, 228, 229

Standard error of estimate ($s_{y \cdot x}$), 295, 296, 303

Standard error of the difference between two population means, 178–181, 201, 203, 204, 212

Standard error of the difference between two population proportions, 178–181, 202–203, 207, 212, 216

Standard error of the mean, 94–95, 100–101, 104, 109, 201

Standard error of the mean difference ($s_{\bar{d}}$), 203, 204, 212

Standard error of the proportion, 96–97, 105

Standardized score (z scores), 50–51, 159–50

Statistical independence, chi square test of, 235, 243–246

Statistical significance, 34–38, 40, 99, 134–138, 147, 158–159, 190–191, 234–235, 228, 266

Success/failure, 33, 46, 80–82

Sum of squares (SS), 33, 37–38, 267–268, 271, 272, 275

    SS between groups (SS$_B$, treatment SS), 267–268, 271, 275

    SS total (SS$_T$), 267–268, 271, 275

    SS within groups (SS$_W$, error SS), 267–268, 272, 275

$s_{y.x}$, standard error of estimate, 295, 296, 303

t and confidence level, 100–101, 180–182, 205, 295–296

t-critical, 149–150, 193, 212, 288, 290, 291, 301

t curves, distributions, 91, 97–101, 108–109, 181

Table A: Areas Under the Normal Curve, 52–54, 58, 59, 148, 149, 334

Table B: Selected Binomial Probabilities, 79–80, 338

Table C: Critical Values of t, 99–101, 148–149, 343 343

Table D: Critical Values of U for Mann-Whitney Test, 232–233, 344

Table E: Critical Values of Chi Square, 234, 237, 346

Table F: Critical Values of the F Ratio for ANOVA, 266, 269, 276, 348

Table G: Studentized Range Q Values for Tukey's HSD Test, 269, 272, 276, 351

Table H: Critical Values for Pearson's r_{xy}, 290–291, 353

Table I: Critical Values for Spearman's r_s, 305–306, 354

Text overview, 2–5

Tied ranks , 228–229, 305

Total variation, 266–267, 297, 298

Treatment groups, 265–266

Trials, independent, 62

True limits, upper and lower, 66–67

t table, features of, 99–101

t table, Table C, 148–149, 99–101, 343

t test, 148–150, 190–191, 212, 287–291, 296, 301

Tukey's HSD, 267, 269–270, 272–274

    Degrees of freedom for, 269, 272

    Interpretation of, 272–273

    With unequal sample sizes, 270

Two tailed test (*See also* Hypothesis testing), 136–137, 140, 142–144, 155–159, 190–192, 193, 216, 229, 287–288, 289

Two-way chi square test (χ^2 test of independence), 144–148, 243–253

Type I error (*See* Alpha error)

Type II error (*See* Beta error)

Unbiased estimator, 35, 99, 105–106

Unexplained variation, 297, 298, 299

Upper tailed test, 135–136, 137, 193–194, 213, 230, 287–288, 305–306

Valid percentage, 8, 10–12

Variables, 5–18

    Compared to constants, 5, 179–181, 186

Variance, 33–34, 40, 180–181

Variation, measures of, 2, 21, 29–41, 266

Variations on chi square test, 247–253

Wilcoxon rank-sum test, 227, 228–231

Within-group variation (error/random variation), 266–268

X, 6, 28

$\bar{X}$, sample average, 26–27, 28, 93, 105–106, 150–151

X axis, 48, 94, 106, 292, 294, 296, 298

$\hat{Y}$, Y predicted (*See also* Correlation and regression), 292–296, 299

Y axis, 292, 294, 296, 298

$\hat{Y}$ line (*See also* Line of best fit, Regression line), 292–294, 295, 297, 299

Yates' correction for continuity (*See also* Chi square test), 247–253

    Theory, logic, and rationale, 247–248

With expected frequencies less than 5, 247–249

With small (1 d.f.) tables, 249–251

2x1 tables, 250–251

2x2 tables, revised χ^2_O formula, 250–252

z-critical, 148–150, 158, 191, 193, 196, 212, 216, 218, 229–230, 288

z-observed, 148–150, 190–192, 203, 212, 229, 288

z scores, 46, 48–54, 95, 100, 148–150, 283, 284

z coefficients and confidence levels, 113–114, 117

z test, 100–102, 148–150, 158, 195–196, 202, 216, 218–219, 228–230, 296, 305

About the Book

Is it possible to demystify statistics? Can math phobia be overcome? Perhaps surprisingly, the answer is yes. *Learning to Live with Statistics,* based on years of teaching experience, explains basic statistical concepts and procedures in a straightforward, digestible way.

Using familiar examples that highlight the relevance of the subject to everyday life, David Asquith provides clear guidelines for defining statistical problems and choosing the right tools for solving them. The result is a student-friendly text that explains how to do statistics, and how to understand the results.

Practice exercises illustrate each of the techniques covered, and answers are provided for more than half of the exercises. The text also includes reference tables and a glossary.

David Asquith is professor of sociology at San Jose State University.